BIOLOGICAL AND MEDICAL PHYSICS, BIOMEDICAL ENGINEERING

BIOLOGICAL AND MEDICAL PHYSICS, BIOMEDICAL ENGINEERING

The fields of biological and medical physics and biomedical engineering are broad, multidisciplinary and dynamic. They lie at the crossroads of frontier research in physics, biology, chemistry, and medicine. The Biological and Medical Physics, Biomedical Engineering Series is intended to be comprehensive, covering a broad range of topics important to the study of the physical, chemical and biological sciences. Its goal is to provide scientists and engineers with textbooks, monographs, and reference works to address the growing need for information.

Books in the series emphasize established and emergent areas of science including molecular, membrane, and mathematical biophysics; photosynthetic energy harvesting and conversion; information processing; physical principles of genetics; sensory communications; automata networks, neural networks, and cellular automata. Equally important will be coverage of applied aspects of biological and medical physics and biomedical engineering such as molecular electronic components and devices, biosensors, medicine, imaging, physical principles of renewable energy production, advanced prostheses, and environmental control and engineering.

For other titles published in this series, go to http://www.springer.com/series/3740

Geoff Dougherty
Editor

Medical Image Processing

Techniques and Applications

Springer

Editor
Geoff Dougherty
Applied Physics and Medical Imaging
California State University Channel Islands
One University Drive
93012 Camarillo
USA
Geoff.Dougherty@csuci.edu

ISSN 1618-7210
ISBN 978-1-4419-9769-2 e-ISBN 978-1-4419-9779-1
DOI 10.1007/978-1-4419-9779-1
Springer New York Dordrecht Heidelberg London

Library of Congress Control Number: 2011931857

Printed on acid-free paper

Springer is part of Springer Science+Business Media (www.springer.com)

To my mother, Adeline Maud Dougherty, and my father, Harry Dougherty (who left us on 17th November 2009)

Preface

The field of medical imaging advances so rapidly that all of those working in it, scientists, engineers, physicians, educators, and others, need to frequently update their knowledge to stay abreast of developments. While journals and periodicals play a crucial role in this, more extensive, integrative publications that connect fundamental principles and new advances in algorithms and techniques to practical applications are essential. Such publications have an extended life and form durable links in connecting past procedures to the present, and present procedures to the future. This book aims to meet this challenge and provide an enduring bridge in the ever expanding field of medical imaging.

This book is designed for end users in the field of medical imaging, who wish to update their skills and understanding with the latest techniques in image analysis. The book emphasizes the conceptual framework of image analysis and the effective use of image processing tools. It is designed to assist cross-disciplinary dialog both at graduate and at specialist levels, between all those involved in the multidisciplinary area of digital image processing, with a bias toward medical applications. Its aim is to enable new end users to draw on the expertise of experts across the specialization gap.

To accomplish this, the book uses applications in a variety of fields to demonstrate and consolidate both specific and general concepts, and to build intuition, insight, and understanding. It presents a detailed approach to each application while emphasizing the applicability of techniques to other research areas. Although the chapters are essentially self-contained, they reference other chapters to form an integrated whole. Each chapter uses a pedagogical approach to ensure conceptual learning before introducing specific techniques and "tricks of the trade".

The book is unified by the theme foreshadowed in the title "Medical Image Processing: Techniques and Applications." It consists of a collection of specialized topics, each presented by a specialist in the field. Each chapter is split into sections and subsections, and begins with an introduction to the topic, method, or technology. Emphasis is placed not only on the background theory but also on the practical aspects of the method, the details necessary to implement the technique, and limits of applicability. The chapter then introduces selected more advanced applications of the topic, method, or technology, leading toward recent achievements and unresolved questions in a manner that can be understood by a reader with no specialist knowledge in that area.

Chapter 1, by Dougherty, presents a brief overview of medical image processing. He outlines a number of challenges and highlights opportunities for further development.

A number of image analysis packages exist, both commercial and free, which make use of libraries of routines that can be assembled/mobilized/concatenated to automate an image analysis task. Chapter 2, by Luengo, Malm, and Bengtsson, introduces one such package, DIPimage, which is a toolbox for MatLab that incorporates a GUI for automatic image display and a convenient drop-down menu of common image analysis functions. The chapter demonstrates how one can quickly develop a solution to automate a common assessment task such as counting cancerous cells in a Pap smear.

Segmentation is one of the key tools in medical image analysis. The main application of segmentation is in delineating an organ reliably, quickly, and effectively. Chapter 3, by Couprie, Najman and Talbot, presents very recent approaches that unify popular discrete segmentation methods.

Deformable models are a promising method to handle the difficulties in segmenting images that are contaminated by noise and sampling artifact. The model is represented by an initial curve (or surface in three dimensions (3D)) in the image which evolves under the influence of internal energy, derived from the model geometry, and an external force, defined from the image data. Segmentation is then achieved by minimizing the sum of these energies, which usually results in a smooth contour. In Chapter 4, Alfiansyah presents a review of different deformable models and issues related to their implementations. He presents some examples of the different models used with noisy medical images.

Over the past two decades, many authors have investigated the use of MRI for the analysis of body fat and its distribution. However, when performed manually, accurate isolation of fat in MR images can be an arduous task. In order to alleviate this burden, numerous segmentation algorithms have been developed for the quantification of fat in MR images. These include a number of automated and semi-automated segmentation algorithms. In Chapter 5, Costello and Kenny discuss some of the techniques and models used in these algorithms, with a particular emphasis on their application and implementation. The potential impact of artifacts such as intensity inhomogeneities, partial volume effect (PVE), and chemical shift artifacts on image segmentation are also discussed.

An increasing portion of medical imaging problems concern thin objects, and particularly vessel filtering, segmentation, and classification. Example applications include vascular tree analysis in the brain, the heart, or the liver, the detection of aneurysms, stenoses, and arteriovenous malformations in the brain, and coronal tree analysis in relation to the prevention of myocardial infarction. Thin, vessel-like objects are more difficult to process in general than most images features, precisely because they are thin. They are prone to disappear when using many common image analysis operators, particularly in 3D. Chapter 6, by Tankyevych, Talbot, Passat, Musacchio, and Lagneau, introduces the problem of cerebral vessel filtering and detection in 3D and describes the state of the art from filtering to segmentation, using local orientation, enhancement, local topology, and scale selection. They apply both linear and nonlinear operators to atlas creation.

Automated detection of linear structures is a common challenge in many computer vision applications. Where such structures occur in medical images, their measurement and interpretation are important steps in certain clinical decision-making tasks. In Chapter 7, Dabbah, Graham, Malik, and Efron discuss some of the well-known linear structure detection methods used in medical imaging. They describe a quantitative method for evaluating the performance of these algorithms in comparison with their newly developed method for detecting nerve fibers in images obtained using *in vivo* corneal confocal microscopy (CCM).

Advances in linear feature detection have enabled new applications where the reliable tracing of line-like structures is critical. This includes neurite identification in images of brain cells, the characterization of blood vessels, the delineation of cell membranes, and the segmentation of bacteria under high resolution phase contrast microscopy. Linear features represent fundamental image analysis primitives. In Chapter 8, Domanski, Sun, Lagerstrom, Wang, Bischof, Payne, and Vallotton introduce the algorithms for linear feature detection, consider the preprocessing and speed options, and show how such processing can be implemented conveniently using a graphical user interface called HCA-Vision. The chapter demonstrates how third parties can exploit these new capabilities as informed users.

Osteoporosis is a degenerative disease of the bone. The averaging nature of bone mineral density measurement does not take into account the microarchitectural deterioration within the bone. In Chapter 9, Haidekker and Dougherty consider methods that allow the degree of microarchitectural deterioration of trabecular bone to be quantified. These have the potential to predict the load-bearing capability of bone.

In Chapter 10, Adam and Dougherty describe the application of medical image processing to the assessment and treatment of spinal deformity, with a focus on the surgical treatment of idiopathic scoliosis. The natural history of spinal deformity and current approaches to surgical and nonsurgical treatment are briefly described, followed by an overview of current clinically used imaging modalities. The key metrics currently used to assess the severity and progression of spinal deformities from medical images are presented, followed by a discussion of the errors and uncertainties involved in manual measurements. This provides the context for an analysis of automated and semi-automated image processing approaches to measure spinal curve shape and severity in two and three dimensions.

In Chapter 11, Cree and Jelinek outline the methods for acquiring and pre-processing of retinal images. They show how morphological, wavelet, and fractal methods can be used to detect lesions and indicate the future directions of research in this area.

The appearance of the retinal blood vessels is an important diagnostic indicator for much systemic pathology. In Chapter 12, Iorga and Dougherty show that the tortuosity of retinal vessels in patients with diabetic retinopathy correlates with the number of detected microaneurysms and can be used as an alternative indicator of the severity of the disease. The tortuosity of retinal vessels can be readily measured in a semi-automated fashion and avoids the segmentation problems inherent in detecting microaneurysms.

With the increasing availability of highly resolved isotropic 3D medical image datasets, from sources such as MRI, CT, and ultrasound, volumetric image rendering techniques have increased in importance. Unfortunately, volume rendering is computationally demanding, and the ever increasing size of medical image datasets has meant that direct approaches are unsuitable for interactive clinical use. In Chapter 13, Zhang, Peters, and Eagleson describe volumetric visualization pipelines and provide a comprehensive explanation of novel rendering and classification algorithms, anatomical feature and visual enhancement techniques, dynamic multimodality rendering and manipulation. They compare their strategies with those from the published literatures and address the advantages and drawbacks of each in terms of image quality and speed of interaction.

In Chapter 14, Bones and Wu describe the background motivation for adopting sparse sampling in MRI and show evidence of the sparse nature of biological image data sets. They briefly present the theory behind parallel MRI reconstruction, compressive sampling, and the application of various forms of prior knowledge to image reconstruction. They summarize the work of other groups in applying these concepts to MRI and then describe their own contributions. They finish with a brief conjecture on the possibilities for future development in the area.

In Chapter 15, Momot, Pope, and Wellard discuss the fundamentals of diffusion tensor imaging (DTI) in avascular tissues and the key elements of digital processing and visualization of the diffusion data. They present examples of the application of DTI in two types of avascular tissue: articular cartilage and eye lens. Diffusion tensor maps present a convenient way to visualize the ordered microstructure of these tissues. The direction of the principal eigenvector of the diffusion tensor reports on the predominant alignment of collagen fibers in both tissues.

Contents

Contributors

Clayton Adam Queensland University of Technology, Brisbane, Australia, c.adam@qut.edu.au

Agung Alfiansyah Surya Research and Education Center, Tangerang, Indonesia, agung.alfiansyah@gmail.com

Ewert Bengtsson Swedish University of Agricultural Sciences, Uppsala, Sweden
Uppsala University, Uppsala, Sweden, ewart.bengtsson@cb.uu.se

Leanne Bischof CSIRO (Commonwealth Scientific and Industrial Research Organisation), North Ryde, Australia, leanne.bischof@csiro.au

Philip J. Bones University of Canterbury, Christchurch, New Zealand, phil.bones@canterbury.ac.nz

David P. Costello Mater Misericordiae University Hospital and University Collage Dublin, Ireland, dcostello@mater.ie

Camille Couprie Université Paris-Est, Paris, France, c.couprie@esiee.fr

Michael J. Cree University of Waikato, Hamilton, New Zealand, cree@waikato.ac.nz

Mohammad A. Dabbah The University of Manchester, Manchester, England, m.a.dabbah@manchester.ac.uk

Luke Domanski CSIRO (Commonwealth Scientific and Industrial Research Organisation), North Ryde, Australia, Luke.Domanski@csiro.au

Geoff Dougherty California State University Channel Islands, Camarillo, CA, USA, geoff.dougherty@csuci.edu

Roy Eagleson The University of Western Ontario, London, ON, Canada, eagleson@uwo.ca

Nathan Efron Institute of Health and Biomedical Innovation, Queensland University of Technology, Brisbane, Australia, n.efron@qut.edu.au

James Graham The University of Manchester, Manchester, England, jim.graham@manchester.ac.uk

Mark A. Haidekker University of Georgia, Athens, Georgia, mhaidekker.uga@gmail.com

Cris L. Luengo Hendriks Uppsala University, Uppsala, Sweden, cris@cb.uu.se

Michael Iorga NPHS, Thousand Oaks, CA, USA, michael.iorga@yahoo.com

Herbert F. Jelinek Charles Stuart University, Albury, Australia, hjelinek@csu.edu.au

Patrick A. Kenny Mater Misericordiae University Hospital and University College Dublin, Ireland, pkenny@mater.ie

Ryan Lagerstrom CSIRO (Commonwealth Scientific and Industrial Research Organisation), North Ryde, Australia, Ryan.Lagerstrom@csiro.au

Michel Lagneau Hôpital Louis-Pasteur, Colmar, France, michel.lagneau@ch-colmar.rss.fr

Rayaz A. Malik The University of Manchester, Manchester, England, Rayaz.A.Malik@manchester.ac.uk

Patrik Malm Swedish University of Agricultural Sciences, Uppsala, Sweden

Uppsala University, Uppsala, Sweden, patrik@cb.uu.se

Konstantin I. Momot Queensland University of Technology, Brisbane, Australia, k.momot@qut.edu.au

Mariano Musacchio Hôpital Louis-Pasteur, Colmar, France, mariano_musacchio@yahoo.fr

Laurent Najman Université Paris-Est, Paris, France, l.najman@esiee.fr

Nicholas Passat Université de Strasbourg, Strasbourg, France, passat@dpt-info.u-strasbg.fr

Matthew Payne CSIRO (Commonwealth Scientific and Industrial Research Organisation), North Ryde, Australia, matthew.payne@csiro.au

Terry M. Peters Robarts Research Institute, University of Western Ontario, London, ON, Canada, tpeters@robarts.ca

James M. Pope Queensland University of Technology, Brisbane, Australia, j.pope@qut.edu.au

Changming Sun CSIRO (Commonwealth Scientific and Industrial Research Organisation), North Ryde, Australia, changmin.sun@csiro.au

Hugues Talbot Université Paris-Est, Paris, France, h.talbot@esiee.fr

Olena Tankyevych Université Paris-Est, Paris, France, tankyevych@gmail.com

Pascal Vallotton CSIRO (Commonwealth Scientific and Industrial Research Organisation), North Ryde, Australia, Pascal.Vallotton@csiro.au

Dadong Wang CSIRO (Commonwealth Scientific and Industrial Research Organisation), North Ryde, Australia, dadong.wang@csiro.au

R. Mark Wellard Queensland University of Technology, Brisbane, Australia, m.wellard@qut.edu.au

Bing Wu Duke University, Durham, NC, USA, contactbing@gmail.com

Qi Zhang Robarts Research Institute, University of Western Ontario, London, ON, Canada, Qi.Zhang@nrc-cnrc.gc.ca

Chapter 1
Introduction

Geoff Dougherty

1.1 Medical Image Processing

Modern three-dimensional (3-D) medical imaging offers the potential and promise for major advances in science and medicine as higher fidelity images are produced. It has developed into one of the most important fields within scientific imaging due to the rapid and continuing progress in computerized medical image visualization and advances in analysis methods and computer-aided diagnosis [1], and is now, for example, a vital part of the early detection, diagnosis, and treatment of cancer. The challenge is to effectively process and analyze the images in order to effectively extract, quantify, and interpret this information to gain understanding and insight into the structure and function of the organs being imaged. The general goal is to understand the information and put it to practical use.

A multitude of diagnostic medical imaging systems are used to probe the human body. They comprise both microscopic (viz. cellular level) and macroscopic (viz. organ and systems level) modalities. Interpretation of the resulting images requires sophisticated image processing methods that enhance visual interpretation and image analysis methods that provide automated or semi-automated tissue detection, measurement, and characterization [2–4]. In general, multiple transformations will be needed in order to extract the data of interest from an image, and a hierarchy in the processing steps will be evident, e.g., enhancement will precede restoration, which will precede analysis, feature extraction, and classification [5]. Often, these are performed sequentially, but more sophisticated tasks will require feedback of parameters to preceding steps so that the processing includes a number of iterative loops.

G. Dougherty (✉)
California State University Channel Islands, Camarillo, CA, USA
e-mail: Geoff.Dougherty@csuci.edu

G. Dougherty (ed.), *Medical Image Processing: Techniques and Applications*, Biological and Medical Physics, Biomedical Engineering, DOI 10.1007/978-1-4419-9779-1_1,

There are a number of specific challenges in medical image processing:

1. Image enhancement and restoration
2. Automated and accurate segmentation of features of interest
3. Automated and accurate registration and fusion of multimodality images
4. Classification of image features, namely characterization and typing of structures
5. Quantitative measurement of image features and an interpretation of the measurements
6. Development of integrated systems for the clinical sector

Design, implementation, and validation of complex medical systems require not only medical expertise but also a strong collaboration between physicians and biologists on the one hand, and engineers, physicists, and computer scientists on the other.

Noise, artifacts and weak contrast are the principal causes of poor image quality and make the interpretation of medical images very difficult. They are responsible for the limited success of conventional or traditional detection and analysis algorithms. Poor image quality invariably leads to problematic and unreliable feature extraction, analysis and recognition in many medical applications. Research efforts are geared towards improving the quality of the images, and finding more robust techniques to successfully handle images of compromised quality in various applications.

1.2 Techniques

The major strength in the application of computers to medical imaging lies in the use of image processing techniques for quantitative analysis. Medical images are primarily visual in nature; however, visual analysis by human observers is usually associated with limitations caused by interobserver variations and errors due to fatigue, distractions, and limited experience. While the interpretation of an image by an expert draws from his/her experience and expertise, there is almost always a subjective element. Computer analysis, if performed with the appropriate care and logic, can potentially add objective strength to the interpretation of the expert. Thus, it becomes possible to improve the diagnostic accuracy and confidence of even an expert with many years of experience.

Imaging science has expanded primarily along three distinct but related lines of investigation: segmentation, registration and visualization [6]. Segmentation, particularly in three dimensions, remains the holy grail of imaging science. It is the important yet elusive capability to accurately recognize and delineate all the individual objects in an image scene. Registration involves finding the transformation that brings different images of the same object into strict spatial (and/or temporal) congruence. And visualization involves the display, manipulation, and measurement of image data. Important advances in these three areas will be outlined in the various chapters in this book.

A common theme throughout this book is the differentiation and integration of images. On the one hand, automatic segmentation and classification of tissues provide the required differentiation, and on the other the fusion of complementary images provides the integration required to advance our understanding of life processes and disease. Measurement of both form and function, of the whole image and at the individual pixel level, and the ways to display and manipulate digital images are the keys to extracting the clinical information contained in biomedical images. The need for new techniques becomes more pressing as improvements in imaging technologies enable more complex objects to be imaged and simulated.

1.3 Applications

The approach required is primarily that of problem solving. However, the understanding of the problem can often require a significant amount of preparatory work. The applications chosen for this book are typical of those in medical imaging; they are meant to be exemplary, not exclusive. Indeed, it is hoped that many of the solutions presented will be transferable to other problems. Each application begins with a statement of the problem, and includes illustrations with real-life images. Image processing techniques are presented, starting with relatively simple generic methods, followed by more sophisticated approaches directed at that specific problem. The benefits and challenges in the transition from research to clinical solution are also addressed.

Biomedical imaging is primarily an applied science, where the principles of imaging science are applied to diagnose and treat disease, and to gain basic insights into the processes of life. The development of such capabilities in the research laboratory is a time-honored tradition. The challenge is to make new techniques available outside the specific laboratory that developed them, so that others can use and adapt them to different applications. The ideas, skills and talents of specific developers can then be shared with a wider community and this will hopefully facilitate the transition of successful research technique into routine clinical use.

1.4 The Contribution of This Book

Computer hardware and software have developed to the point where large images can be analyzed quickly and at moderate cost. Automated processes, including pattern recognition and computer-assisted diagnosis (CAD), can be effectively implemented on a personal computer. This chapter has outlined some of the successes in the field, and brought attention to some of the remaining problems. The following chapters will describe in detail some of the techniques and applications that are currently being used by experts in the field.

Medical imaging is very visual. Although the formalism of the techniques and algorithms is mathematical, we understand the advantages offered through visualization. Therefore, this book offers many images and diagrams. Some are for pedagogical purposes, to assist with the exposition, and others are motivational, to reveal interesting features of particular applications.

The book is a collection of chapters, written by experts in the field of image analysis, in a style to build intuition, insight, and understanding. Each chapter represents the state-of-the-art wisdom in a particular subfield, the result of ongoing, world-wide collaborative efforts over a period of time. Although the chapters are essentially self-contained they reference other chapters to form an integrated whole. Each chapter employs a pedagogical approach to ensure conceptual learning before introducing specific techniques and "tricks of the trade." The book aims to address recent methodological advances in imaging techniques by demonstrating how they can be applied to a selection of topical applications. It is hoped that this will empower the reader to experiment with and use the techniques in his/her own research area and beyond.

Chapter 2 describes an intuitive toolbox for MatLab, called DipImage, and demonstrates how it can be used to count cancerous cells in a Pap smear.

Chapter 3 introduces new approaches that unify discrete segmentation techniques, Chapter 4 shows how deformable models can be used with noisy images, and Chapter 5 applies a number of automated and semi-automated segmentation methods to MRI images.

Automated detection of linear structures is a common challenge in many computer vision applications. Chapters 6–8 describe state-of-the-art techniques and apply them to a number of biomedical systems.

Image processing methods are applied to osteoporosis in Chapter 9, to idiopathic scoliosis in Chapter 10, and to retinal pathologies in Chapters 11 and 12.

Novel volume rendering algorithms are discussed in Chapter 13.

Sparse sampling algorithms in MRI are presented in Chapter 14, and the visualization of diffusion tensor images of avascular tissues is discussed in Chapter 14.

References

1. Dougherty, G.: Image analysis in medical imaging: recent advances in selected examples. Biomed. Imaging Interv. J. 6(3), e32 (2010)
2. Beutel, J., Kundel, H.L., Van Metter, R.L.: Handbook of Medical Imaging, vol. 1. SPIE, Bellingham, Washington (2000)
3. Rangayyan, R.M.: Biomedical Image Analysis. CRC, Boca Raton, FL (2005)
4. Meyer-Base, A.: Pattern Recognition for Medical Imaging. Elsevier Academic, San Diego, CA (2004)
5. Dougherty, G.: Digital Image Processing for Medical Applications. Cambridge University Press, Cambridge (2009)
6. Robb, R.A.: Biomedical Imaging, Visualization and Analysis. Wiley-Liss, New York (2000)

Chapter 2
Rapid Prototyping of Image Analysis Applications

Cris L. Luengo Hendriks, Patrik Malm, and Ewert Bengtsson

2.1 Introduction

When developing a program to automate an image analysis task, one does not start with a blank slate. Far from it. Many useful algorithms have been described in the literature, and implemented countless times. When developing an image analysis program, experience points the programmer to one or several of these algorithms. The programmer then needs to try out various possible combinations of algorithms before finding a satisfactory solution. Having to implement these algorithms just to see if they work for this one particular application does not make much sense. This is the reason programmers and researches build up libraries of routines that they have implemented in the past, and draw on these libraries to be able to quickly string together a few algorithms and see how they work on the current application. Several image analysis packages exist, both commercial and free, and they can be used as a basis for building up such a library. None of these packages will contain all the necessary algorithms, but they should provide at least the most basic ones. This chapter introduces you to one such package, DIPimage, and demonstrates how one can proceed to quickly develop a solution to automate a routine medical task. As an illustrative example we use some of the approaches taken over the years to solve the long-standing classical medical image analysis problem of assessing a Pap smear. To make best use of this chapter, you should have MATLAB and DIPimage running on your computer, and try out the command sequences given.

C.L. Luengo Hendriks (✉)
Centre for Image Analysis, Swedish University of Agricultural Sciences,
Box 337, SE-751 05 Uppsala, Sweden
e-mail: cris@cb.uu.se

G. Dougherty (ed.), *Medical Image Processing: Techniques and Applications*, Biological and Medical Physics, Biomedical Engineering, DOI 10.1007/978-1-4419-9779-1_2,

2.2 MATLAB and DIPimage

2.2.1 *The Basics*

DIPimage is built on MATLAB (The MathWorks, Natick, MA, USA), which provides a powerful and intuitive programming language, publication-quality graphing, and a very extensive set of algorithms and tools. DIPimage adds to this a large collection of image processing and analysis algorithms, easy computation with images, and interactive graphical tools to examine images. It is designed for ease of use, using MATLAB's simple command syntax and several graphical user interfaces, and is accessible to novices but fast and powerful enough for the most advanced research projects. It is available free of charge for academic and other noncommercial purposes from its website: http://www.diplib.org/.

DIPimage extends the MATLAB language with a new data type. Natively, MATLAB knows about arrays of numbers, characters, structures or cells (the latter can contain any other data type). With this toolbox installed, images are added to this list. Even though images can be seen simply as an array of numbers, there are several advantages to this new type: indexing works differently than in an array, the toolbox can alter the way operations are performed depending on the pixel representation, and images can be automatically displayed. This latter point is significant, for it greatly enhances accessibility to novices and significantly increases the interactivity in the design phase of an image analysis application.

In MATLAB, assigning the value `1` to a variable `a` is accomplished with:

```
a = 1;
```

Additionally, if the semicolon is left off this statement, MATLAB will reply by displaying the new value of the variable:

```
a = 1
a =
     1
```

Similarly, when leaving the semicolon off a DIPimage statement that assigns an image into a variable, MATLAB will reply by displaying that image in a figure window. For example, the next statement reads in the Pap smear image in file "papsmear.tif"[1] and assigns it to variable `a`.

```
a = readim('papsmear.tif')
Displayed in figure 10
```

Depending on the chosen configuration, the image will be displayed to a new window or an existing window. To suppress automatic display, all that is thus needed is to add a semicolon at the end of all statements.

[1]You can obtain this file from http://www.cb.uu.se/~cris/Images/papsmear.tif

DIPimage features a graphical user interface (GUI) that gives access to the most commonly used functions in the toolbox (if the GUI does not appear by default in your installation of DIPimage, run the command `dipimage` in the MATLAB command window). The GUI's menus list all available functions. Selecting one of these functions changes the area below the menus to allow parameter selection and execution of the function. For example, the function used above, `readim`, is available under the "File" menu. Selecting it brings up a control to select the file to read and choose a name for the variable that will hold the image. Pressing the "Execute" button will read the selected file and put its contents into the chosen variable. Additionally, the result of the command is displayed.

2.2.2 *Interactive Examination of an Image*

The figure windows in which the images are automatically displayed have four menus. The second and third ones allow the user to change the way the image is displayed. Note that some menu options are only present when applicable. For example, the "Mappings" menu has options to choose the slicing direction in 3D and 4D images, which are not visible with 1D or 2D images; two- or higher-dimensional, gray-value images have options to select a color map, which are hidden for color images. The fourth menu, "Actions," contains all the interactive tools that a user can use to examine the image in the display. The "Action" enabled by default is "Pixel testing," which allows the user to hold the left mouse button down to get a reading of the values of the pixel under the cursor. The title bar of the figure window shows the coordinates and either the gray value or the RGB values of the pixel. Holding down the right mouse button allows the user to measure distances. The "Zoom" and "Pan" modes allow closer examination of large images. For 3D and 4D images, several additional options exist. "Step through slices" is to use the mouse to change the slice of the image shown (it is also possible to do this with the keyboard, without changing the mode). Most interestingly, "Link displays" can be used to link various windows displaying 3D or 4D images. These windows will then all show the same slice at all times. This is very useful when, for example, comparing the output of various filters. We encourage the reader to explore these options and read more about them in the user manual [1].

2.2.3 *Filtering and Measuring*

A large selection of filters and analysis tools are available through the DIPimage GUI. Many more are accessible only from the command line, and are consequently hidden. Typing

```
help dipimage
```

gives an overview of (almost) all functions in DIPimage. For more information on any one function, use the `help` command with the function's name. We will stick

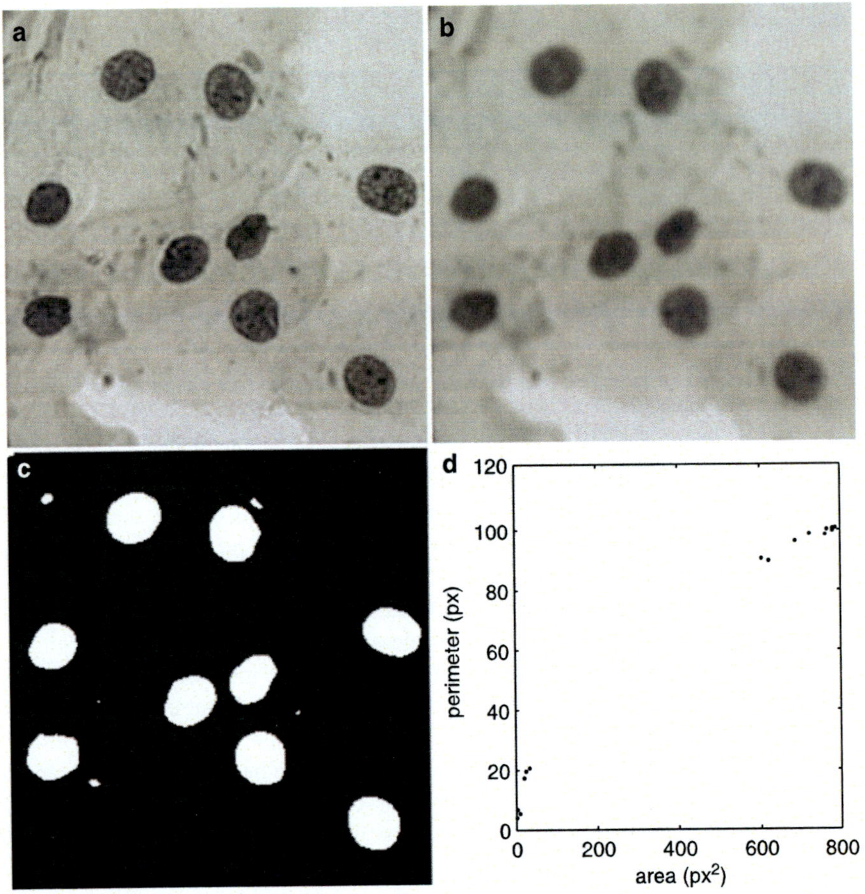

Fig. 2.1 A first look at DIPimage: (**a**) input image, (**b**) result of gaussf, (**c**) result of threshold, and (**d**) plot of measured area vs. perimeter

to the functions in the GUI for now. Let us assume that we still have the Pap smear image loaded in variable `a`. We select the "Gaussian filter" from the "Filters" menu. For the first parameter we select variable a, for the second parameter we enter 2, and for output image we type `b`. After clicking "Execute," we see

```
b = gaussf(a,2,'best')
Displayed in figure 11
```

on the command line, and the filtered image is shown in a window (Fig. 2.1b). Typing the above command would have produced the same result, without the need for the GUI. Because `'best'` is the default value for the third parameter, the same result would also have been accomplished typing only

```
b = gaussf(a,2)
```

Next we will select the "Threshold" function from the "Segmentation" menu, enter `-b` for input image and `c` for output image, and execute the command. We now have a binarized image, where the nuclei are marked as objects (red) and the rest as background (Fig. 2.1c). If we had left the minus sign out of the input to the threshold, the output would have been inverted. Finally, we select the "Measure" function from the "Analysis" menu, use `c` as the first input image, select several measurements by clicking on the "Select..." button (for example: "size," "center" and "perimeter," hold the control key down while clicking the options to select more than one), and execute the command. On the command window we will now see the result of the measurements. We can generate a plot of surface area ("size") vs. perimeter (Fig. 2.1d) by typing

```
figure, plot(msr.size, msr.perimeter,'.')
```

Note several small objects are found that are obviously not nuclei. Based on the size measure these can be discarded. We will see more of this type of logic in Sect. 2.4.

2.2.4 Scripting

If you collect a sequence of commands in a plain text file, and save that file with a ".m" extension, you have created a script. This script can be run simply by typing its name (without the extension) at the MATLAB command prompt. For example, if we create a file "analyse.m" with the following content:

```
a = readim('papsmear.tif');
b = smooth(a,2);
c = threshold(-b);
msr = measure(c,[],{'Size','Center','Perimeter'});
figure, plot(msr.size,msr.perimeter,'.')
```

then we can execute the whole analysis in this section by just typing

```
analyse
```

It is fairly easy to collect a sequence of commands to solve an application in such a file, execute the script to see how well it works, and modify the script incrementally. Because the GUI prints the executed command to the command line, it is possible to copy and paste the command to the script. The script works as both a record of the sequence of commands performed, and a way to reuse solutions. Often, when trying to solve a problem, one will start with the working solution to an old problem. Furthermore, it is easier to write programs that require loops and complex logic in a text editor than directly at the command prompt.

If you are a programmer, then such a script is obvious. However, compared to many other languages that require a compilation step, the advantage with MATLAB is that you can select one or a few commands and execute them independently of the rest of the script. You can execute the program line by line, and if the result of one line is not as expected, modify that line and execute it again, without having to run the whole script anew. This leads to huge time savings while developing new algorithms, especially if the input images are large and the analysis takes a lot of time.

2.3 Cervical Cancer and the Pap Smear

Cervical cancer is one of the most common cancers for women, killing about a quarter million women world-wide every year. In the 1940s, Papanicolaou discovered that vaginal smears can be used to detect the disease at an early, curable stage [2]. Such smears have since then commonly been referred to as Pap smears. Screening for cervical cancer has drastically reduced the death rate for this disease in the parts of the world where it has been applied [3]. Mass screens are possible because obtaining the samples is relatively simple and painless, and the equipment needed is inexpensive.

The Pap smear is obtained by collecting cells from the cervix surface (typically using a spatula), and spreading them thinly (by smearing) on a microscope slide (Fig. 2.2). The sample is then stained and analyzed under the microscope by a cytotechnologist. This person needs to scan the whole slide looking for abnormal cells, which is a tedious task because a few thousand microscopic fields of view need to be scrutinized, looking for the potentially few abnormal cells among the

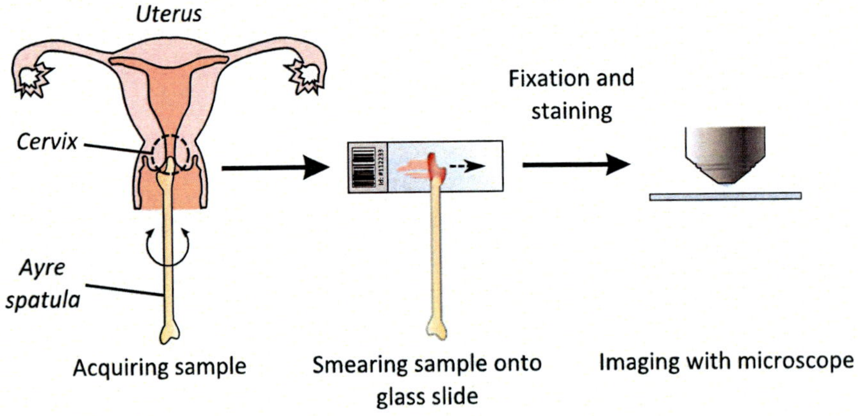

Fig. 2.2 A Pap smear is obtained by thinly smearing collected cells onto a glass microscope slide

several hundred thousand cells that a slide typically contains. This work is made even more difficult due to numerous artifacts: overlapping cells, mucus, blood, etc. The desire to automate the screening has always been great, for all the obvious reasons: trained personnel are expensive, they get tired, their evaluation criteria change over time, etc. The large number of images to analyze for a single slide, together with the artifacts, has made automation a very difficult problem that has occupied numerous image analysis researchers over the decades.

2.4 An Interactive, Partial History of Automated Cervical Cytology

This section presents an "interactive history," meaning that the description of methods is augmented with bits of code that you, the reader, can try out for yourself. This both makes the descriptions easier to follow, and illustrates the use of DIPimage to quickly and easily implement a method from the literature. This section is only a partial history, meaning that we show the highlights but do not attempt to cover everything; we simplify methods to their essence, and focus only on the image analysis techniques, ignoring imaging, classification, etc. For a somewhat more extensive description of the history of this field see for instance the paper by Bengtsson [4].

2.4.1 The 1950s

The first attempt at automation of Pap smear assessment was based on the observation that cancer cells are typically bigger, with a greater amount of stained material, than normal cells. Some studies showed that all samples from a patient with cancer had at least some cells with a diameter greater than 12 μm, while no normal cells were that large. And thus a system, the cytoanalyzer, was developed that thresholded the image at a fixed level (Fig. 2.3a), and measured the area (counted the pixels) and the integrated optical density (summed gray values) for each connected component [5]. To replicate this is rather straightforward, and very similar to what we did in Sect. 2.2.3:

```
a = readim('papsmear.tif');
b = a<128;
msr = measure(b,a,{'Size','Sum'});
```

As you can see, this only works for very carefully prepared samples. Places where multiple cytoplasms overlap result in improper segmentation, creating false large regions that would be identified as cancerous. Furthermore, the threshold value of 128 that we selected for this image is not necessarily valid for other images. This

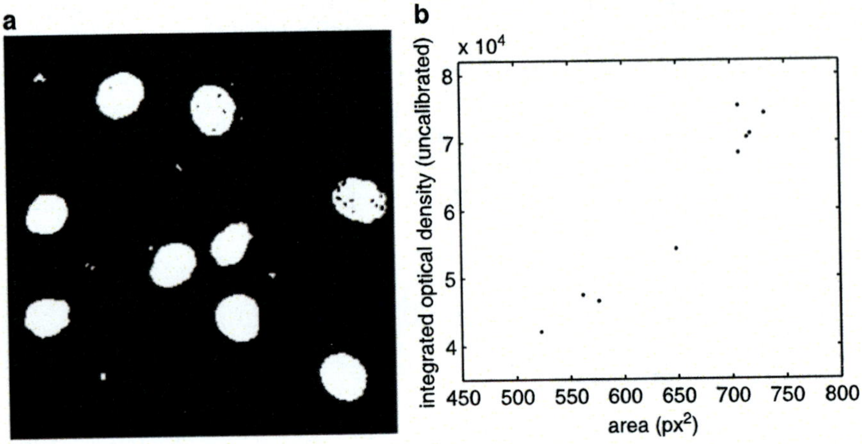

Fig. 2.3 (**a**) Segmented Pap-smear image and (**b**) plot of measured area vs. integrated optical density

requires strong control over the sample preparation and imaging to make sure the intensities across images are constant.

The pixel size in this machine was about 2 μm, meaning that it looked for segmented regions with a diameter above 6 pixels. For our image this would be about 45 pixels. The resulting data was analyzed as 2D scatter plots and if signals fell in the appropriate region of the plot the specimen was called abnormal (Fig. 2.3b):

```
figure, plot(msr.size,msr.sum,'.')
```

All the processing was done in hardwired, analog, video processing circuits. The machine could have worked if the specimens only contained well-preserved, single, free-lying cells. But the true signal was swamped by false signals from small clumps of cells, blood cells, and other debris [6].

2.4.2 The 1960s

One of the limitations of the cytoanalyzer was the fixed thresholding; it was very sensitive to proper staining and proper system setup. Judith Prewitt (known for her local gradient operator) did careful studies of digitized cell images and came up with the idea of looking at the histogram of the cell image [7]. Although this work was focused on the identification of red blood cells, the method found application in all other kinds of (cell) image analysis, including Pap smear assessment.

Assuming three regions with different intensity (nucleus, cytoplasm, and background), we would expect three peaks in the histogram (Fig. 2.4a). For simple shapes, we expect fewer pixels on the border between the regions than in the

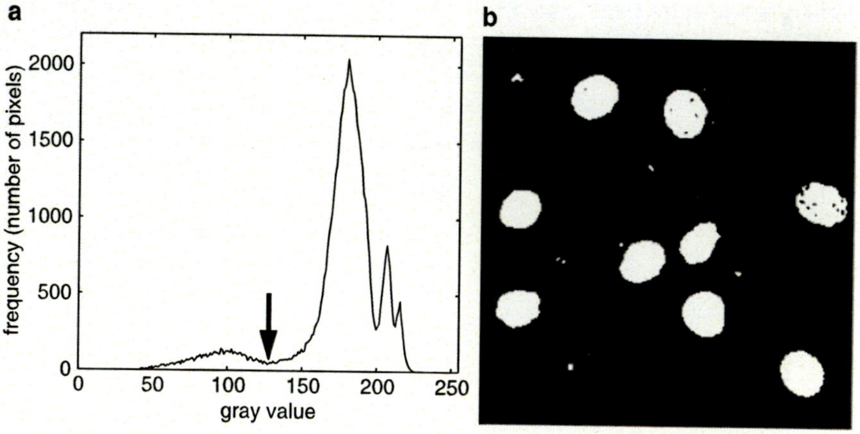

Fig. 2.4 (**a**) Histogram of Pap-smear image, with calculated threshold and (**b**) segmented image

regions themselves, meaning that there would be two local minima in between these three peaks, corresponding to the gray values of the pixels forming the borders. The two gray values corresponding to these two local minima are therefore good candidates for thresholding the image, thereby classifying each pixel into one of the three classes (Fig. 2.4b). Detecting these two local minima requires simplifying the histogram slightly, for example by a low-pass filter, to remove all the local minima caused by noise:

```
a = readim('papsmear.tif');
h = diphist(a);        % obtain a histogram
h = gaussf(h,3);       % smooth the histogram
t = minima(h);         % detect the local minima
t(h==0) = 0;           % mask out the minima at the tails
t = find(t)            % get coordinates of minima
a < t(1)               % threshold the image at the first
                         local minimum
```

Basically, this method substitutes the fixed threshold of the cytoanalyzer with a smoothing parameter for the histogram. If this smoothing value is taken too small, we find many more than two local minima; if it is too large, we do not find any minima. However, the results are less sensitive to the exact value of this parameter, because a whole range of smoothing values allows the detection of the two minima, and the resulting threshold levels are not affected too much by the smoothing. And, of course, it is possible to write a simple algorithm that finds a smoothing value such that there are exactly two local minima:

```
h = diphist(a);        % obtain a histogram
t = [0,0,0];           % initialize threshold array
while length(t)>2      % repeat until we have 2 local
                         minima
```

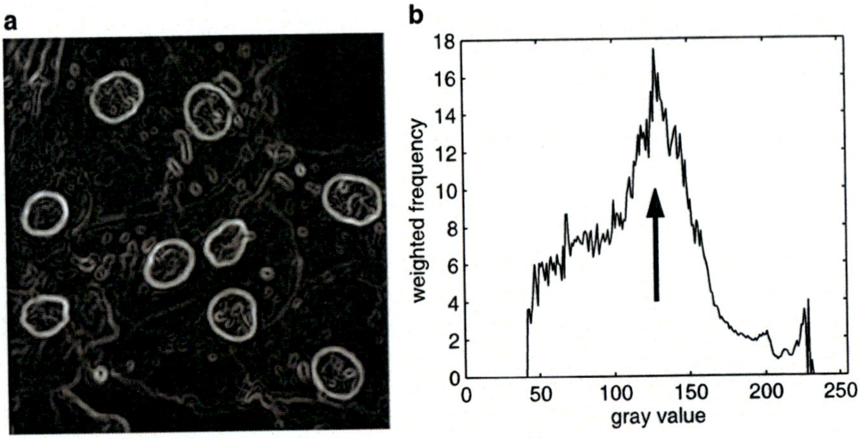

Fig. 2.5 (**a**) Gradient magnitude of Pap-smear image and (**b**) differential histogram computed from the Pap-smear image and its gradient magnitude

```
        h = gaussf(h,1); % (the code inside the loop is
                           identical to that used above)
        t = minima(h);
        t(h==0)=0;
        t = find(t);
    end
```

The loop then repeats the original code, smoothing the histogram more and more, until at most two local minima are found.

2.4.3 The 1970s

In the late 1960s, a group at Toshiba, in Japan, started working on developing a Pap smear screening machine they called CYBEST. They used a differential histogram approach for the automated thresholding [8,9]. That is, they computed a histogram weighted by a measure of edge strength; pixels on edges contribute more strongly to this histogram than pixels in flat areas. Peaks in this histogram indicate gray values that occur often on edges (Fig. 2.5). Though CYBEST used a different scheme to compute edge strength, we will simply use the Gaussian gradient magnitude (`gradmag`).

```
a = readim('papsmear.tif');
b = gradmag(a);
h = zeros(255,1);     % initialize array
for i = 1:255
    t = a==i;     % t is a binary mask
    n = sum(t); % n counts number of pixels with value i
```

```
      if n>0
      h(i) = sum(b(t))/n;      % average gradient at pixels
                                 with value i
      end
   end
   h = gaussf(h,2);      % smooth differential histogram
   [~,t] = max(h);       % find location of maximum
   a < t    % threshold
```

A second peak in this histogram gives a threshold to distinguish cytoplasm from background, much like in Prewitt's method.

This group studied which features were useful for analyzing the slides and ended up using four features [10]: nuclear area, nuclear density, cytoplasmic area, and nuclear/cytoplasmic ratio. These measures can be easily obtained with the `measure` function as shown before. They also realized that nuclear shape and chromatin pattern were useful parameters but were not able to reliably measure these features automatically, mainly because the automatic focusing was unable to consistently produce images with all the cell nuclei in perfect focus. Nuclear shape was determined as the square of the boundary length divided by the surface area. Determining the boundary length is even more sensitive to a correct segmentation than surface area. The `measure` function can measure the boundary length (`'perimeter'`), as well as the *shape factor* (`'p2a'`). The shape factor, computed by $perimeter^2/(4\pi area)$, is identical to CYBEST's nuclear shape measure, except it is normalized to be 1 for a perfect circle. The chromatin pattern measure that was proposed by this group and implemented in CYBEST Model 4 is simply the number of blobs within the nuclear region [11]. For example (using the `a` and `t` from above):

```
m = gaussf(a) < t;      % detect nuclei
m = label(m)==3;        % pick one nucleus
m = (a < mean(a(m))) & m;    % detect regions within
                               nucleus
max(label(m))       % count number of regions
```

Here, we just used the average gray value within the nucleus as the threshold, and counted the connected components (Fig. 2.6). The procedure used in CYBEST was more complex, but not well described in the literature.

The CYBEST system was developed in four generations over two decades, and tested extensively, even in full scale clinical trials, but was not commercially successful.

2.4.4 The 1980s

In the late 1970s and 1980s, several groups in Europe were working on developing systems similar to CYBEST, all based on the so-called "rare event model," that is,

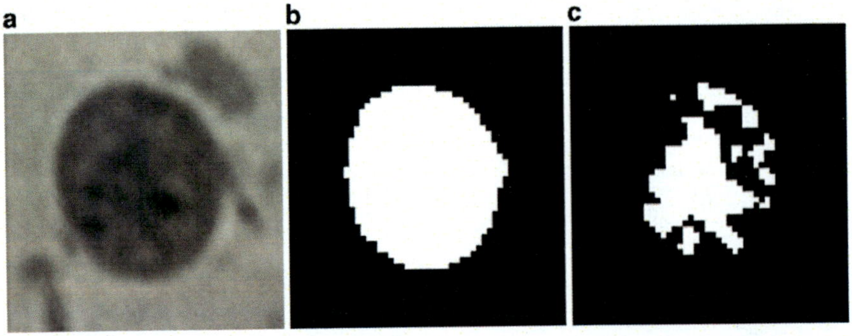

Fig. 2.6 A simple chromatin pattern measure: (**a**) the nucleus, (**b**) the nucleus mask, and (**c**) the high-chromatin region mask within that nucleus

looking for the few large, dark, diagnostic cells among the few hundred thousand normal cells. And these systems had to do this sufficiently fast, while avoiding being swamped by false alarms due to misclassifications of overlapping cells and small clumps of various kinds.

As a side effect of the experimentation on feature extraction and classification, carried out as part of this research effort, a new concept emerged. Several groups working in the field observed that even "normal" cells on smears from patients with cancer had statistically significant shifts in their features towards the abnormal cells. Even though these shifts were not strong enough to be useful on the individual cell level, it made it possible to detect abnormal specimens through a statistical analysis of the feature distributions of a small population, a few hundred cells, provided these features were extracted very accurately. This phenomenon came to be known as MAC, malignancy associated changes [12]. The effect was clearly most prominent in the chromatin pattern in the cell nuclei. The CYBEST group had earlier noted that it was very difficult to extract features describing the chromatin pattern reliably in an automated system. A group at the British Colombia Cancer Research Centre in Vancouver took up this idea and developed some very careful cell segmentation and chromatin feature extraction algorithms.

To accurately measure the chromatin pattern, one first needs an accurate delineation of the nucleus. Instead of using a single, global threshold to determine the nuclear boundary, a group in British Colombia used the gradient magnitude to accurately place the object boundary [13]. They start with a rough segmentation, and selected a band around the boundary of the object in which the real boundary must be (Fig. 2.7a):

```
a = readim('papsmear.tif');
b = gaussf(a,2)<128;     % quick-and-dirty threshold
c = b-berosion(b,1);     % border pixels
c = bdilation(c,3);  % broader region around border
```

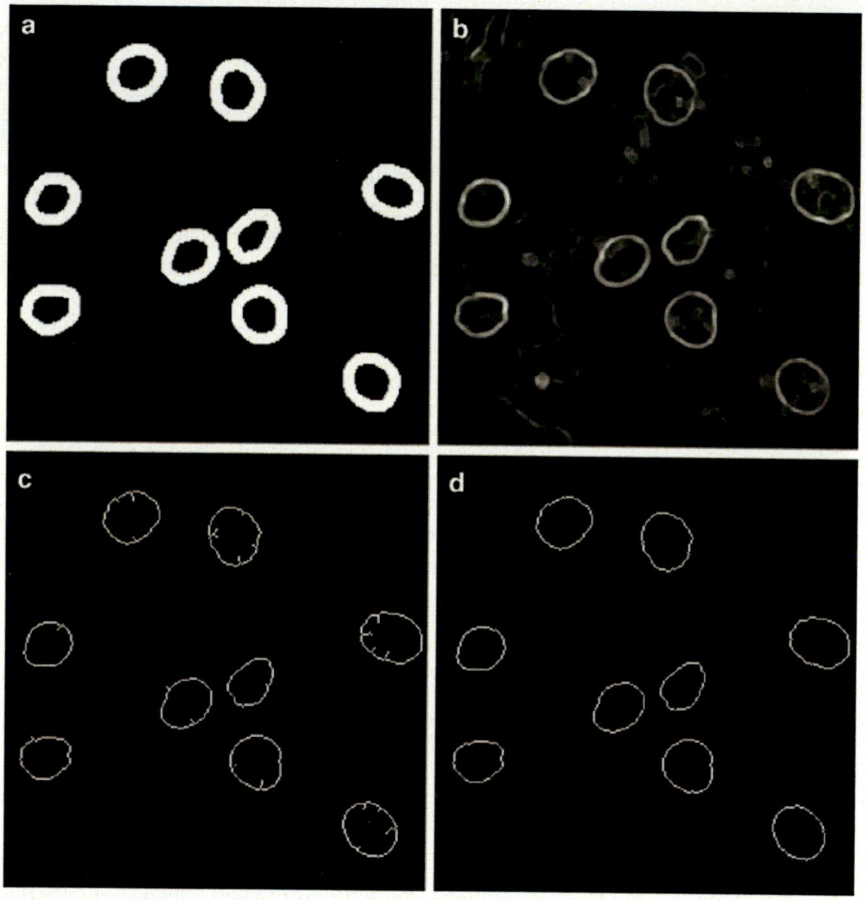

Fig. 2.7 Accurate delineation of the nucleus: (**a**) boundary regions, (**b**) gradient magnitude, (**c**) upper skeleton, and (**d**) accurate boundaries

Now comes the interesting part: a conditional erosion that is topology preserving (like the binary skeleton), but processes pixels in order of the gray value of the gradient magnitude (Fig. 2.7b), low gray values first. This implies that the skeleton will lie on the ridges of the gradient magnitude image, rather than on the medial axis of the binary shape. This operation is identical to the upper skeleton or upper thinning, the gray-value equivalent of the skeleton operation [14], except that the latter does not prune terminal branches nor isolated pixels (Fig. 2.7c). We can prune these elements with two additional commands (Fig. 2.7d):

```
g = gradmag(a);
g = closing(g,5);    % reducing number of local minima
                       in g
d = dip_upperskeleton2d(g*c);
      % compute upper skeleton in border region only
```

```
d = bskeleton(d,0,'looseendsaway');
     % prune terminal branches
d = d -getsinglepixel(d);    % prune isolated pixels
```

It is interesting to note, the upper skeleton is related to the watershed in that both find the ridges of the gray value image. We could have used the function `watershed` to obtain the exact same result.

We now have an accurate delineation of the contour. The following commands create seeds from the initial segmentation, and grow them to fill the detected contours:

```
e = b & ~c;
e = bpropagation(e,~d,0,1,0);
```

Because the pixels comprising the contours are exactly on the object edge, we need an additional step to assign each of these pixels to either the background or the foreground. In the paper, the authors suggest two methods based on the gray value of the pixel, but do not say which one is better. We will use option 1: compare the border pixel's value to the average for all the nuclei and the average for all the background, and assign it to whichever class it is closest:

```
gv_nuc = mean(a(e));      % average nuclear gray value
gv_bgr = mean(a(~(d|e)));     % average background gray
                                value
t = a < (gv_nuc+gv_bgr)/2;    % threshold halfway between
                                the two averages
e(d) = t(d);     % reassign border pixels only
```

The other option is to compare each border pixel with background and foreground pixels in the neighborhood, and will likely yield a slightly better result for most cells.

A very large group of features were proposed to describe each nucleus. Based on the outline alone, one can use the mean and maximum radius, sphericity, eccentricity, compactness, elongation, etc., as well as Fourier descriptors [15]. Simple statistics of gray values within one nucleus are maximum, minimum, mean, variance, skewness, and kurtosis. Texture features included contour analysis and region count after thresholding the nucleus into areas of high, medium and low chromatin content; statistics on the co-occurrence matrix [16] and run lengths [17]; and the fractal dimension[18]. The fractal dimension is computed from the fractal area measured at different resolutions, and gives an indication of how the image behavior changes with scale. The fractal area is calculated with:

```
fs = 1 + abs(dip_finitedifference(a,0,'m110')) + ...
     abs(dip_finitedifference(a,1,'m110'));
m = label(e)==4;     % pick one nucleus
sum(fs(m))    % sum values of fs within nucleus
```

The function dip_finitedifference calculates the difference between neighboring pixels, and is equivalent to MATLAB's function diff, except it returns an image of the same size as the input.

Multivariate statistical methods were finally used to select the best combination of features to determine whether the cell population was normal or from a slide influenced by cancer.

2.4.5 The 1990s

In the 1990s, research finally lead to successful commercial systems being introduced: AutoPap [19] and PAPNET [20]. They built on much of the earlier research, but two concepts were extensively used in both of these commercial systems: mathematical morphology and neural networks. In short, what the PAPNET systems did was detect the location of nuclei of interest, extract a square region of fixed size around this object, and use that as input to a neural network that classified the object as debris/benign/malignant [21]. Using such a system, these machines avoided the issues of difficulty in segmentation, careful delineation, and accurate measurement. Instead, the neural network does all the work. It is trained with a large collection of nuclei that are manually classified, and is then able to assign new nuclei to one of the classes it was trained for. However, the neural network is a "black box" of sorts, in that it is not possible to know what features of the nucleus it is looking at to make the decision [22]. Slightly simplified, the method to extract fixed-sized image regions containing a nucleus is as follows:

```
a = readim('papsmear.tif');
b = gaussf(a); % slight smoothing of the image
b = closing(b,50)-b;    % top-hat, max. diameter is 50
                          pixels
c = threshold(b);
```

The closing minus the input image is a top-hat, a morphological operation that eliminates large dark regions. The result, c, is a mask where all the large objects (nuclei clusters, for example) have been removed. But we also want to make sure we only look at objects that are dark in the original image:

```
c = c & ~threshold(a);
```

c now contains only objects that are dark and small (Fig. 2.8a). The next step is to remove the smallest objects and any object without a well-defined edge:

```
d = gradmag(a,3);
d = threshold(d);     % detect strong edges
d = brmedgeobjs(~d);    % find inner regions
```

The first step finds regions of large gradient magnitude. In the second step we invert that mask and remove the part that is connected to the image border.

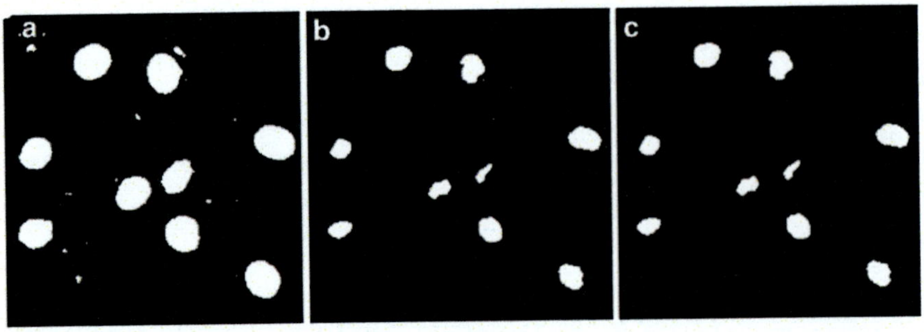

Fig. 2.8 Detecting the location of possible nuclei: (**a**) all dark, small regions, (**b**) all regions surrounded by strong edges, and (**c**) the combination of the two

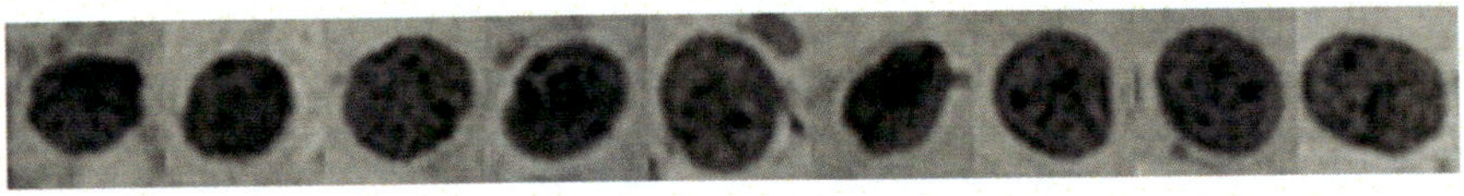

Fig. 2.9 Extracted regions around potential nuclei. These subimages are the input to a neural network

The regions that remain are all surrounded by strong edges (Fig. 2.8b). The combination of the two partial results,

```
e = c & d;
```

contains markers for all the medium-sized objects with a strong edge (Fig. 2.8c). These are the objects we want to pass on to the neural network for classification. We shrink these markers to single-pixel dots and extract an area around each dot (Fig. 2.9):

```
e = bskeleton(e,0,'looseendsaway');
    % reduce each region to a single dot
coords = findcoord(e);    %get the coordinates for each
                              dot
N = size(coords,1); % number of dots
reg = cell(N); % we'll store the little regions in here
for ii=1:N
    x = coords(ii,1);
    y = coords(ii,2);
    reg{ii} = a(x-20:x+20,y-20:y+20);
        % the indexing cuts a region from the image
end
reg = cat(1,reg{:}) % glue regions together for display
```

That last command just glues all the small regions together for display. Note that, for simplicity, we did not check the `coords` array to avoid out-of-bounds indexing when extracting the regions. One would probably want to exclude regions that fall partially outside the image.

The PAPNET system recorded the 64 most "malignant looking" regions of the slide for human inspection and verification, in a similar way to what we did just for our single image field.

2.4.6 *The 2000s*

There was a reduced academic research activity in the field after the commercial developments took over in the 1990s. But there was clearly room for improvements, so some groups continued basic research. This period is marked by the departure from the "black box" solutions, and a return to accurate measurements of specific cellular and nuclear features. One particularly active group was located in Brisbane, Australia [23]. They applied several more modern concepts to the Pap smears and demonstrated that improved results could be achieved. For instance, they took the dynamic contour concept (better known as the "snake," see Chapter 4) and applied it to cell segmentation [24]. Their snake algorithm is rather complex, since they used a method to find the optimal solution to the equation, rather than the iterative approach usually associated with snakes, which can get stuck in local minima. Using "normal" snakes, one can refine nuclear boundary thus:

```
a = readim('papsmear.tif')
b = bopening(threshold(-a),5);  % quick-and-dirty
   %segmentation
c = label(b);     % label the nuclei
N = max(c);       % number of nuclei
s = cell(N,1);   % this will hold all snakes
vf = vfc(gradmag(a));     % this is the snake's ''external
                                 force''
for ii = 1:N    % we compute the snake for each nucleus
                   separately
   ini = im2snake(c==ii);    % initial snake given by
                                 segmentation
   s{ii} = snakeminimize(ini,vf,0.1,2,1,0,10);
       % move snake so its energy is minimized
   snakedraw(s{ii}) % overlays the snake on the image
end
```

The snake is initialized by a rough segmentation of the nuclei (Fig. 2.10a), then refined by an iterative energy minimization procedure (`snakeminimize`, Fig. 2.10b). Note that we used the function `vfc` to compute the *external force*, the image that drives the snake towards the edges of the objects. This VFC (vector field

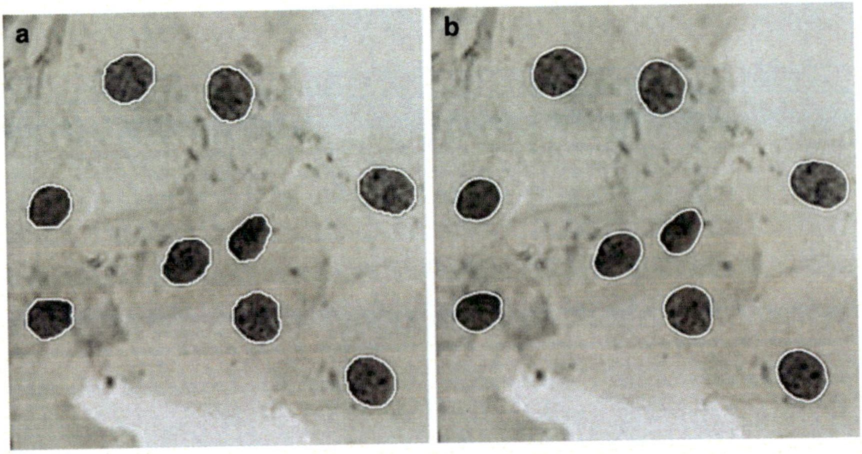

Fig. 2.10 (**a**) Initial snake and (**b**) final snake in active contour method to accurately delineate nuclei

convolution) approach is a recent improvement to the traditional snake [25]. For this example image, the results are rather similar when using the traditional gradient, because the initial snake position is close to its optimal. Also note the large number of parameters used as input to the function `snakeminimize`, the function that moves the control points of the snake to minimize the snake's energy function. This large number of parameters (corresponding to the various weights in the energy function) indicates a potential problem with this type of approach: many values need to be set correctly for the method to work optimally, and thus this particular program is only applicable to images obtained under specific circumstances.

2.5 The Future of Automated Cytology

An interesting observation that can be made from the brief samples of the long history of automated cervical cytology that has been presented in this chapter is that the focus of the research in the field has been moving around the world about once every decade – Eastern USA, Japan, Europe, Western USA/Canada, and Australia (Fig. 2.11) – although there are, of course, outliers to this pattern. This pattern might have arisen because of the apparent ease of the problem: when researchers in one region have been making strong promises of progress for too long, without being able to deliver on these promises, it becomes increasingly difficult to obtain more research funds in that region; researchers in a different region in the world are then able to take the lead.

One significant development that we have not discussed so far is the efforts of producing cleaner specimens that are more easy to analyze than the smears, which can be very uneven in thickness and general presentation of the cells. These

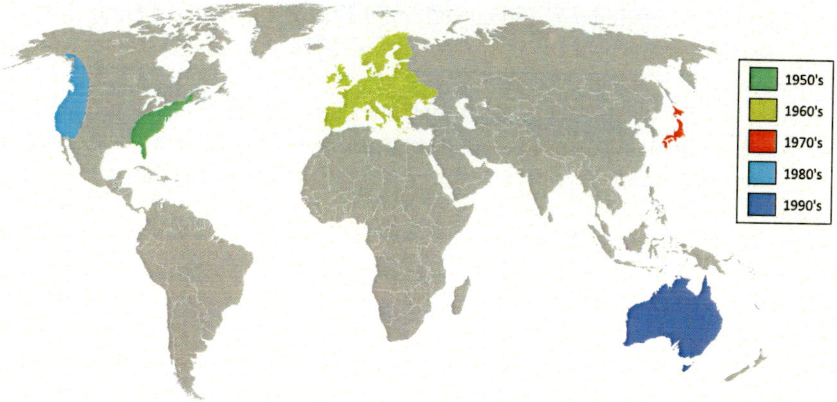

Fig. 2.11 A map of the world showing the location of major Pap smear analysis automation research over the decades

efforts led to two commercial liquid cytology systems in the 1990s [26, 27]. The companies behind these sample preparation devices have also developed dedicated image analysis systems. These devices work well, but the modified kits for preparing the specimens are so expensive that most of the economic gain from automation disappears.

The cost of automation is an important issue. One of the original motivations for developing automation was the high cost of visual screening. Still, the first generation automated systems were very complex and expensive machines costing about as much to operate as visual screening. The need for modified sample preparation for some systems added to these costs. A third aspect was that the combined effect of visual and machine screening gives a higher probability of actually detecting a lesion than either one alone, making it hard in some countries, due to legal liability reasons, to use automation alone even if it is comparable in performance to visual screening. All of this has made the impact of automation very limited in the poorer parts of the world, so cervical cancer is still killing a quarter million women each year. Most of these deaths could be prevented by a globally functioning screening program.

A significant challenge for the future, therefore, is to come up with a screening system that is significantly cheaper than the present generation. How can this be achieved? Looking at the history we can see two approaches.

One is the "rare event" approach. Modern whole-slide scanners are much more robust, cheaper, and easier to operate than earlier generations of robot microscopes. With software for such systems, a competitive screening system should be possible, perhaps utilizing a somewhat modified specimen preparation approach that gives cleaner specimens without the high cost of the present liquid-based preparations.

The alternative approach is to use the MAC concept. The obstacle there is to achieve sufficiently robust imaging to consistently detect the subtle changes of chromatin texture between normal and malignancy-influenced cells in an automated system. A malignancy-influenced cell looks too much like a normal cell when it is slightly out of focus or poorly segmented. Here, modern 3D scanning concepts may make a significant difference.

So perhaps the next generation systems will be developed in third-world countries, to solve their great need for systems that are better and more economical than the systems that have been developed in the richer parts of the world.

2.6 Conclusions

As we have seen, the availability of a toolbox of image analysis routines greatly simplifies the process of "quickly trying out" an idea. Some of the algorithms that we approximated using only two or three lines of code would require many hundreds of lines of code without such a toolbox.

Another benefit to using an environment like MATLAB is the availability of many other tools not directly related to images that are very useful in developing novel algorithms. For example, in this chapter we have created graphs with some of the intermediate results. Being able to graphically see the numbers obtained is invaluable. In Sect. 2.4.5, we implemented only the first stage of the PAPNET system, but with equal ease we could have constructed the rest, using any of the neural network toolboxes that exist for MATLAB.

These two benefits are augmented in the MATLAB/DIPimage environment with an interpreted language, allowing interactive examination of intermediate results, and a high-level syntax, allowing easy expression of mathematical operations. The downside is that certain algorithms can be two orders of magnitude faster when expressed in C than in MATLAB, and algorithms written in MATLAB are more difficult to deploy. A common approach is to develop the algorithms in MATLAB, and translate them to C, C++, or Java when the experimentation phase is over. Though it is not always trivial to translate MATLAB code to C, MATLAB code that uses DIPimage has a big advantage: most of the image processing and analysis routines are implemented in a C library, DIPlib, that can be distributed independently of the DIPimage toolbox and MATLAB. All these things considered, even having to do the programming a second time in C, one can save large amounts of time when doing the first development in an environment such as that described here.

References

1. Luengo Hendriks, C.L., van Vliet, L.J., Rieger, B., van Ginkel, M., Ligteringen, R.: DIPimage User Manual. Quantitative Imaging Group, Delft University of Technology, Delft, The Netherlands (1999–2010)

2. Traut, H.F., Papanicolaou, G.N.: Cancer of the uterus: the vaginal smear in its diagnosis. Cal. West. Med. 59(2), 121–122 (1943)
3. Christopherson, W.M., Parker, J.E., Mendez, W.M., Lundin Jr., F.E.: Cervix cancer death rates and mass cytologic screening. Cancer **26**(4), 808–811 (1970)
4. Bengtsson, E.: Fifty years of attempts to automate screening for cervical cancer. Med. Imaging Technol. **17**(3), 203–210 (1999)
5. Tolles, W.E., Bostrom, R.C.: Automatic screening of cytological smears for cancer: the instrumentation. Ann. N. Y. Acad. Sci. **63**, 1211–1218 (1956)
6. Spencer, C.C., Bostrom, R.C.: Performance of the Cytoanalyzer in recent clinical trials. J. Natl. Cancer Inst. **29**, 267–276 (1962)
7. Prewitt, J.M.S., Mendelsohn, M.L.: The analysis of cell images. Ann. N. Y. Acad. Sci. 128(3), 1035–1053 (1965)
8. Watanabe, S., the CYBEST group: An automated apparatus for cancer prescreening: CYBEST. Comput. Graph. Image Process. **3**(4), 350–358 (1974)
9. Tanaka, N., Ikeda, H., Ueno, T., Watanabe, S., Imasato, Y.: Fundamental study of automatic cyto-screening for uterine cancer. II. Segmentation of cells and computer simulation. Acta Cytol. **21**(1), 79–84 (1977)
10. Tanaka, N., Ikeda, H., Ueno, T., Takahashi, M., Imasato, Y.: Fundamental study of automatic cyto-screening for uterine cancer. I. Feature evaluation for the pattern recognition system. Acta Cytol. **21**(1), 72–78 (1977)
11. Tanaka, N., Ueno, T., Ikeda, H., Ishikawa, A., Yamauchi, K., Okamoto, Y., Hosoi, S.: CYBEST model 4: automated cytologic screening system for uterine cancer utilizing image analysis processing. Anal. Quant. Cytol. Histol. **9**(5), 449–453 (1987)
12. Burger, G., Jutting, U., Rodenacker, K.: Changes in benign cell populations in cases of cervical cancer and its precursors. Anal. Quant. Cytol. **3**(4), 261–271 (1981)
13. MacAulay, C., Palcic, B.: An edge relocation segmentation algorithm. Anal. Quant. Cytol. Histol. **12**(3), 165–171 (1990)
14. Serra, J.: Image Analysis and Mathematical Morphology. Academic, London (1982)
15. Kuhl, F.P., Giardina, C.R.: Elliptic Fourier features of a closed contour. Comput. Graph. Image Process. **18**(3), 236–258 (1982)
16. Haralick, R.M., Shanmugam, K., Dinstein, I.: Textural features for image classification. IEEE Trans. Syst. Man Cybern. **3**(6), 610–621 (1973)
17. Galloway, M.M.: Texture analysis using gray level run lengths. Comput. Graph. Image Process. **4**(2), 172–179 (1975)
18. MacAulay, C., Palcic, B.: Fractal texture features based on optical density surface area. Use in image analysis of cervical cells. Anal. Quant. Cytol. Histol. **12**(6), 394–398 (1990)
19. Lee, J., Nelson, A., Wilbur, D.C., Patten, S.F.: The development of an automated Papanicolaou smear screening system. Cancer **81**, 332–336 (1998)
20. DeCresce, R.P., Lifshitz, M.S.: PAPNET cytological screening system. Lab Med. **22**, 276–280 (1991)
21. Luck, R.L., Scott, R.: Morphological classification system and method. US Patent 5,257,182, 1993
22. Duda, R.O., Hart, P.E., Stork, D.G.: Pattern Classification, 2nd edn. Wiley, New York (2001)
23. Mehnert, A.J.H.: Image Analysis for the Study of Chromatin Distribution in Cell Nuclei. Ph.D. Thesis, University of Queensland, Brisbane, Australia, 2003
24. Bamford, P., Lovell, B.: Unsupervised cell nucleus segmentation with active contours. Signal Process. **71**(2), 203–213 (1998)
25. Li, B., Acton, S.T.: Active contour external force using vector field convolution for image segmentation. IEEE Trans. Image Process. **16**(8), 2096–2106 (2007)
26. Hutchinson, M.L., Cassin, C.M., Ball, H.G.: The efficacy of an automated preparation device for cervical cytology. Am. J. Clin. Pathol. **96**(3), 300–305 (1991)
27. Howell, L.P., Davis, R.L., Belk, T.I., Agdigos, R., Lowe, J.: The AutoCyte preparation system for gynaecologic cytology. Acta Cytol. 42(1), 171–177 (1998)

Chapter 3
Seeded Segmentation Methods for Medical Image Analysis

Camille Couprie, Laurent Najman, and Hugues Talbot

Segmentation is one of the key tools in medical image analysis. The objective of segmentation is to provide reliable, fast, and effective organ delineation. While traditionally, particularly in computer vision, segmentation is seen as an early vision tool used for subsequent recognition, in medical imaging the opposite is often true. Recognition can be performed interactively by clinicians or automatically using robust techniques, while the objective of segmentation is to precisely delineate contours and surfaces. This can lead to effective techniques known as "intelligent scissors" in 2D and their equivalent in 3D.

This chapter is divided as follows. Section 3.1 starts off with a more "philosophical" section setting the background for this study. We argue for a segmentation context where high-level knowledge, object information, and segmentation method are all separate.

In Sect. 3.2, we survey in some detail a number of segmentation methods that are well-suited to image analysis, in particular of medical images. We illustrate this, make some comparisons and some recommendations.

In Sect. 3.3, we introduce very recent methods that unify many popular discrete segmentation methods and we introduce a new technique. In Sect. 3.4, we give some remarks about recent advances in seeded, globally optimal active contour methods that are of interest for this study.

In Sect. 3.5, we compare all presented methods qualitatively. We then conclude and give some indications for future work.

H. Talbot (✉)
Université Paris-Est, Paris, France

G. Dougherty (ed.), *Medical Image Processing: Techniques and Applications*, Biological and Medical Physics, Biomedical Engineering, DOI 10.1007/978-1-4419-9779-1_3,

3.1 The Need for Seed-Driven Segmentation

Segmentation is a fundamental operation in computer vision and image analysis. It consists of identifying regions of interests in images that are semantically consistent. Practically, this may mean finding individual white blood cells amongst red blood cells; identifying tumors in lungs; computing the 4D hyper-surface of a beating heart, and so on.

Applications of segmentation methods are numerous. Being able to reliably and readily characterize organs and objects allows practitioners to measure them, count them and identify them. Many images analysis problems begin by a segmentation step, and so this step conditions the quality of the end results. Speed and ease of use are essential to clinical practice.

This has been known for quite some time, and so *numerous* segmentation methods have been proposed in the literature [57]. However, segmentation is a difficult problem. It usually requires high-level knowledge about the objects under study. In fact, semantically consistent, high-quality segmentation, in general, is a problem that is indistinguishable from strong Artificial Intelligence and has probably no exact or even generally agreeable solution. In medical imaging, experts often disagree amongst themselves on the placement of the 2D contours of normal organs, not to mention lesions. In 3D, obtaining expert opinion is typically difficult, and almost impossible if the object under study is thin, noisy and convoluted, such as in the case of vascular systems. At any rate, segmentation is, even for humans, a difficult, time-consuming and error-prone procedure.

3.1.1 Image Analysis and Computer Vision

Segmentation can be studied from many angles. In computer vision, the segmentation task is often seen as a low-level operation, which consists of separating an arbitrary scene into reasonably alike components (such as regions that are consistent in terms of color, texture and so on). The task of grouping such component into semantic objects is considered a different task altogether. In contrast, in image analysis, segmentation is a high-level task that embeds high-level knowledge about the object.

This methodological difference is due to the application field. In computer vision, the objective of segmentation (and grouping) is to recognize objects in an arbitrary scene, such as persons, walls, doors, sky, etc. This is obviously extremely difficult for a computer, because of the generality of the context, although humans do generally manage it quite well. In contrast, in image analysis, the task is often to *precisely* delineate some objects sought in a particular setting known in advance. It might be for instance to find the contours of lungs in an X-ray photograph.

The segmentation task in image analysis is still a difficult problem, but not to the same extent as in the general vision case. In contrast to the vision case, experts might agree that a lesion is present on a person's skin, but may disagree on its exact contours [45]. Here, the problem is that the boundary between normal skin and lesion might be objectively difficult to specify. In addition, sometimes there does exist an object with a definite physical contour (such as the inner volume of the left ventricle of the heart). However, imaging modalities may be corrupted by noise and partial volume effects to an extent that delineating the precise contours of this physical object in an image is also objectively difficult.

3.1.2 Objects Are Semantically Consistent

However, in spite of these difficulty, we may assume that, up to some level of ambiguity, an object (organ, lesion, etc) may still be specified somehow. This means that semantically, an object possess some consistency. When we point at a particular area on an image, we expect to be, again with some fuzziness, either inside or outside the object

This leads us to the realize that there must exist some mathematical indicator function, that denotes whether we are inside or outside of the object with high probability. This indicator function can be considered like a series of constraints, or labels. They are sometimes called *seeds* or *markers*, as they provide starting points for segmentation procedures, and they mark where objects are and are not.

In addition, a *metric* that expresses the consistency of the object is likely to exist. A gradient on this metric may therefore provide object contour information. Contours may be weak in places where there is some uncertainty, but we assume they are not weak everywhere (else we have an ambiguity problem, and our segmentation cannot be precise). The metric may simply be the image intensity or color, but it may express other information like consistency of texture for instance. Even though this metric may contain many descriptive elements (as a vector of descriptors for instance), we assume that we are still able to compute a gradient on this metric [61].

This is the reason why many segmentation methods focus on contours, which are essentially discontinuities in the metric. Those that focus on regions do so by defining and utilizing some consistency metric, which is the same problem expressed differently.

The next and final step for segmentation is the actual contour placement, which is equivalent to object delineation. This step can be considered as an optimization problem, and this is the step on which segmentation methods in the literature focus the most. We will say more about this in Sect. 3.2 listing some image segmentation categories.

3.1.3 A Separation of Powers

In summary, to achieve segmentation in the analysis framework, we need three ingredients: (1) an indicator function that denotes whether we are inside or outside of the object of interest; (2) a metric from which we may derive contour information, and (3) an optimization method for placing the contour accurately.

To achieve accuracy, we need flexibility and robustness. Some have argued that it is useful to treat these three steps separately. This was first described in [47]) as the *morphological* method, but is also called by others *interactive* or *seeded* segmentation [31]. In this context, this does not mean that user interaction is required, only that object identification is provided by some means, and contour extraction separately by a segmentation operator.

The first ingredient, the object identification, or our indicator function, is of course essential and it is frustrating to be obliged to only write here "by some means". Accurate content identification can simplify the requirements on the segmentation operator greatly. Unfortunately, the means in question for contents identification are problem-dependent and sometimes difficult to publish, because they are often seen as *ad hoc* and of limited interest beyond their immediate use in the problem at hand. Fortunately, some journals accept such publications, such as the *Journal of Image Analysis and Stereology* and applications journals (e.g. *Journal of Microscopy*, materials, etc). There are also a few recent books on the matter [23,52]. Software libraries are also important but not many are freely available for training, although the situation is improving.

Also, whereas in computer vision a fully automated solution is required, in medical imaging a semi-automated method might be sufficient. In biomedical imaging, a large number of objects are typically measured (such as cells, organelles, etc.), and a fully automated method is often desirable. However, in medical imaging, typically a relatively small number of patients is being monitored, treated or surveyed, and so human-guided segmentation can be sufficient. The objective of the segmentation method in this context is to provide reasonable contours quickly, which can be adjusted easily by an operator.

In this variety of contexts, is it possible to define precisely the segmentation problem? The answer is probably no, at this stage at least in image analysis research. However, it is possible to provide *formulations* of the problem. While this may sound strange or even suspicious, the reason is that there exists a real need for automated or semi-automated segmentation procedures for both image analysis and computer vision, and so solutions have been proposed. They can still be explained, compared and evaluated.

3.1.4 Desirable Properties of Seeded Segmentation Methods

We come to the first conclusion that to provide reliable and accurate results, we must rely on a segmentation procedure and not just an operator. Object identification and constraints analysis will set us in good stead to achieve our results, but not all segmentation operators are equivalent. We can list here some desirable properties of interactive segmentation operators.

- It is useful if the operator can be expressed in an energy or cost optimization formulation. It is then amenable to existing optimization methods, and this entails a number of benefits. Lowering the cost or the energy of the formulation can be done in several ways (e.g. continuous or discrete optimization), which results in different characteristics and compromises, say between memory resources and time. Optimization methods improve all the time through the work of researchers, and so our formulations will benefit too.
- It is desirable if the optimization formulation can provide a solution that is at least locally optimal, and if possible globally optimal, otherwise noise will almost certainly corrupt the result.
- The operator should be fast, and provide guaranteed convergence, because it will be most likely restarted several times, in order to adjust parameters. Together with this requirement, the ability to segment many objects at once is also desirable, otherwise the operator will need to be restarted as many time as there are objects in the image. This may not be a big problem if objects do not overlap and if bounding boxes can be drawn around them, because the operator can then be run only within the bounding box, but this is not the general case.
- The operator should be bias-free: e.g. with respect to objects size or to the discretization grid or with respect to initialization.
- The operator should be flexible: it is useful if it can be coupled with topology information for instance, or with multi-scale information.
- It should be generic, not tied to particular data or image types.
- It should be easy to use. This in practice means possessing as few parameters as possible. Of course one can view constraints setting as an enormous parameter list, but this is the reason why we consider this step as separate.

Such a method certainly does not yet exist to our knowledge, although some might be considered to come close. We describe some of them in the next section.

3.2 A Review of Segmentation Techniques

Here, we list and detail some segmentation categories that are compatible with the image analysis viewpoint, although cannot hope to present a complete description of this field.

3.2.1 Pixel Selection

Pixel selection is likely the oldest segmentation method. It consists of selecting pixels solely based on their values and irrespective of their spatial neighborhood.

The simplest pixel selection method is humble thresholding, where we select pixels that have a gray-level value greater or smaller than some threshold value. This particular method is of course very crude, but is used frequently nonetheless. Multiple thresholding uses several values instead of a single value; color and multi-spectral thresholding using vectors of values and not just scalars. By definition all histogram-based methods for finding the parameters of the thresholding, including those that optimize a metric to achieve this [54], are pixel selection methods. Statistical methods (e.g. spectral classification methods) that include no spatial regularization fall into this category as well. This is therefore a veritable plethora of methods that we are including here, and research is still active in this domain.

Of course, thresholding and related methods are usually very fast and easily made interactive, which is why they are still used so much. By properly pre-processing noisy, unevenly illuminated images, or by other transforms, it is surprising how many problems can be solved by interactive or automated thresholding. However, this is of course not always the case, hence the need for more sophisticated methods.

3.2.2 Contour Tracking

It was realized early on that (1) human vision is sensitive to contours and (2) there is a duality between simple closed contours and objects. A simple closed contour (or surface) is one that is closed and does not self-intersect. By the Jordan theorem, in the Euclidean space, any such contour or surface delineates a single object of finite extent. There are some classical difficulties with the Jordan theorem in the discrete setting [52], but they can be solved by selecting proper object/background connectivities, or by using a suitable graph, for instance, the 6-connected hexagonal grid or the Khalimsky topology [22, 40].

A contour can be defined locally (it is a frontier separating two objects (or an object and its background in the binary case)), while an object usually cannot (an object can have an arbitrary extent). A gradient (first derivative) or a Laplacian (second derivative) operator can be used to define an object border in many cases, and gradients are less sensitive to illumination conditions than pixel values. As a result, contour detection through the use of gradient or Laplacian operators became popular, and eventually led to the Marr–Hildreth theory [44].

Given this, it is only natural that most segmentation method use contour information directly in some ways, and we will revisit this shortly. Early methods used *only* this information to detect contours and then tried to combine them in some way. By far the most popular and successful version of this approach is the Canny edge detector [9]. In his classical paper, Canny proposed a closed-form optimal 1D edge detector assuming the presence of additive white Gaussian noise, and successfully proposed a 2D extension involving edge tracking using non-maxima suppression with hysteresis.

One problem with this approach is that there is no optimality condition in 2D, no topology or connectivity constraints and no way to impose markers in the final result. All we get is a series of contours, which may or may not be helpful. Finding a suitable combination of detected contours (which can be incomplete) to define objects is then a combinatorial problem of high complexity. Finally, this approach extends even less to 3D.

Overall, in practical terms, these contour tracking methods have been superseded by more recent methods and should not be used without good reasons. For instance, more recent minimal-path methods can be applied to contour tracking methods, although they are much more sophisticated in principle [3, 14]. In this class of methods belongs also the "intelligent scissors" types. There were many attempts in previous decades to provide automated delineating tools in various image processing software packages, but a useful contribution was provided relatively recently by Mortensen [48]. This method is strictly interactive, in the sense that it is designed for human interaction and feedback. "Intelligent scissor" methods are useful to clinicians for providing ground truth data for instance. Such methods are still strictly 2D. As far as we know, no really satisfying 3D live-wire/intelligent scissor method is in broad use today [5]. However, minimal surfaces methods, which we will describe shortly in Sect. 3.4.3, in some ways do perform this extension to nD [30].

3.2.3 Statistical Methods

The opposite approach to contour detection is to work on the objects, or regions themselves. An early and intuitive approach has been to try to divide (the *splitting* step) an image into uniform regions, for example using a hierarchical representation of an image in the form of quadtrees (in 2D) and octrees (in 3D). Uniformity can be defined by statistical parameters and/or tests. Subsequently, a *merging* step considering neighboring and statistical region information is performed [36]. Initial considered statistics were color and intensity, but other region descriptors can be used as well, for instance including texture, motion and so on. In this approach, even though regions statistics are used, they are inevitably derived at the pixel level. The split and merge approach consists of acquiring all the statistics first and basing a decision on them.

A different approach, which is also productive, consists of building a *model* first. One way is to consider an image as a 2D or 3D graph of pixels, to start from a vast over-segmentation at the pixel level, and to evolve cliques of pixels (e.g. sets of one, two or more pixels that are fully-connected, respectively called unary, binary or higher-level cliques) to fit that model. This is the *Markov Random Field* (MRF) model, named in this way by comparison to classical one-dimensional Markov chains, for which only immediate neighboring relationships matter. Models that can be written using these cliques turn out to corresponds to energies featuring weighted finite sums with as many terms as there are different kinds of cliques. In [26] Geman and Geman proposed to optimize these sums using Gibbs sampling (a form of Monte-Carlo Markov Chain algorithm) and simulated annealing. This was first used for image restoration, but can be readily applied to segmentation as well. This approach was very successful because it is very flexible. Markers and texture terms can be added in, and many algorithmic improvement were proposed over the years. However, it remains a relatively costly and slow approach. Even though Geman and Geman showed that their simulated annealing strategy converges, it only does so under conditions that make the algorithm extremely slow, and so usually only a non-converged or approximate result is used. More recently, it was realized that Graph-Cut (GC) methods were well-suited to optimized some MRF energies very efficiently. We will give more details in the corresponding section.

MRFs belong to the larger class of Bayesian methods. Information-theoretic perspectives and formulations, such as following the Minimum Description Length principle, also exist. These frameworks are also very flexible, allowing for example region competition [69]. However, the corresponding models might be complicated both to understand and run, and sometimes possess many parameters that are not obvious to tune. Well-designed methods are guaranteed to converge to at least a local minimum.

In general, when dealing with regions that have complex content (for instance, textures, or multispectral content), statistical methods can be a very good choice although they cannot be recommended for general work, since simpler and faster methods often are sufficient.

3.2.4 *Continuous Optimization Methods*

In the late 1980s, it was realized that contour tracking methods were too limited for practical use. Indeed, getting closed contours around objects were difficult to obtain with contour tracking. This meant that detecting actual objects was difficult except in the simplest cases.

3.2.4.1 Active Contours

Researchers, therefore, proposed to start from already-closed loops, and to make them evolve in such a way that they would converge towards the true contours of the image. Thus were introduced *active contours*, or *snakes* [39]. The formulation of snakes takes the following continuous-domain shape:

$$E_{\text{snake}} = \int_0^1 \{E_{\text{internal}}(\mathbf{v}(s)) + E_{\text{data}}(\mathbf{v}(s)) + E_{\text{constraints}}(\mathbf{v}(s))\}\, \mathrm{d}s. \tag{3.1}$$

where $\mathbf{v}(s)$ is a parametric representation of the contour.

This model is very flexible. It contains internal terms, image data terms and constraints terms (see Chapter 4 for more details):

- The first term, the internal energy, contains a curvature term and a "rubber band" energy. The former tends to smooth the resulting contour following a thin plate, while the latter tends to make it shrink around features of interest. Other terms such as kinetic energy can be added too, which makes it possible for the snake to avoid noisy zones and flat areas.
- The second term, the data energy, attracts the active contours towards points of interest in the image: typically, image contours (zones of high gradient), lines or termination points.
- The last term, the constraint term, is optional, but allows interaction with the snake by defining zones of attraction and repulsion.

To solve this equation, the Euler–Lagrange of (3.1) is worked out (typically in closed form), and a gradient descent algorithm is used. All the terms are combined in a linear combination, allowing them to be balanced according to the needs of the user. Due to its flexibility, the active contour model was very popular in the literature as well as in applications. It fits very well into the interactive segmentation paradigm because constraints can be added very easily, and it can be quite fast because it uses a so-called Lagrangian framework. The contour itself is discretized at regular interval points and evolves according to (3.1). Convergence towards a local minimum of the energy is guaranteed, but may require many iterations.

In practice, there are some difficulties: the snake energy is flexible but difficult to tune. Because of the contour evolution, points along the contour tend to spread out or bunch up, requiring regular and frequent resampling. There can also be topological difficulties, for instance causing the snake to self-intersect. The snake is also sensitive to its parametrization and to initialization. Finally, even though a local optimum is guaranteed, in practice, it may not be of good quality due to noise sensitivity.

One major difficulty with snakes is that they can be extended to 3D via triangulation, but such extensions can be complicated, and topological problems plaguing snakes in 2D are usually more difficult to avoid in 3D. However, 3D active surfaces are still widely used, because they make it easy to improve or regularize a triangulated surface obtained by other means. For instance, the brain segmentation software FreeSurfer includes such a method. To distinguish them from other models we are going to introduce now, snake-like active contours or surfaces are sometimes called *parametric deformable models*.

3.2.4.2 Level Sets

One way to avoid altogether some of the problems brought about by the way parametric deformable models are discretized, is to embed the contour into a higher-dimensional manifold. This idea gave rise to *level sets*, proposed by Osher and Sethian in 1988 [53]. Remarkably, this is around the same time when active contours

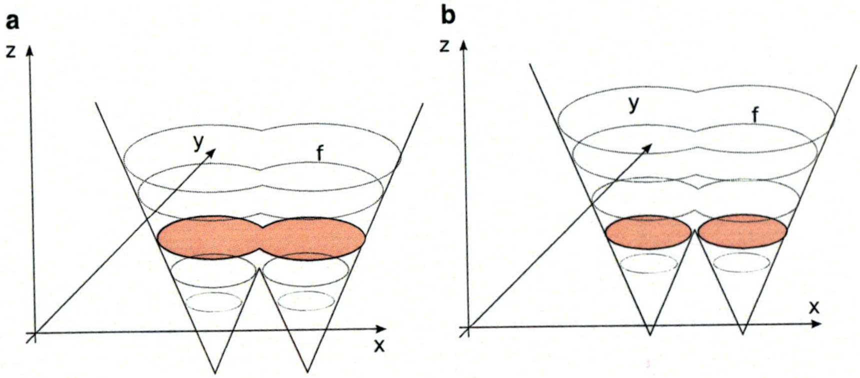

Fig. 3.1 Embedding and evolving a curve as a level set of a higher-dimension function. The zero-level of function ψ is shown in color, representing a 2D contour. To evolve the contour, the whole function evolves. Note that topology changes can occur in the contour, while the embedding surface shows no such effect

were proposed. However level sets were initially proposed for computational fluid dynamics and numerical simulations. They were applied to imaging somewhat later [43, 62]. A contour is represented on the surface S of an evolving regular function ψ by its zero level-set, which is simply the threshold of the function ψ at zero. By using sufficiently regular embedding functions ψ, namely signed distance transforms from an initial contour, it was possible to propose effective evolution equations to solve similar problems to Lagrangian active contours.

The main advantages of the level-sets method were that contour resampling was no longer necessary, and contour self-intersection (shock solutions) was avoided because level sets were able to change topology easily (see Fig. 3.1b). This means practically that it was possible at least in theory to initialize a segmentation by drawing a box around a series of object of interest, and the level set could find a contour around each of them. This was seen as a major benefit by the vision community. The level set Eulerian formulation (where the whole space is discretized) is thought to offer better theoretical guarantees than the Lagrangian framework of previous non-embedded formulations, and the simulation of function evolution is a well-researched topic with many usable and interesting results. Finally, the formulation is dimension independent. Level sets work virtually unchanged in 3D or more, which is a major benefit.

There are also a number of drawbacks. First, the level set formulation is more expensive than earlier active contour formulations. It requires the iterative solving of PDEs in the whole space, which is expensive. In practice, it is possible to limit the computation in a narrow band around the contour, but this is still more costly than if they were limited to the contour itself, and requires the resampling that was sought to be avoided. The surface S of function ψ is implicitly represented by the function itself, but it requires more space than the contour. In 3D or more, this may be prohibitive. Some contour motions are not representable (e.g. contour rotation), but this is a minor problem. More importantly, the fact that level-sets can undergo topology changes is actually a problem in image analysis, where it is useful to know that a contour initialized somewhere will converge to a single simple closed contour. In some cases, a contour can split or even disappear completely, leading to undesirable results.

Nonetheless, level-set formulations are even more flexible than active contours, and very complex energies solving equally complex problems have been proposed in the literature. Solving problem involving texture, motion, competing surfaces and so on is relatively easy to formulate in this context [55, 56]. For this reason, they were and remain popular. Complex level-set formulation tend to be sensitive to noise and can converge to a poor locally optimal solution. On the other hand, more robust, closer to convex solutions can now be solved via other means. An example of relatively simple PDE that can be solved by level sets is the following:

$$\psi_t + F|\nabla \psi_t| = 0, \tag{3.2}$$

where F is the so-called speed function. Malladi and Sethian proposed the following for F:

$$F = \frac{1-\varepsilon\kappa}{1+|\nabla I|} + \beta(\nabla\psi.\nabla|\nabla I|). \tag{3.3}$$

The first part of the equation is a term driving the embedding function ψ towards contours of the image with some regularity and smoothing controlled by the curvature κ. The amount of smoothing is controlled by the parameter ε. The second term is a "balloon" force that tends to expand the contour. It is expected that the contour initially be placed inside the object of interest, and that this balloon force should be reduced or eliminated after some iterations, controlled by the parameter β. We see here that even though this model is relatively simple for a level-set one, it already has a few parameters that are not obvious to set or optimize.

3.2.4.3 Geodesic Active Contours

An interesting attempt to solve some of the problems posed by overly general level sets was to go back and simplify the problem, arguing for consistency and a geometric interpretation of the contour obtained. The result was the geodesic active contour (GAC), proposed by Caselles et al. in 1997 [10]. The level set formulation is the following:

$$\psi_t = |\nabla\psi|\,\mathrm{div}\left(g(I)\frac{\nabla\psi}{|\nabla\psi|}\right). \tag{3.4}$$

This equation is virtually parameter-free, with only a g function required. This function is a *metric* and has a simple interpretation: it defines at point x the cost of a contour going through x. This metric is expected to be positive definite, and in most cases is set to be a scalar functional with values in $\mathbb{R}^+$. In other words, the GAC equation finds the solution of:

$$\mathrm{argmin}_C \int_C g(s)\mathrm{d}s, \tag{3.5}$$

where C is a closed contour or surface. This is the minimal closed path or minimal closed surface problem, i.e. finding the closed contour (or surface) with minimum weight defined by g. In addition to simplified understanding and improved consistency, (3.4) has the required form for Weickert's PDE operator splitting [28, 68], allowed PDEs to be solved using separated semi-implicit schemes for improved efficiency. These advances made GAC a reference method for segmentation, which is now widely used and implemented in many software packages such as ITK. The GAC is an important interactive segmentation method due to the importance of initial contour placement, as with all level-sets methods. Constraints such as forbidden or attracting zones can all be set through the control of function g, which has an easy interpretation.

As an example, to attract the GAC towards zones of actual image contours, we could set

$$g \equiv \frac{1}{1+|\nabla I|^p}, \tag{3.6}$$

With $p = 1$ or 2. We see that for this function, g is small (costs little) for zones where the gradient is high. Many other functions, monotonically decreasing for increasing values for ∇I, can be used instead. One point to note, is that GAC has a so-called *shrinking bias*, due to the fact that the globally optimal solution for (3.5) is simply the null contour (the energy is then zero). In practice, this can be avoided with balloon forces but the model is again non-geometric. Because GAC can only find a local optimum, this is not a serious problem, but this does mean that contours are biased towards smaller solutions.

3.2.5 *Graph-Based Methods*

The solution to (3.5) proposed in the previous section was in fact inspired by preexisting discrete solutions to the same problem. On computers, talking about continuous-form solutions is a bit of a misnomer. Only the mathematical formulation is continuous, the computations and the algorithms are all necessarily discrete to be computable. The idea behind discrete algorithm is to embrace this constraint and embed the discrete nature of numerical images in the formulation itself.

3.2.5.1 Graph Cuts

We consider an image as a graph $\Gamma(\mathscr{V},\mathscr{E})$ composed of n vertices $\mathscr{V}$ and m edges $\mathscr{E}$. For instance, a 2D $N \times N$ 4-connected square grid image will have $n = N^2$ vertices and $m = 2 \times N \times (N-1)$ edges.[1] We assume that both the edges and the vertices are weighted. The vertices will typically hold image pixel values and the edge values relate to the gradient between their corresponding adjacent pixels, but this is not necessary. We assume furthermore that a segmentation of the graph can be represented as a graph *partition*, i.e:

$$V = \bigcup_{V_i \in \Gamma} V_i; \forall i \neq j, V_j \cap V_i = \emptyset. \tag{3.7}$$

Then $E^\star$ is the set of edges that are such that their corresponding vertices are in different partitions.

$$E^\star = \{e = \{p_i, p_j\} \in E, p_i \in V_i; p_j \in V_j, i \neq j\}. \tag{3.8}$$

The set $E^\star$ is called the *cut*, and the cost of the cut is the sum of the edge weights that belong to the cut:

$$C(E^\star) = \sum_{e \in E^\star} w_e, \tag{3.9}$$

where w_e is the weight of individual edge e. We assume these weights to be positive. Reinterpreting these weights as *capacities*, and specifying a set of vertices as connected to a *source* s and a distinct set connected to a *sink* t, the celebrated 1962 Ford and Fulkerson result [25] is the following:

Theorem 3.1. *Let P be a path in Γ from s to t. A flow through that path is a quantity which is constrained by the minimum capacity along the path. The edges with this capacity are said to be saturated, i.e. the flow that goes through them is equal to their capacity. For a finite graph, there exists a maximum flow that can go through the whole graph Γ. This maximum flow saturates a set of edges E^s. This set of edges define a cut between s and t, and this cut has minimal weight.*

This theorem is illustrated in Fig. 3.2.

In 2D and if Γ is planar, this duality essentially says that the Ford and Fulkerson minimum cut can be interpreted as a shortest path in a suitable dual graph to Γ [2]. In arbitrary dimension, the maxflow – mincut duality allows us to compute discrete minimal hypersurfaces by optimizing a discrete version of (3.4).

There exist many algorithms that can be used to compute the maximum flow in a graph (also called network in this framework), but none with a linear complexity. Augmenting paths algorithms [7] are effective in 2D where the number of vertices is relatively high compared to the number of edges. In 3D and above, where the reverse is true, push-relabel algorithms [27] are more efficient. These algorithms can only be used when there is one source and one sink. The case where there are multiple sources or sinks is known to be NP-hard. To compute energies comprising several sources or sinks and leading to multi-label segmentation, approximations

[1] This particular computation is left as an exercise to the reader.

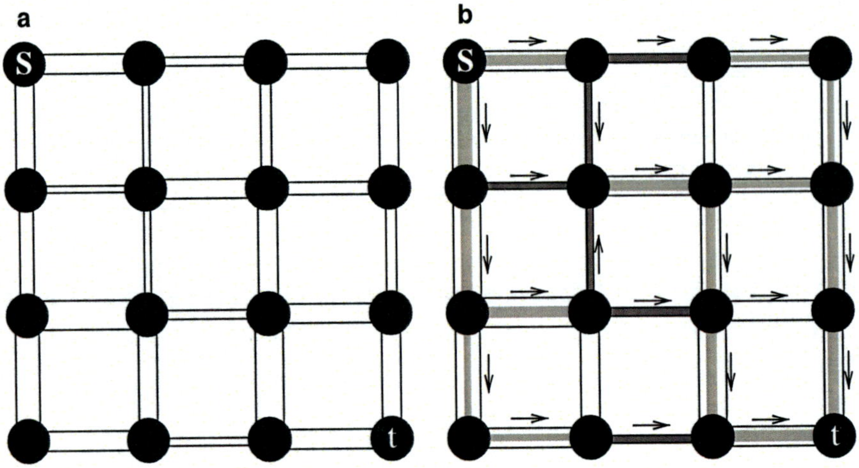

Fig. 3.2 (**a**) A graph with edge weights interpreted as capacities, shown as varying diameters in this case. (**b**) A maximum flow on this graph. We see that the saturated vertices (in *black*) separate s from t, and they form a cut of minimum weight

can be used, such as α-expansions. These can be used to formulate and optimize complex discrete energies with MRF interpretations [8, 66], but the solution is only approximate. Under some conditions, the result is not necessarily a local minimum of the energy, but can be guaranteed not to be too far from the globally optimal energy (within a known factor, often 2).

In the last 10 years, GC methods have become extremely popular due to their ability to solve a large number of problems in computer vision, particularly in stereo-vision and image restoration. In image analysis, their ability to form a globally optimal binary partition with a geometric interpretation is very useful. However, GC do have some drawbacks. They are not easy to parallelize, they are not very efficient in 3D, they have a so-called *shrinking bias*, just as GAC and continuous maxflow have as well. In addition, they have a *grid bias*, meaning that they tend to find contours and surfaces that follow the principal directions of the underlying graph. This results in "blocky" artifacts, which may or may not be problematic.

Due to their relationship with sources and sinks, which can be seen as internal and external markers, as well as their ability to modify the weights in the graph to select or exclude zones, GC are at least as interactive as the continuous methods of previous sections.

3.2.5.2 Random Walkers

In order to correct some of the problems inherent to graph cuts, Grady introduced the Random Walker (RW) in 2004 [29, 32]. We set ourselves in the same framework as in the Graph Cuts case with a weighted graph, but we consider from the start a multilabel problem, and, without loss of generality, we assume that the edge weights are all normalized between 0 and 1. This way, they represent the probability that a random particle may cross a particular edge to move from a vertex to a neighboring one. Given a set of starting points on this graph for each label, the algorithm considers the probability for a particle moving freely and randomly on this weighted graph to reach any arbitrary unlabelled vertex in the graph before any other coming from the other labels. A vector of probabilities, one for each label, is therefore computed at each unlabeled vertex. The algorithm considers the computed probabilities at each vertex and assigns the label of the highest probability to that vertex.

Intuitively, if close to a label starting point the edge weights are close to 1, then its corresponding “random walker” will indeed walk around freely, and the probability to encounter it will be high. So the label is likely to spread unless some other labels are nearby. Conversely, if somewhere edge weights are low, then the RW will have trouble crossing these edges. To relate these observations to segmentation, let us assume that edge weights are high within objects and low near edge boundaries. Furthermore, suppose that a label starting point is set within an object of interest while some other labels are set outside of it. In this situation, the RW is likely to assign the same label to the entire object and no further, because it spreads quickly within the object but is essentially stopped a the boundary. Conversely, the RW spreads the other labels outside the object, which are also stopped at the boundary. Eventually, the whole image is labeled with the object of interest consistently labeled with a single value.

This process is similar in some way to classical segmentation procedures like seeded region growing [1], but has some interesting differentiating properties and characteristics. First, even though the RW explanation sounds stochastic, in reality the probability computations are deterministic. Indeed, there is a strong relationship between random walks on discrete graphs and various physical interpretations. For instance, if we equate an edge weight with an electrical resistance with the same value, thereby forming a resistance lattice, and if we set a starting label at 1 V and all the other labels to zero volt, then the probability of the RW to reach a particular vertex will be the same as its voltage calculated by the classical Kirchhoff’s laws on the resistance lattice [24]. The problem of computing these voltages or probability is also the same as solving the discrete Dirichlet problem for the Laplace equation, i.e. the equivalent of solving $\nabla^2 \varphi = 0$ in the continuous domain with some suitable boundary conditions [38]. To solve the discrete version of this equation, discrete calculus can be used [33], which in this case boils down to inverting the graph Laplacian matrix. This is not too costly as it is large but very sparse. Typically calculating the RW is less costly and more easily parallelizable than GC, as it exploits the many advances realized in numerical analysis and linear algebra over the past few decades.

The RW method has some interesting properties with respect to segmentation. It is quite robust to noise and can cope well with weak boundaries (see Fig. 3.3). Remarkably, in spite of the RW being a purely discrete process, it exhibits no grid bias. This is due to the fact that level lines of the resistance distance (i.e. the

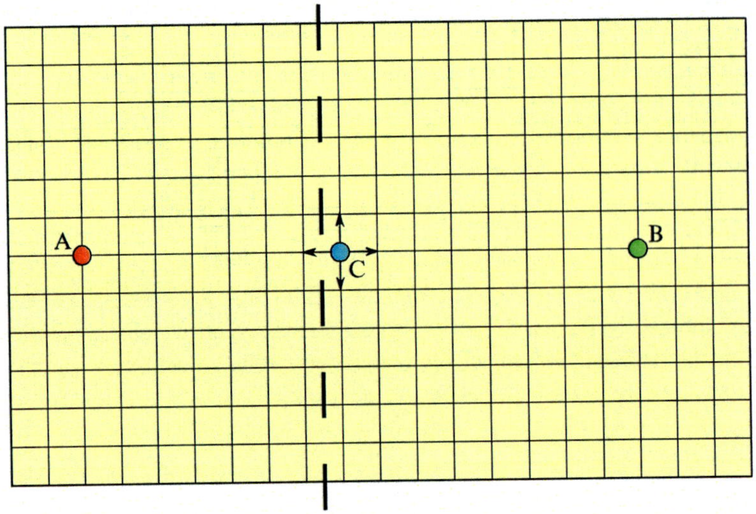

Fig. 3.3 An intuitive explanation of why the Random Walker copes well with weak boundaries. We assuming constant, high probabilities everywhere on this graph, except where *thick vertical lines* cross an edge, where the probabilities are low. A and B represent labels, and we estimate the probability of a random walker in C to move to the left as opposed to all the other directions (north, south, or east). We see that locally the probabilities are identical, but globally, there are many ways for a random walker to come from B to the north, east or south position from C. However, there is only one way to move to the west of C, and that is to go through C. Therefore, Random walker probabilities must be high up to C, and then drop precipitously. Since the situation is symmetrical with respect to A, it is likely that the region left of he thick lines will be labelled with A, and the region right to it are going to be labelled with B. This is in spite of the fact that the boundary defined by the *thick vertical lines* is weak and closer to A than B

resistance between a fixed node and all the others) in an infinite graph with constant edge weights are asymptotically isotropic [21]. RW exhibit a shrinking bias but not as strong as GC.

3.2.5.3 Watershed

While there are many variations on discrete segmentation methods, we will consider one last method: the Watershed Transform (WT). It was introduced in 1979 by Beucher and Lantuéjoul [6] by analogy to the topography feature in geography. It can be explained intuitively in the following manner: consider a gray-level image to be a 3D topographical surface or terrain. A drop of water falling onto this surface would follow a descending path towards a local minimum of the terrain. The set of points, such that drops falling onto them would flow into the same minimum, is called a *catchment basin*. The set of points that separate catchment basins form the *watershed line*. Finally, the transform that takes an image as input and produces its set of watershed lines is called the *Watershed Transform*. To use

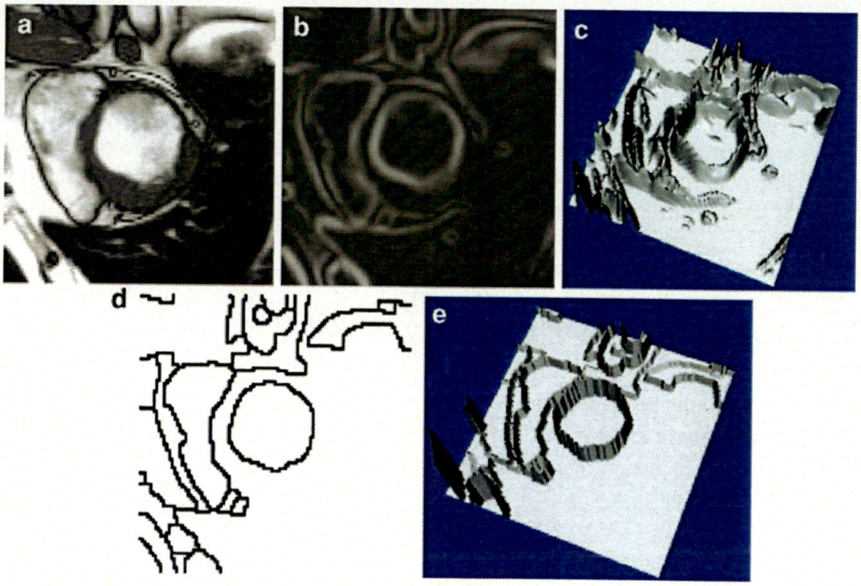

Fig. 3.4 Watershed segmentation: (**a**) an MRI image of the heart, (**b**) its smoothed gradient, (**c**) the gradient seen as a topographic surface, (**d**) Watershed of the gradient, and (**e**) topographical view of the watershed

this transform in practical segmentation settings, we must reverse the point of view somewhat. Assume now that labels are represented by lakes on this terrain and that by some flooding process, the water level rises evenly. The set of points that are such that waters from different lakes meet is also called the watershed line. Now this watershed line is more constrained, because there are only as many lines as necessary to separate all the lakes.

This intuitive presentation is useful but does not explain why the WT is useful for segmentation. As the "terrain", it is useful to consider the magnitude of the gradient of the image. On this gradient image, interior objects will have values close to zero and will be surrounded by zones of high values: the contours of the objects. They can therefore be assimilated to catchment basins, and the WT can delineate them well (see Fig. 3.4).

The WT is a seeded segmentation method, and has many interesting interpretations. If we consider the image again as a graph as in the GC setting, then on this graph the set of watershed lines from the WT form a graph cut. The edges of this tree can be weighted with a functional derived from a gradient exactly as in the GC case. Computing the WT can be performed in many efficient ways [46, 67], but an interesting one is to consider the Maximum Spanning Forest (MSF) algorithm [19]. In this algorithm, the classical graph algorithm for maximum spanning tree (MST) is run on the graph of the image, following for instance Kruskal's algorithm [41,59], with the following difference: when an edge selected by the MST algorithm is connected with a seed, then all vertices that are connected with it become also labeled with this seed, and so on recursively. However, when an edge selected by the MST algorithm would be connecting two different seeds, the connection is simply not performed. It is easy to show that (1) eventually all edges of the graph are labeled with this algorithm; (2) the set of edge that are left connected form a graph cut separating all the seeds; and (3) the labels are connected to the seeds by subtrees. The result is a MSF, and the set of unconnected edges form a watershed line. The MSF algorithm can be run in quasi-linear time [20].

3.2.6 *Generic Models for Segmentation*

Even though seeded models are the focus of this chapter, we say here a few words about generic models that are not seeded by default, because they contain powerful ideas for the future of seeded models.

3.2.6.1 Continuous Models

Over the years, several now widely cited formulations of the segmentation problem have been proposed, including for instance the Mumford–Shah functional [49] or the Chan–Vese active contour without edges (AWE) [13]. They generally seek to solve the segmentation problem in the vision setting, and can be used for image restoration as well (denoising, inpainting, etc).

In particular, the Mumford–Shah functional is the following:

$$E(\mathbf{f},C) = \beta \int_{\Omega} (\mathbf{f}-\mathbf{g})^2 dA + \alpha \int_{\Omega \setminus C} |\nabla \mathbf{f}|^2 dA + \gamma \int_C ds. \tag{3.10}$$

This formulation is very interesting because it has been an inspiration to many. In this expression, $\mathbf{g}$ is the original image, $\mathbf{f}$ a piecewise smooth approximation of $\mathbf{g}$ and C a collection of contours where $\mathbf{f}$ is discontinuous. In essence, C represents the segmentation of $\mathfrak{F}$ and $\mathbf{f}$ is a restored (denoised, etc) model of $\mathfrak{F}$. The first term in (3.10) is a data fidelity term; the second is a total variation term (TV), and the last optimizes an unweighted contour length.

Both MS and AWE initially were solved using level-sets methods, but more recently convex methods have been used. The MS functional is NP-hard in general, but convex relaxations are computable, and can be exact in the binary case. In particular, the ROF model is convex, and correspond to the MS model without the last term [12]. From the image analysis point of view, these models are not readily usable, because they correspond to simplistic models of vision, and if markers or shape constraints are added, they tend to dominate the model, which then does not help very much.

3.2.6.2 Hierarchical Models

Hierarchies of segmentations are a powerful way to deal with the multi-resolution inherent to nature. Many images contain objects at different scales. In medical imaging a vascular network is a typical example. It is very difficult to come up with a seeded strategy to solve this case. One general idea is to perform many segmentations at once or in sequence, taking into account various scales. This is not as easy to do as it sounds, because simply repeating a segmentation procedure with different parameters will not yield compatible segmentations, in the sense that contours are not likely to remain stable as the scale increases or decreases. One way of dealing with this is to offer a measure to the strength of a particular piece of contour, and as the scale increase, remove pieces of contours with weak strength first. This *saliency* idea was proposed by Najman and Schmitt in [51] in the context of watershed segmentation, but more work has been done on this idea since, for example, on ultrametric watershed and connections [50, 63]. A saliency map or ultrametric watershed is an interactive segmentation because edge strength can be selected by interactive thresholding for instance, but it is not always obvious how to combine this with seeded segmentation.

Hierarchical methods do offer some other benefits, such as the ability to efficiently optimize Mumford–Shah-like functionals on a saliency map [35]. Other functionals are also possible, such as optimizing minimum ratio costs [34]. There are some drawbacks as well, such as decreased speed and extra memory requirement, and again the question of compatibility with other constraints. This is at present a very interesting area of research.

3.2.6.3 Combinations

Many segmentation algorithms can be combined to provide different sets of compromises or extensions. For instance, Yuille proposed an interesting model combining Bayesian methods with level-sets [69]. An active area of research today are so-called *turbopixels*, where a first-level over-segmentation is performed in order to group pixels into consistent regions of similar size. Then these regions are linked in a graph and a discrete segmentation is performed over them [42]. This two-level segmentation procedure has some advantages in terms of speed and resource allocations. Final segmentation can still be precise if the first-order grouping is done well, and these methods are compatible with seeded segmentation. However, segmentation quality may be poor in the presence of weak edges [64].

3.3 A Unifying Framework for Discrete Seeded Segmentation

In many early segmentation methods, the focus was on the values of the pixels themselves, or in graph terms the values of the vertices. Since the advent of GC methods, it was realized that focusing instead on the edges was useful. In particular, defining a gradient function on the edges is easy. Let p and q be two vertices in the graph $\Gamma(\mathscr{V},\mathscr{E})$ of image I, that we have been using so far (see Sect. 3.2.5), then we can set as weight $w_{p,q}$ for the edge linking p and q any value depending on the discrete gradient $I_q - I_p$, where I_q represents the value of I at vertex q. For instance, we can use $w_{p,q} = \exp(-\beta|I_q - I_p|^2)$, with β a positive scalar parameter. This is a monotonically decreasing function of the gradient, recommended by several authors. In addition, there are topological advantages, as a cut in such a graph obeys the Jordan property in arbitrary dimension. In addition, there is a fundamental difference between regions, formed of uniformly labeled vertices, and cuts formed of edges. In former pixel-based segmentation procedures, the contours were themselves made of pixels, which created problems [19]. The only significant drawback is that storing edge weights rather than pixels costs roughly twice as much memory in 2D, or three times as much in 3D for the simplest nearest-neighbor connectivity. This extra cost increases with the connectivity, and may be indeed be a problem in some applications.

3.3.1 *Discrete Optimization*

Assuming then this simple model of discrete images, the segmentation problem can be viewed as an optimization problem over cliques of one or two pixels, like in the MRF setting. For instance, classical graph cut can optimize the following problem exactly:

$$\operatorname{argmin}_x E(x) = \sum_{u\in\mathscr{V}} w_u|x_u - y_u| + \sum_{(u,v)\in\mathscr{E}} w_{u,v}|x_u - x_v|, \tag{3.11}$$

in the case where x is a binary vertex labeling, y a reference binary image that can, for instance, represent seeds, and w_u and $w_{u,v}$ positive unary weights and binary weights respectively. The seeded segmentation case corresponds to an image y containing some vertices labelled with 0, others with 1 (the seeds) and unlabelled ones as well. The w_u for the labelled vertices in y have infinite weights, and the unlabeled one zero. Using the same notation, the Random Walker optimizes the following energy:

$$\operatorname{argmin}_x E(x) = \sum_{u\in\mathscr{V}} w_u(x_u - y_u)^2 + \sum_{(u,v)\in\mathscr{E}} w_{u,v}(x_u - x_v)^2. \tag{3.12}$$

In this case, the optimal labelling $x^\star$ is not binary even if y is binary. It expresses the probability of a vertex belonging to label 0 or label 1. To reach a unique solution, we must threshold the result:

$$s_u = 0 \text{ if } x_u < \frac{1}{2}, s_u = 1 \text{ otherwise.} \tag{3.13}$$

In this case, the binary result s represents the segmentation. There is a striking similarity between (3.11) and (3.12), which leads us to propose a unifying framework.

Table 3.1 Our generalized scheme for image segmentation includes several popular segmentation algorithms as special cases of the parameters p and q. The power watershed are previously unknown in the literature, but may be optimized efficiently with a MSF calculation

q\p	0	Finite	∞
1	Collapse to seeds	Graph cuts	Power watershed $q = 1$
2	ℓ_2 norm Voronoi	Random walker	Power watershed $q = 2$
∞	ℓ_1 norm Voronoi	ℓ_1 norm Voronoi	Shortest path forest

3.3.2 A Unifying Framework

We propose to optimize the following general discrete energy:

$$\text{argmin}_x E(x) = \sum_{u \in \mathscr{V}} w_u^p |x_u - y_u|^q + \sum_{(u,v) \in \mathscr{E}} w_{u,v}^p |x_u - x_v|^q, \tag{3.14}$$

The p and q terms are integer exponents. In cases where the optimal $x^\star$ is not binary, we threshold it in the end as in (3.13). An analysis of the influence of p and q provides us with Table 3.1.

In this table, we find some well-known algorithms, such as previously mentioned GR and RW, in addition to the Shortest Path Forests algorithm [20], that uses forests of shortest path leading to seeds as segmentation criteria. Most of the other cases are not interesting (Voronoi diagrams, for instance), but the case $q = 1$ or 2 and $p \longrightarrow \infty$ is novel and interesting: this is the Power Watershed algorithm [15].

3.3.3 Power Watershed

Among the drawbacks of traditional watershed as described in Sect. 3.2.5.3 are the following: (1) watershed has no energy interpretation and is purely a segmentation algorithm; (2) watershed segmentations are not unique: for the same seed placement and edge weights, the same definition can provide different results; (3) watershed results tend to leak in the presence of weak boundaries. We intend to solve all three problems.

An analysis of the convergence of (3.14) in the case $q = 1$ or 2 and $p \longrightarrow \infty$ led us to the algorithm shown below.

This algorithm is illustrated in Fig. 3.5. We also show some pictorial results in Fig. 3.6, where we compare qualitatively the results of PW with the other classical discrete segmentation algorithms, namely GC, RW, SPF and the classical WT in the form of a MSF.

More details on the Power Watershed algorithm can be found in [16]. We show the PW algorithm performs very well in terms of quantitative results, that qualitatively PW is devoid of size bias and grid artifacts, while being only slightly

Algorithm: power watershed algorithm, optimizing $p \rightarrow \infty, q \geq 1$

Data: A weighted graph $\Gamma(\mathscr{V}, \mathscr{E})$ and a reference image y containing seed information

Result: A **potential function** x and a labeling s associating a label to each vertex.

Set x values as unknown except seed values.
Sort the edges of E by decreasing order of weight.
while *any node has an unknown potential* **do**
  Find an edge (or a plateau) E_{MAX} in E of maximal weight; denote by S the set of nodes connected by E_{MAX}.
  if *S contains any nodes with known potential* **then**
    Find x_S minimizing (3.14) (using the input value of q) on the subset S with the weights in E_{MAX} set to $w_{ij} = 1$, all other weights set to $w_{ij} = 0$ and the known values of x within S fixed to their known values.
    Consider all x_S values produced by this operation as known.
  else
    Merge all of the nodes in S into a single node, such that when the value of x for this merged node becomes known, all merged nodes are assigned the same value of x and considered known.
Set $s_i = 1$ if $x_i \geq \frac{1}{2}$ and $s_i = 0$ otherwise.

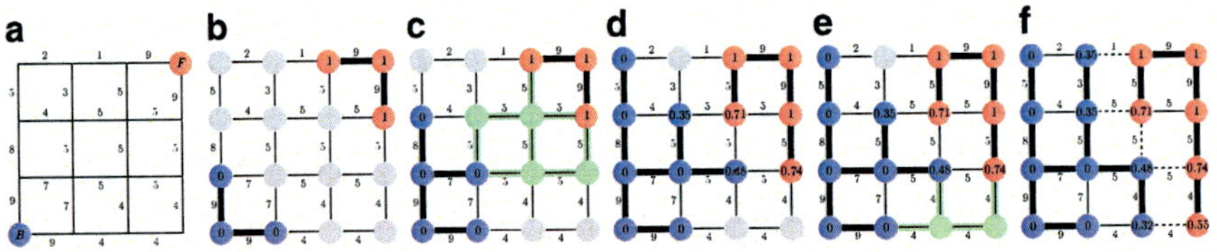

Fig. 3.5 Illustration of the different steps for Algorithm in the case $q = 2$. The values on the nodes correspond to x, their color to s. The bold edges represents edges belonging to a Maximum Spanning Forest. (**a**) A weighted graph with two seeds, all maxima of the weight function are seeded, (**b**) First step, the edges of maximum weight are added to the forest, (**c**) After several steps, the next largest edge set belongs to a plateau connected to two labeled trees, (**d**) Minimize (3.14) on the subset (considering the merged nodes as a unique node) with $q = 2$ (i.e., solution of the Random Walker problem), (**e**) Another plateau connected to three labeled vertices is encountered, and (**f**) Final solutions x and s obtained after few more steps. The q-cut, which is also an MSF cut, is represented in *dashed lines*

slower than standard watershed and much faster than either GC or RW, particularly in 3D. The PW algorithm provides a unique unambiguous result, and an energy interpretation for watershed, which allows it to be used in wider contexts as a solver, for instance in filtering [17] and surface reconstruction. One chief advantage of PW with respect with GC for instance, is its ability to compute a globally optimal result in the presence of multiple labels. When segmenting multiple objects this can be important.

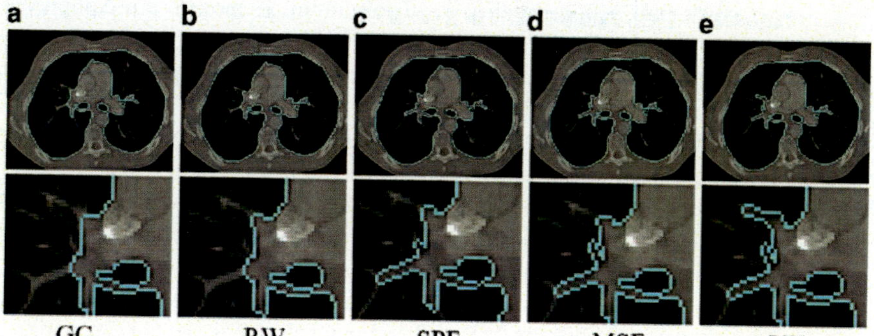

Fig. 3.6 Slides of a 3D lung segmentation. The foreground seed used for this image is a small rectangle in one slice of each lung, and the background seed is the frame of the image (**a**) GC (**b**) RW (**c**) SPF (**d**) MSF (**e**) PW

3.4 Globally Optimum Continuous Segmentation Methods

Here, we provide some arguments for globally optimal segmentation in the context of continuous-domain optimization.

3.4.1 *Dealing with Noise and Artifacts*

Even assuming we can construct a contents metric as explained in the first section, there are several sources of artifacts in segmentation: (1) weak edges cause uncertainty in the result; (2) noise tends to corrupt boundaries, and for some methods tend to lead to wrong results; (3) method artifacts, such as a size bias or blockiness artifacts can cause undesirable results. Of course all these artifacts are linked and essentially due to the contents metric, reflecting insufficient knowledge about the content, but it is precisely to solve this problem that we require segmentation.

Weak edges are a fact of life in medical imaging. Most often in CT for example it is difficult to delineate a lesion because it has a similar radiation absorption profile to surrounding tissues. In this case, it is better to use methods that interpolate contours and surfaces well. The GAC is very useful in this context because of its geometric formulation and shortest path/minimal surface interpretation. In addition, it is straightforward to add simple shape information, such as elliptical or spherical shape priors.

Many iterative methods do not cope well with noise. One reason might be that the formulation of the corresponding energy is not convex, which implies that it would probably not have a single global optimum. This is unfortunately the case with most active contours and level set formulations, including the classic formulation of GACs. In addition, these methods make it easy to add terms to the energy and make it look like it can be optimized. The reality is that in most cases, these methods get stuck into a poor quality local minimum. If models are complex, tweaking their parameters is difficult and error-prone. This is the reason why most recent segmentation models feature few parameters and tend to propose formulations that can be optimized globally.

Finally, all segmentation methods present artifacts. Graph Cuts for instance tend to both produce blocky results (grid bias) and favour small objects (shrinking bias). They can be coped with by augmenting the connectivity of the graph and by metric manipulation knowing the position of the seeds. However, it is preferable to use formulations that are isotropic in nature, such as continuous-domain ones.

These are some of the reasons that motivate us to mention continuous, isotropic, efficient formulations for finding the global solution to the GAC equation exactly.

3.4.2 Globally Optimal Geodesic Active Contour

In spite of advances in GAC optimization, more efficient ways of solving equation (3.5) do exist. In particular, in 2D, this equation can be solved by a continuous-domain, non point-convex circular shortest path [3]. The solution, called the globally optimal geodesic active contour (GOGAC) is globally optimal and extremely efficient [4], although it can only find a single contour at a time. The GOGAC solution is as flexible as the original GAC, but due to its formulation and algorithm, it is significantly less affected by noise.

This GOGAC has no shrinking bias and no grid bias, however, it tends to favor circular boundaries due to its polar coordinate equivalence. This may be desirable in some applications, but not in others. This can be avoided by using a different weighting than the $1/r$ given in the original article. A flat weighting can be used if small solutions are forbidden for instance.

3.4.3 Maximal Continuous Flows and Total Variation

The GOGAC solution is extremely efficient but does not extend to 3D and higher, but in 2006, Appleton and Talbot proposed a continuous maximum flow (CMF) solution to solve this problem. Their solution, inspired by local solutions for discrete graph cuts, consists of simulating a flow originating from a source s and ending in a sink t, and a pressure field, linked by a PDE system forming a propagation equation and constrained by the metric g:

$$\begin{aligned} \frac{\partial \vec{F}}{\partial t} &= -\nabla P \\ \frac{\partial P}{\partial t} &= -\operatorname{div} \vec{F} \\ \|\vec{F}\| &\leq g. \end{aligned} \tag{3.15}$$

This unusual system, at convergence, produces a scalar field P that acts as an indicator function for the interior of the contour C of (3.5). It solves the closed minimal surface problem exactly and efficiently, and so this represents a better way to solve it than (3.4). The result in 2D is exactly the same as that obtained with GOGAC. This CMF result provides a direct algorithm for solving the problem posed by Iri and Strang in [37, 65]. Interestingly, researchers in image restoration had proposed over the years solutions to Strang's dual problem, that of minimizing the *total variation* (TV) of a functional. Initial solutions used level-set formulations [60], and later ones convex optimization methods [11, 58]. Nonetheless, it is thought that primal maximum flow methods are better suited to segmentation than TV formulations [18]. Note that CMF are also biased towards small contours, and because they find a global optimum, this is a more serious problem than with standard GAC. However, there exist ways to remove this shrinking bias for an arbitrary collection of sources and sinks [2], and the bias is less strong in 3D and can be ignored, as long as small solutions are forbidden, using for instance large enough inner seeds. CMFs are about as fast as GC, but can be parallelized easily.

3.5 Comparison and Discussion

In the space of a single chapter it is not possible to present a thorough, quantitative assessment of the most popular segmentation methods. However, in Table 3.2, we present a qualitative comparison.

In this table, we have presented all the methods discussed in the chapter. A score of 1 indicates a low, undesirable score and the highest score is 5. These scores are potentially controversial and represent experience and opinion rather than hard fact. We have ranked all methods according to some desirable features. In the following discussion, we present robustness as the ability of a method to cope with noise and weak edges. Flexibility denotes the ability of a method to be used in different contexts: seeded or non-seeded segmentation, and the possibility to optimize different models, for instance with texture.

Taking the methods in order, we see that (1) Pixel selection uses low resources but is extremely simplistic; (2) Contour tracking has some speed and flexibility advantages but is limited to 2D; (3) Split-and-merge methods generally have high scores but are not robust and not flexible; (4) MRFs and Bayesian methods optimized by simulated annealing feature a lot of flexibility but are very slow; (5) Active contours are fast and flexible but not robust, they find only one object at a time, and cannot be extended easily to 3D; (6) Level sets (LS) are similar in some ways but are quite slow, require lots of resources and are not robust. They do extend to 3D readily; (7) GAC are a particular case of LS methods, which are popular in medical imaging because they are faster and more robust but less flexible. However standard GAC is slow compared to many other methods and still not robust enough; (8) Graph cuts are a very popular recent method, which feature relatively high scores across the board, in particular they are among the most flexible and

Table 3.2 A qualitative assessment of many popular segmentation methods. See text for details

	High Speed	Low memory	Multi-label	Flexibility	Robustness	No bias	3D and more	Parallelizable	Multi-resolution	score
Pixel selection	5	5	1	1	1	5	5	5	1	29
Contour tracking	5	5	1	4	3	4	1	1	2	26
Split and merge	4	4	5	2	2	3	4	3	5	32
MRF – SA	1	3	4	4	3	3	3	5	2	28
Active contours	4	4	1	5	2	2	2	2	2	24
Level sets	1	2	2	5	2	3	5	4	3	27
GAC	2	2	2	3	3	3	5	4	3	27
Graph cuts	2	3	2	4	5	2	4	2	3	27
Watershed	4	4	5	2	3	5	5	3	4	35
Random Walker	3	3	5	3	4	4	4	4	3	33
GOGAC	5	3	1	1	5	3	1	1	2	22
CMF	3	2	1	2	5	3	5	4	3	28
Power watershed	4	3	5	3	4	5	5	2	4	35

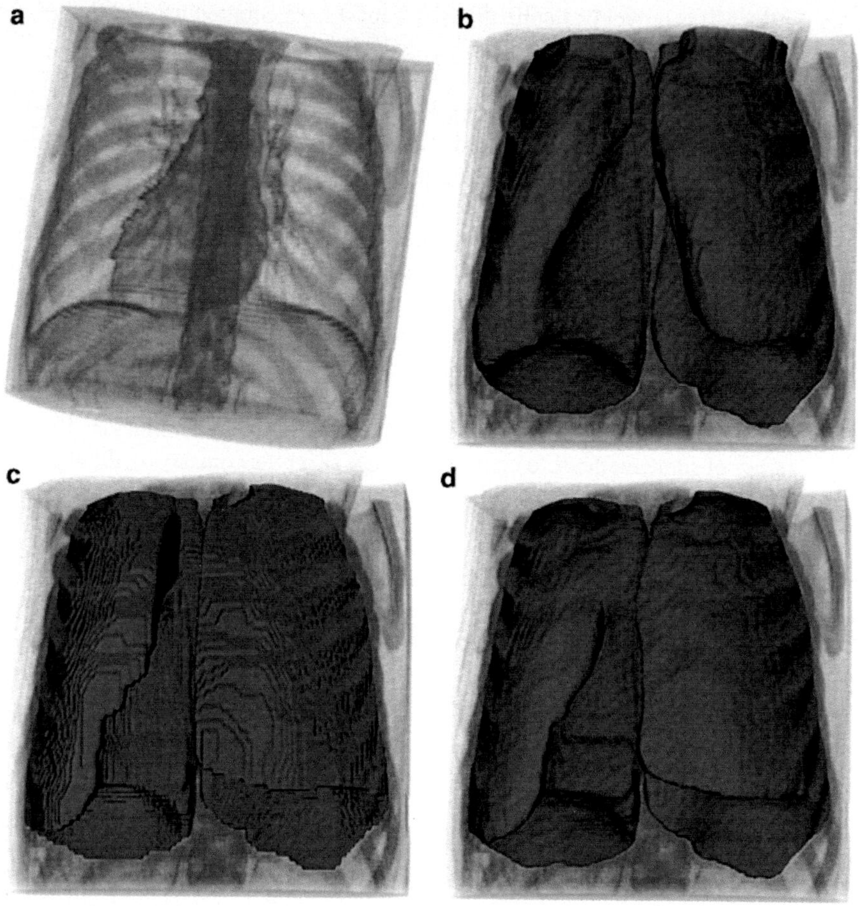

Fig. 3.7 Segmentation of the lungs in a chest CT image. (**a**) The CT image. (**b**) Segmentation using 3D standard Geodesic Active Contours. The surfaces fail to fill the base of the lung. (**c**) Segmentation using a discrete maximal flow algorithm. Observe the directional bias due to the grid. (**d**) Segmentation from identical input using continuous maximal flows

robust methods. However, they are slow, particularly in 3D, not parallelizable easily and feature much bias; (9) Watershed is an old method but has definite advantages for segmentation: it is fast, bias-free and multi-label (it can segment many objects at once). However, it is not flexible or very robust. Watershed can be extended readily for multi-resolution, and due to its age, many parallel implementations exist, including hardware ones; (10) The Random Walker is a recent method which is similar in some ways to Watershed, but is significantly more robust. It requires more resources however.

Among the newer methods presented in this chapter, (11) GOGAC solves GAC exactly and quickly in 2D, and so provides a quick robust solution, which is good for 2D interactive segmentation of single objects. However, is not flexible in its model;

(12) CMF is probably among the most robust segmentation method in the literature for 3D segmentation, but it segments only one object at a time, is not very flexible, and has no grid bias but does feature a shrinking bias. Finally, (13) Power watershed fits in between standard watershed and random walker. It is significantly more flexible and robust than standard watershed. Its speed is also comparable, but it uses more memory, and is less parallelizable.

The global score is probably even more subject to controversy than the individual ones, but it would tend to show that active contour methods should not be tried as a first choice method. For medical imaging, Random Walker and watershed-based methods are probably a good first choice, particularly for ease of use. It is comforting to realize that more modern methods suitable for 3D medical imaging (GC, RW, PW and CMF) are all very robust.

Many advantages presented in the literature, such as purported sub-pixel accuracy of segmentation, are not listed here because they are an illusion. The reported ability of some methods to control topology or on the contrary to allow it to change is not necessarily a drawback or advantage either way, so we do not include it as well.

3.6 Conclusion and Future Work

In conclusion, we argue that seeded or interactive segmentation is useful in medical imaging. Compared with model-based segmentation, seeded segmentation is more robust in actual image analysis applications, as opposed to computer vision. The ability to separate seeds/markers, use contour information, and perform contour optimization are very useful, as these elements generally result in a higher likelihood of good results. From this point of view, we argue that segmentation is a process and not merely an operator.

In general, the literature focuses on contour placement optimization at the expense of the other two components, with some rare exceptions. This is unfortunate but understandable with respect to seeds/markers, because they are highly application dependent. The choice of methods for obtaining contour information is also limited, and this is probably a good area for future research. One conclusion of this study is that contour placement optimization methods are important. More recent methods focus on optimization robustness, which is important. For someone not yet experienced in medical segmentation, simpler, more robust methods should be preferred over complex ones. Among those, power-watershed is a good candidate because of its combination of speed, relative robustness, ability to cope with multiple labels, absence of bias and availability (the code is easily available online). The random walker is also a very good solution, but is not generally and freely available.

We have not surveyed or compared methods that encompass shape constraints. We recognize that this is important in some medical segmentation methods, but this would require another study altogether.

Finally, at present there exists a dichotomy between the way discrete and continuous-domain work and are presented. In the near future, it is likely we will see methods unifying both aspects to great advantage.

References

1. Adams, R., Bischof, L.: Seeded region growing. IEEE Trans. Pattern Anal. Mach. Intell. **16**(6), 641–647 (1994)
2. Appleton, B.: Globally minimal contours and surfaces for image segmentation. Ph.D. thesis, University of Queensland (2004). Http://espace.library.uq.edu.au/eserv/UQ:9759/ba_thesis.pdf
3. Appleton, B., Sun, C.: Circular shortest paths by branch and bound. Pattern Recognit. **36**(11), 2513–2520 (2003)
4. Appleton, B., Talbot, H.: Globally optimal geodesic active contours. J. Math. Imaging Vis. **23**, 67–86 (2005)
5. Ardon, R., Cohen, L.: Fast constrained surface extraction by minimal paths. Int. J. Comput. Vis. **69**(1), 127–136 (2006)
6. Beucher, S., Lantuéjoul, C.: Use of watersheds in contour detection. In: International Workshop on Image Processing. CCETT/IRISA, Rennes, France (1979)
7. Boykov, Y., Kolmogorov, V.: An experimental comparison of min-cut/max- flow algorithms for energy minimization in vision. PAMI **26**(9), 1124–1137 (2004)
8. Boykov, Y., Veksler, O., Zabih, R.: Fast approximate energy minimization via graph cuts. IEEE Trans. Pattern Anal. Mach. Intell. **23**(11), 1222–1239 (2001)
9. Canny, J.: A computational approach to edge detection. IEEE Trans. Pattern Anal. Mach. Intell. **8**(6), 679–698 (1986)
10. Caselles, V., Kimmel, R., Sapiro, G.: Geodesic active contours. Int. J. Comput. Vis. **22**(1), 61–79 (1997)
11. Chambolle, A.: An algorithm for total variation minimization and applications. J. Math. Imaging Vis. **20**(1–2), 89–97 (2004)
12. Chan, T., Bresson, X.: Continuous convex relaxation methods for image processing. In: Proceedings of ICIP 2010 (2010). Keynote talk, http://www.icip2010.org/file/Keynote/ICIP
13. Chan, T., Vese, L.: Active contours without edges. IEEE Trans. Image Process. **10**(2), 266–277 (2001)
14. Cohen, L.D., Kimmel, R.: Global minimum for active contour models: A minimal path approach. Int. J. Comput. Vis. **24**(1), 57–78 (1997). URL citeseer.nj.nec.com/cohen97global.html
15. Couprie, C., Grady, L., Najman, L., Talbot, H.: Power watersheds: A new image segmentation framework extending graph cuts, random walker and optimal spanning forest. In: Proceedings of ICCV 2009, pp. 731–738. IEEE, Kyoto, Japan (2009)
16. Couprie, C., Grady, L., Najman, L., Talbot, H.: Power watersheds: A unifying graph-based optimization framework. IEEE Transactions on Pattern Analysis and Machine Intelligence, **33**(7), 1384–1399 (2011)
17. Couprie, C., Grady, L., Talbot, H., Najman, L.: Anisotropic diffusion using power watersheds. In: Proceedings of the International Conference on Image Processing (ICIP), pp. 4153–4156. Honk-Kong (2010)
18. Couprie, C., Grady, L., Talbot, H., Najman, L.: Combinatorial continuous maximum flows. SIAM J. Imaging Sci. (2010). URL http://arxiv.org/abs/1010.2733. In revision
19. Cousty, J., Bertrand, G., Najman, L., Couprie, M.: Watershed cuts: Minimum spanning forests and the drop of water principle. In: IEEE Transactions on Pattern Analysis and Machine Intelligence, pp. 1362–1374. (2008)

20. Cousty, J., Bertrand, G., Najman, L., Couprie, M.: Watershed cuts: Thinnings, shortest-path forests and topological watersheds. IEEE Trans. Pattern Anal. Mach. Intell. **32**(5), 925–939 (2010)
21. Cserti, J.: Application of the lattice Green's function for calculating the resistance of an infinite network of resistors. Am. J. Phys. **68**, 896 (2000)
22. Daragon, X., Couprie, M., Bertrand, G.: Marching chains algorithm for Alexandroff-Khalimsky spaces. In: SPIE Vision Geometry XI, vol. 4794, pp. 51–62 (2002)
23. Dougherty, E., Lotufo, R.: Hands-on Morphological Image Processing. SPIE press, Bellingham (2003)
24. Doyle, P., Snell, J.: Random Walks and Electric Networks. Carus Mathematical Monographs, vol. 22, p. 52. Mathematical Association of America, Washington, DC (1984)
25. Ford, J.L.R., Fulkerson, D.R.: Flows in Networks. Princeton University Press, Princeton, NJ (1962)
26. Geman, S., Geman, D.: Stochastic relaxation, gibbs distributions, and the bayesian restoration of images. PAMI **6**, 721–741 (1984)
27. Goldberg, A., Tarjan, R.: A new approach to the maximum-flow problem. J. ACM **35**, 921–940 (1988)
28. Goldenberg, R., Kimmel, R., Rivlin, E., Rudzsky, M.: Fast geodesic active contours. IEEE Trans. Image Process. **10**(10), 1467–1475 (2001)
29. Grady, L.: Multilabel random walker image segmentation using prior models. In: Computer Vision and Pattern Recognition, IEEE Computer Society Conference, vol. 1, pp. 763–770 (2005). DOI http://doi.ieeecomputersociety.org/10.1109/CVPR.2005.239
30. Grady, L.: Computing exact discrete minimal surfaces: Extending and solving the shortest path problem in 3D with application to segmentation. In: Computer Vision and Pattern Recognition, 2006 IEEE Computer Society Conference, vol. 1, pp. 69–78. IEEE (2006)
31. Grady, L.: Random walks for image segmentation. IEEE Trans. Pattern Anal. Mach. Intell. **28**(11), 1768–1783 (2006)
32. Grady, L., Funka-Lea, G.: Multi-label image segmentation for medical applications based on graph-theoretic electrical potentials. In: Computer Vision and Mathematical Methods in Medical and Biomedical Image Analysis, pp. 230–245. (2004)
33. Grady, L., Polimeni, J.: Discrete Calculus: Applied Analysis on Graphs for Computational Science. Springer Publishing Company, Incorporated, New York (2010)
34. Grady, L., Schwartz, E.: Isoperimetric graph partitioning for image segmentation. Pattern Anal. Mach. Intell. IEEE Trans. **28**(3), 469–475 (2006)
35. Guigues, L., Cocquerez, J., Le Men, H.: Scale-sets image analysis. Int. J. Comput. Vis. **68**(3), 289–317 (2006)
36. Horowitz, S., Pavlidis, T.: Picture segmentation by a directed split-and-merge procedure. In: Proceedings of the Second International Joint Conference on Pattern Recognition, vol. 424, p. 433 (1974)
37. Iri, M.: Survey of Mathematical Programming. North-Holland, Amsterdam (1979)
38. Kakutani, S.: Markov processes and the Dirichlet problem. In: Proceedings of the Japan Academy, vol. 21, pp. 227–233 (1945)
39. Kass, M., Witkin, A., Terzopoulos, D.: Snakes: Active contour models. Int. J. Comput. Vis. **1**, 321–331 (1988)
40. Khalimsky, E., Kopperman, R., Meyer, P.: Computer graphics and connected topologies on finite ordered sets. Topol. Appl. **36**(1), 1–17 (1990)
41. Kruskal, J.J.: On the shortest spanning subtree of a graph and the travelling salesman problem. Proc. AMS **7**(1) (1956)
42. Levinshtein, A., Stere, A., Kutulakos, K., Fleet, D., Dickinson, S., Siddiqi, K.: Turbopixels: Fast superpixels using geometric flows. Pattern Anal. Mach. Intell. IEEE Trans. **31**(12), 2290–2297 (2009)
43. Malladi, R., Sethian, J., Vemuri, B.: Shape modelling with front propagation: A level set approach. IEEE Trans. Pattern Anal. Mach. Intell. **17**(2), 158–175 (1995)

44. Marr, D., Hildreth, E.: Theory of edge detection. Proc. R. Soc. Lond. Ser. B Biol. Sci. **207**, 187–217 (1980)
45. Menzies, S.W., Crotty, K.A., Ingvar, C., McCarthy, W.H.: An Atlas of Surface Microscopy of Pigmented Skin Lesions. McGraw-Hill, Roseville, Australia (1996). ISBN 0 07 470206 8
46. Meyer, F.: Topographic distance and watershed lines. Signal Process. **38**(1), 113–125 (1994)
47. Meyer, F., Beucher, S.: Morphological segmentation. J. Vis. Commun. Image Represent. **1**(1), 21–46 (1990)
48. Mortensen, E., Barrett, W.: Interactive segmentation with intelligent scissors. Graph. Models Image Process. **60**(5), 349–384 (1998)
49. Mumford, D., Shah, J.: Optimal approximations by piecewise smooth functions and associated variational problems. Commun. Pure Appl. Math. **42**(5), 577–685 (1989)
50. Najman, L.: On the equivalence between hierarchical segmentations and ultrametric watersheds. Journal of Mathematical Imaging and Vision, **40**(3), 231–247 (2011)
51. Najman, L., Schmitt, M.: Geodesic saliency of watershed contours and hierarchical segmentation. Pattern Anal. Mach. Intell. IEEE Trans. **18**(12), 1163–1173 (2002)
52. Najman, L., Talbot, H. (eds.): Mathematical Morphology: From theory to applications. ISTE-Wiley, London, UK (2010)
53. Osher, S., Sethian, J.: Fronts propagating with curvature-dependent speed: Algorithms based on Hamilton-Jacobi formulations. J. Comput. Phys. **79**(1), 12–49 (1988)
54. Otsu, N.: A threshold selection method from gray-level histograms. Automatica **11**, 285–296 (1975)
55. Paragios, N., Deriche, R.: Geodesic active contours and level sets for the detection and tracking of moving objects. Pattern Anal. Mach. Intell. IEEE Trans. **22**(3), 266–280 (2002)
56. Paragios, N., Deriche, R.: Geodesic active regions and level set methods for supervised texture segmentation. Int. J. Comput. Vis. **46**(3), 223–247 (2002)
57. Pham, D., Xu, C., Prince, J.: Current methods in medical image segmentation1. Biomed. Eng. **2**(1), 315 (2000)
58. Pock, T., Cremers, D., Bischof, H., Chambolle, A.: An algorithm for minimizing the Mumford-Shah functional. In: 12th International Conference on Computer Vision, pp. 1133–1140. IEEE (2009)
59. Prim, R.: Shortest connection networks and some generalizations. Bell Syst. Techn. J. **36**(6), 1389–1401 (1957)
60. Rudin, L.I., Osher, S., Fatemi, E.: Nonlinear total variation based noise removal algorithms. Phys. D **60**(1-4), 259–268 (1992). DOI http://dx.doi.org/10.1016/0167-2789(92)90242-F
61. Sagiv, C., Sochen, N., Zeevi, Y.: Integrated active contours for texture segmentation. Image Process. IEEE Trans. **15**(6), 1633–1646 (2006)
62. Sethian, J.: Level set methods and fast marching methods. Cambridge University Press, Cambridge (1999). ISBN 0-521-64204-3
63. Soille, P.: Constrained connectivity for hierarchical image partitioning and simplification. IEEE Trans. Pattern Anal. Mach. Intell. **30**(7), 1132–1145 (2008)
64. Stawiaski, J., Decencière, E., Bidault, F.: Computing approximate geodesics and minimal surfaces using watershed and graph cuts. In: Banon, G.J.F., Barrera, J., Braga-Neto, U.d.M., Hirata, N.S.T. (eds.) Proceedings of the 8th International Symposium on Mathematical Morphology, vol. 1, pp. 349–360. Instituto Nacional de Pesquisas Espaciais (INPE) (2007). URL http://urlib.net/dpi.inpe.br/ismm@80/2007/03.19.13.04
65. Strang, G.: Maximal flow through a domain. Math. Program. **26**, 123–143 (1983)
66. Veksler, O.: Efficient graph-based energy minimization. Ph.D. thesis, Cornell University (1999)
67. Vincent, L., Soille, P.: Watersheds in digital spaces: An efficient algorithm based on immersion simulations. IEEE Trans. Pattern Anal. Mach. Intell. **13**(6), 583–598 (1991)
68. Weickert, J., Romeny, B., Viergever, M.: Efficient and reliable schemes for nonlinear diffusion filtering. Image Process. IEEE Trans. **7**(3), 398–410 (2002)
69. Zhu, S., Yuille, A.: Region competition: Unifying snakes, region growing, and Bayes/MDL for multiband image segmentation. Pattern Anal. Mach. Intell. IEEE Trans. **18**(9), 884–900 (2002)

Chapter 4
Deformable Models and Level Sets in Image Segmentation

Agung Alfiansyah

4.1 Introduction

Segmentation is a partitioning process of an image domain into non-overlapping connected regions that correspond to significant anatomical structures. Automated segmentation of medical images is a difficult task. Images are often noisy and usually contain more than a single anatomical structure with narrow distances between organ boundaries. In addition, the organ boundaries may be diffuse. Although medical image segmentation has been an active field of research for several decades, there is no automatic process that can be applied to all imaging modalities and anatomical structures [1].

In general, segmentation techniques can be classified into two main categories: (a) segmentation methods that allow users to explicitly specify the desired feature and (b) algorithms where the specification is implicit. The first segmentation class considers the segmentation as a real-time interaction process between the user and the algorithm. The user is provided with the output and allowed to perform feed-back directly in order to modify the segmentation until he/she gets a satisfactory result. In the extreme case, this framework might degenerate into manual segmentation with the user forcing his/her desirable results. Some examples of this approach are the livewire segmentation algorithms [2–5]. These algorithms produce a piecewise optimal boundary representation of an object, by viewing the image as a weighted graph and finding the shortest path between consecutive specified boundary points of the user. A more recent example of this approach based on the concept of random walks is described in [6].

The majority of segmentation methods belong to the second category, where the desired result is specified implicitly. Segmentation algorithms belonging to this category include: thresholding, various contour-based and region-based segmentation

A. Alfiansyah (✉)
Surya Research and Education Center Tangerang, Tangerang, Indonesia
e-mail: agung.alfiansyah@gmail.com

G. Dougherty (ed.), *Medical Image Processing: Techniques and Applications*, Biological and Medical Physics, Biomedical Engineering, DOI 10.1007/978-1-4419-9779-1_4,

methods, Markov random fields, active contours, model-based, and deformable model methods. As this domain has been studied extensively, there exist many published reviews of medical image segmentation [7, 39, 40], with specialized surveys on deformable models [7,8], vessel extraction [9,10], or brain segmentation [10,11].

This chapter is organized as follows. In Sect. 4.2, we review the main concepts of the deformable model and its role in image segmentation. We present several types of deformable models according to how they are represented, followed by a description of their energy definition and possible optimization methods. In Sect. 4.3, we compare the performance of these deformable models according to properties such as initialization, topological change handling, stability, etc. In Sect. 4.4, we discuss several cases of developed applications of medical image segmentation and conclude in Sect. 4.5.

4.2 Deformable Models

The basic idea of active contours for image segmentation is to embed an initial contour (or surface in the three-dimensional case) into the image, and to subsequently let it evolve while being subjected to various constraints. In order to detect objects in the image, the contour has to stop its evolution on the boundary of the object of interest. Although the term deformable models first appeared in the work by Kass et al. [12] in the late eighties, the idea of deforming a template for extracting image features dates back much further, with work on spring-loaded templates [13] and on the rubber mask technique [14]. In image processing literature, deformable models are also variously known as snakes, active contours or surfaces, balloons, and deformable contours or surfaces. An extensive review of the current research in this area can be found in [1,38].

In the following subsection, we present the different types of deformable models classified according to contour representation, and some energy optimization strategies.

4.2.1 Point-Based Snake

This classical snake was firstly proposed by Kass et al. [12] and represents the contour using discrete points. The behavior of this classical snake is usually modeled by a weighted linear combination of: *internal energy* calculated from the contour that determines the regularity of the curve; *external energy* which attracts the contour towards the significant features in the image; and often an additional *user energy* allowing the operator to better interact with the model.

In the first snake-type, one applies the simplest way to represent the model: a set of discrete points as snake elements $(C(s))$. Using this representation, closed contours can be formed by connecting the last "snaxel" (snake element) to the first one. For segmentation, the snake has to minimize the energy as follows:

$$E(C(s)) = \int_{\Omega} \left\{ \underbrace{E_{\text{contour}}(C(s))}_{\text{internal energy}} + \underbrace{E_{\text{image}}(C(s)) + E_{\text{user}}(C(s))}_{\text{external energy}} \right\} ds. \tag{4.1}$$

Applying discrete-point representation, the internal energy can be approximated by accommodating the elasticity and rigidity terms:

$$\begin{aligned} E_{\text{contour}}(C(s)) &= w_1(s)E_{\text{elasticity}}(C(s)) + w_2(s)E_{\text{rigidity}}(C(s)) \\ &= w_1(s)C_s(s)^2 + w_2(s)C_{ss}(s)^2, \end{aligned} \tag{4.2}$$

where the subscripts s and ss denote the first and second derivatives with respect to the curve parameter. The model behavior is controlled by constants w_1 and w_2, respectively, weighting the curve elasticity and rigidity.

Internal Energy Definition: In the first term in equation (4.2), $C_s(s)$ represents the elastic energy and makes the snake behave like a membrane. The second term $C_{ss}(s)$ represents the contour's bending energy that makes the model act like a thin plate. Decreasing elasticity allows the contour to increase in length, while increasing elasticity increases the tension of the model by reducing its length. Decreasing rigidity allows the active contour model to develop corners, while increasing rigidity makes the model smoother and less flexible. Setting either of the weighting coefficients to zero permits first- and second-order discontinuities, respectively.

This energy equation can then be discretized using the finite difference method as:

$$\mathrm{E}_{\text{elasticity}} \approx (x_s - x_{s-1})^2 + (y_s - y_{s-1})^2. \tag{4.3}$$

This term will minimize the distance between the points on the snake, causing shrinking during optimization of the energy process in the absence of external energy. In a similar way, the rigidity term is discretized as:

$$E_{\text{rigidity}} \approx (x_{s+1} - x_s - x_{s-1})^2 + (y_{s+1} - y_s - y_{s-1})^2. \tag{4.4}$$

The *elasticity* definition using finite differences discretization scheme is valid for the condition that the model's snaxels are evenly spaced [15]. In other cases, a continuity term can be defined that subtracts the average distance of the snaxels, otherwise the energy value will be larger for points which are further apart. This constraint forces the points to be more evenly spaced, and avoids possible contraction of the snake.

External Energy Definition: The external energy term in equation (4.1) represents the image potential derived from image data and guides the contour in finding the desired object. Ideally, this energy has a minimum value at the feature subject to detection. However, due to the presence of noise, there is often a convergence problem at object concavities. The snake can also integrate the constraint energy to interactively guide itself towards or away from particular features. This energy helps the contour to overcome the initialization problem or the sensitivity to noise.

Kass [12] proposed a weighted sum of the following energy terms in order to detect image features:

$$E_{\text{ext}}(C(s)) = \alpha_{\text{line}} \cdot I(s) + \alpha_{\text{edge}} \cdot -\nabla I^2 + \alpha_{\text{term}} \cdot E_{\text{term}}. \tag{4.5}$$

The most common image functional in this model is the image intensity function I. The first term will simply attract the contour to lower or higher intensity values depending on the α_{line} value. Large positive values of α_{line} tend to make the snake align itself with dark regions in the image $I(s)$, whereas large negative values of α_{line} tend to make the snake align itself with bright regions in the image.

The edge energy that attracts the contour towards high gradient values is squared to narrow the edge-gradient response. Similarly, large positive values of α_{edge} tend to make the snake align itself with sharp edges in the image whereas large negative values of α_{edge} make the snake avoid the edges. E_{term} is defined to find termination of line segments and corners. Kass proposed to use the curvature of iso-contours in a Gaussian smoothed image to attract the contours towards line termination.

4.2.1.1 User Constraint Energy

Constraint energy is applied to interactively guide the snake towards or away from particular features. This energy helps the contour to overcome the initialization problem or the sensitivity to noise.

Constraint energy was first proposed for the classical snake by allowing the user to attach springs between points of the contour and fix their position in the image plane. Kass define an energy in terms of *spring* (to attract the snake towards specified points) and *volcano* (to repulse the snake from specified points) within the image (Fig. 4.1).

The *spring* term attracts a contour point towards a spring point in the image plane, with a given constant, *spring constant*. The active contour model is attracted or repelled by the spring depending on this spring constant sign and value. The *volcano term* acts as a repulsive force between a point on the image at a inverse value of the distance from a point on the snake.

Balloon Force: Cohen [16, 17] proposed an additional force that pushes the contour in the direction normal to the contour. Since it can either inflate or deflate the contour, this force is known as a balloon force, defined as:

$$F_{\text{balloon}}(s) = k\vec{N}(s), \tag{4.6}$$

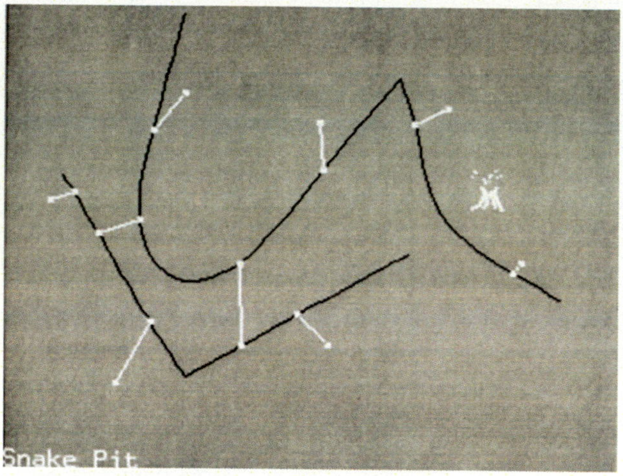

Fig. 4.1 Classical Snake [after [12]]

where $\vec{N}(s)$ is the normal unit vector, k is the weighting parameter representing the strength of the balloon force, and the sign of k determines whether the model inflates or deflates.

Incorporating an additional balloon force helps the user with the classical problem of contour initialization when it is not close enough to the desired solution, as shown in Fig. 4.2b. This force also reduces the model sensitivity to noise, and can also push the model into object concavities. It should be noted that this force only has a single direction for all of the evolved deformable models. Hence, to capture the desired object perfectly the initial contour should be totally inside (or outside) the contour. A snake without any additional force and in the absence of image energy tends to shrink into a point to minimize the internal energy. Figure 4.2c shows this condition.

Gradient Vector Flow: The basic idea of this method [18, 19, 41] is to replace the external force term $E_{\text{ext}}(\boldsymbol{I}) = \nabla P(I)$ with a *gradient vector field* (v), which can be derived from the equilibrium state of the following partial differential equation:

$$v_t = g(|\nabla f|)\nabla^2 v - h(|\nabla f|)(v - \nabla f). \tag{4.7}$$

The first term in (4.7) is referred to as the *smoothing term* since it produces a smoothly varying vector field. The second term involving $(v - \nabla f)$ is referred to as the *data term*, since it encourages the vector field v to be close to ∇f. The weight functions $g(.)$ and $h(.)$ are applied to the smoothing and data terms.

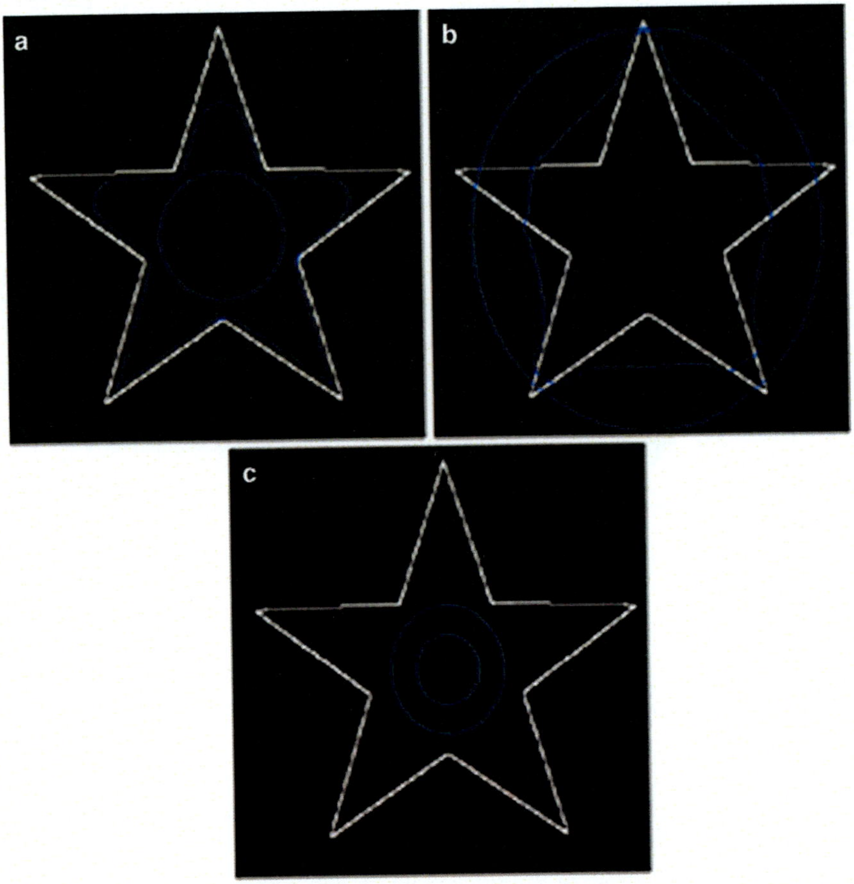

Fig. 4.2 Deformable model with additional balloon force

The authors in [5] proposed the following weight functions:

$$g(\nabla f) = e^{-\frac{|\nabla f|}{K}}$$
$$h(\nabla f) = 1 - e^{-\frac{|\nabla f|}{K}}. \tag{4.8}$$

Using these weight functions, the gradient vector flow field will conform to the distance map gradient near the relevant features, but will vary smoothly away from them. The constant K determines the extent of the field smoothness and the conformity gradient.

Figure 4.3 Illustrates the Gradient Vector Flow performance when a thin concavity or sharp corner is present in the segmented object. The Gradient Vector Flow force is able to attract the snake towards the desired contour and towards concave areas. This would not be the case with conventional external energy, since no external force attracts the snake towards that part of the image.

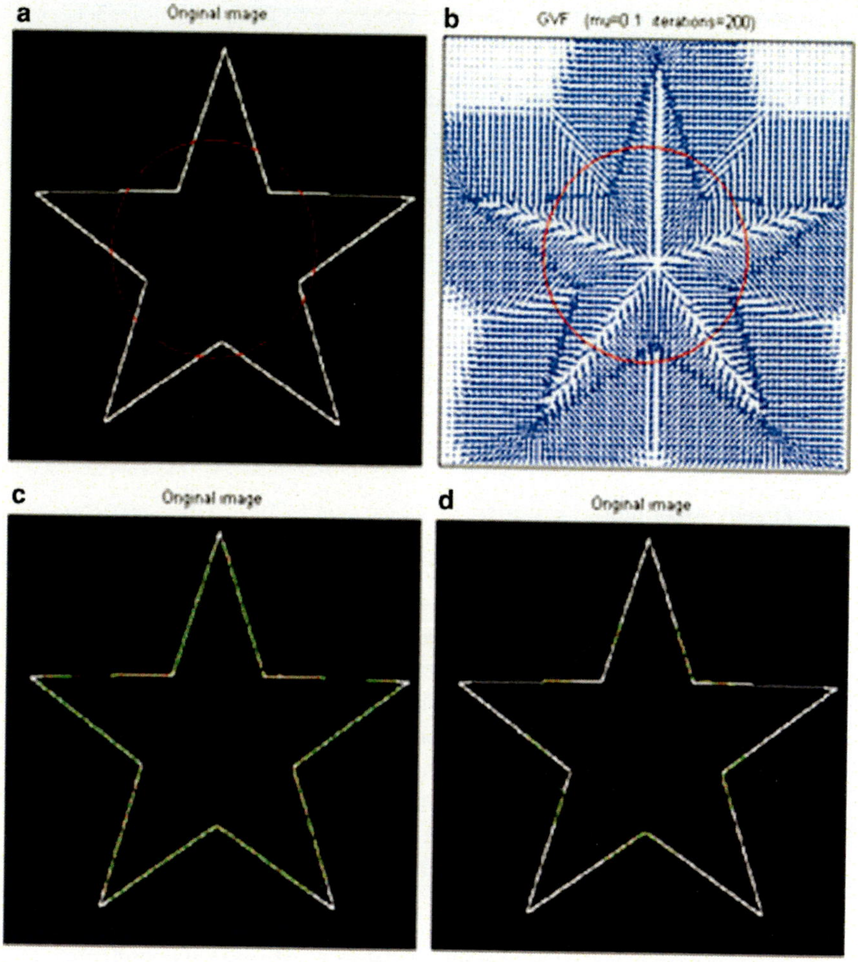

Fig. 4.3 Additional force provided by Gradient Vector Flow for synthetic image segmentation; (**a**) Initial contour defined as a circle in the image; (**b**) Vector flow visualization derived from the image; (**c**) Segmentation result using Gradient Vector Flow: the snake captures the desired object; (**d**) Segmentation result without additional force

4.2.1.2 Snake Optimization Method

Image segmentation using an active contour is formulated as a process of energy minimization that evolves the contour. This minimization controls the model deformation to reach the desired segmentation result. The term “snake” comes from the “slip and slide” movement of the contour during this minimization process.

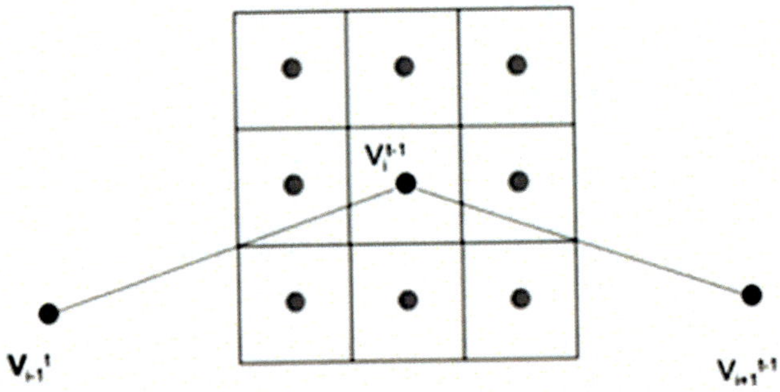

Fig. 4.4 Local neighborhood in optimization using a greedy algorithm. The energy function is evaluated at v_i^{t-1} and each of the neighbors, using v_{i-1}^t and v_{i+1}^{t-1} to compute the internal energy terms. The location with lowest energy is chosen as the new position v_i^t

Greedy Algorithm

Using greedy algorithms optimization finds the solution incrementally by choosing at each step the direction which is locally the most promising for the final result, i.e. which provides larger energy decrease.

Figure 4.4 illustrates the optimization procedure using a greedy algorithm. The energy function at the current point v_i^{t-1} and each of its neighbors is computed under consideration of adjacent contour points v_{i-1}^t and v_{i+1}^{t-1}. The location with the smallest energy value is chosen as the new position of v_i^t. The previous point v_{i-1}^t has already been updated to the new position in the current iteration over the contour, while v_{i+1}^{t-1} will be updated next.

At the first stage of the algorithm, all contour points are sequentially updated within one iteration. At the second stage, the forming of corners is determined by recalculating the rigidity energy terms with the updated points, and also by adjusting the weighting rigidity parameter w_2 for each contour point accordingly.

Dynamic Programming

Dynamic programming minimization was presented by Amini [20] to solve the variational problem in energy minimization of active contour models. Dynamic programming is different from variational optimizations in the sense that it ensures a globally optimal solution with respect to the search space and numerical stability by moving the contour points on a discrete grid without any derivative numerical approximations. The optimization process can be viewed as a discrete multi-stage decision process and is solved by a time-delayed discrete dynamic programming algorithm. Dynamic programming bypasses local minima as it embeds the minimization problem in a neighborhood-related problem. This is achieved by replacing the minimization of the total energy measurement by the minimization of a function of the form:

$$E(v_0,v_1,v_2\dots v_{n-1}) = E_1(v_0,v_1,v_2)+E_2(v_1,v_2,v_3)$$
$$+E_3(v_1,v_2,v_3)+\cdots+E_{n-2}(v_{n-1},v_{n-2},v_{n-1}). \quad (4.9)$$

$$S(v_{i+1},v_1) = \min_{v_1} v_{i-1}S_{i-1}(v_i,v_{i-1})+E_{\text{elast}}(v_{i-1},v_i)+E_{\text{rigid}}(v_{i-1},v_i,v_{i+1})$$
$$+E_{\text{ext}}(v_i). \quad (4.10)$$

Apart from the energy matrix corresponding to the optimal value function S_i, a position matrix is also needed so the value of v_i minimizing equation (4.10) can be stored. The optimal contour of minimum energy $E_{\min}(s)$ can be then found by back-tracking from the end position represented in the matrix.

This process is iterated until the active contour finds an energy that does not change significantly. It consists of a forward pass to determine the minimum energy values for each v_i and a backward pass to find the minimum energy path in the position matrix.

Similar to minimization using the greedy algorithm, dynamic programming allows hard constraints (e.g. minimum distance between snaxels) on the behavior of the global minimum solution directly and naturally. The other advantage of this method comes from the execution time and the required memory. The complexity of this method is $O(mn)$ with $O(mn^2)$ memory required, where n is the number of points on the contour and m is the number of potential locations in the search space to which every point is allowed to move during one optimization step.

Variational Method

This approach is based on the *Euler–Lagrange* condition in order to derive a differential equation that can be solved to minimize the snake energy. For each iteration, an implicit Euler forward step is performed with respect to the internal energy, and an explicit Euler step with respect to the external image and constraints energy terms.

The basic deformable model (in (4.2)) can be rewritten as:

$$E(C(s)) = \int_{\Omega} \frac{1}{2}\left(w_1(s)|C'(s)|^2+w_2(s)|C''(s)|^2\right)\mathrm{d}s+\int_{\Omega} E_{\text{ext}}(C(s))\mathrm{d}s. \quad (4.11)$$

Representing the integrand by $E(s,C',C'')$, the necessary condition for the variation to locally minimize $E(C(s))$ must satisfy the following Euler–Lagrange equation:

$$-(w_1(s)C')'+(w_2(s)C'')''+\nabla E_{\text{ext}}(C(s))=0. \quad (4.12)$$

The equation can be solved by changing C over time t, then C is a function of the contour s and time $t \rightarrow C(s,t)$. When the snake reaches a stable state $C_t(s,t) = 0$, the solution of (4.12) is obtained. Thus, insertion of this term gives:

$$\frac{\partial E_{\text{ext}}}{\partial s} - (w_1(s)C')' + (w_2(s)C'')'' + \nabla E_{\text{ext}}(C(s)) = 0. \tag{4.13}$$

If the sum of all external forces and image energy is $E_{\text{ext}}(s)$, the equation can be solved numerically by using a finite difference approach and represented implicitly in matrix multiplication form such as:

$$\mathbf{F} = \mathbf{A}.\mathbf{V}, \tag{4.14}$$

where A is a penta-diagonal banded matrix. This expression is correct for snakes with fixed-point positions at their extremities or closed snakes (i.e. $C(0)=C(N-1)$).

Generally, we assume that elasticity w_1 and rigidity w_2 are constant for both discretized space and time during curve evolution, thus:

$$a_1 = e_1 = \frac{w_2}{h^4} \quad b_1 = d_1 = \left(\frac{w_1}{h^4} + 4\frac{w_2}{h^4}\right) \quad c = \left(2\frac{w_1}{h^4} + 6\frac{w_2}{h^4}\right). \tag{4.15}$$

To solve (4.14) iteratively, the successive over-relaxation method [21] can be applied. In the two-dimensional image case, the resulting equations for evaluating time t from time $t-1$ can be solved iteratively after matrix inversion using:

$$\mathbf{V}^{\mathrm{t}} = \tau(\mathbf{A} + \mathbf{I})^{-1} \cdot (\mathbf{V}^{\mathrm{t}-1} + \tau\mathbf{F}(\mathrm{x}^{\mathrm{t}-1}, \mathrm{y}^{\mathrm{t}-1}), \tag{4.16}$$

$\mathbf{I}$ is the identity matrix, and $\tau(\mathbf{A} + \mathbf{I})^{-1}$ is also penta-diagonal.

This optimization approach does not guarantee a global minimum solution, and requires estimates of high-order derivatives on the discrete data. Moreover, hard constraints cannot be directly enforced. A desired constraint term like mean or minimum snaxel spacing can only be enforced by increasing the associated weighting term, which will force more effect on this constraint, but at the cost of other terms.

4.2.2 Parametric Deformable Models

These deformable models represent the curve or surface in an explicit parametric form during the model deformation. This representation allows for direct interaction and gives a compact representation for real-time implementation. Parametric deformable models are usually too sensitive to their initial conditions because of the non-convexity of the energy functional and the contraction force which arises from the internal energy term.

B-Spline is often used as a representation of parametric deformable models [42]. In this case, the deformable model is split into segments by knot points. Each curve segment $C(t) = \{x(t), y(t)\}$ is approximated by a piecewise polynomial function, which is obtained by a linear combination of basis functions β_i and a set of control points $v = \{x_i, y\}$. In general, however, representations using smooth basis functions require fewer parameters than point-based approaches and thus result in faster optimization algorithms [22]. Moreover, such curve models have inherent regularity and hence do not require extra constraints to ensure smoothness [22,23].

Both point-based and parametric snakes represent the model in an explicit way, hence it is easier to integrate an a priori shape constraint into the deformable model. Moreover, the user's interaction can be accommodated in a straightforward manner by allowing the user to specify some points through the desired contour evolution. The inconvenience of this model lies in reduced flexibility in accounting for topological changes during the evolution, although much effort has been spent to overcome this limitation.

4.2.2.1 Internal Energy Definition

Similar to the discrete point-based snake, internal energy is responsible for ensuring the smoothness of the contour. Actually, Kass proposed a linear combination of the length of the contour and the integral of the square of the curvature along the contour. Thus, in explicit contour representation, this energy can be defined as:

$$E_{\text{contour}} = w_1 \int_0^M (x'(t)^2 + y'(t)^2)^{\frac{1}{2}} \mathrm{d}t + w_2 \int_0^M \left(\frac{x''(t)y'(t) - x'(t)y''(t)}{(x'(t)^2 + y'(t)^2)^{\frac{3}{2}}} \right)^2 \mathrm{d}t. \quad (4.17)$$

where the second term represents the curvature at point $r(t)$.

4.2.2.2 Image Energy Definition

The most common image energy applied for the snake is defined as the integral of the square of the gradient magnitude along the curve. The main drawback widely known in using this energy is the lack of gradient direction. This information can be used to detect edges, since at the boundary image gradient is usually perpendicular to the curve. This direction should be incorporated into the image energy to make the snake more robust for image segmentation.

Region-based energy. This region-based energy represents the statistical characteristics in a region in the contour and provides snake boundary information. This is indeed very helpful when the contour is far away from the real contour to be detected. For this purpose we assume two regions in the images (which can be expanded two more times) with different probability distributions in which each of these regions has different means and variances. Staib's [24] formulation to determine the region likelihood function can be used for this case:

$$E_{\text{region}} = -\int\limits_{S} \log(P(f(s)|s \in \mathcal{R}))dS - \int\limits_{S'} \log(P(f(s)|s \in \mathcal{R}'))dS. \tag{4.18}$$

Where $\mathcal{R}$ and $\mathcal{R}'$ denote the different regions of the curve and S and S′ indicate the position inside or outside the region, respectively. The energy will be maximum when $\mathcal{R} = S$ and $\mathcal{R}' = S'$, and the regional based energy can be reformulated as:

$$E_{\text{region}} = -\int\limits_{S} \log\left(\frac{P(f(s)|s \in \mathcal{R})}{P(f(s)|s \in \mathcal{R}')}\right) dS. \tag{4.19}$$

4.2.2.3 External Constraint Energy Definition

External constraint, proposed by Kass, can be integrated in such a way that the user might specify a few points which should lie on the contour to be detected. It can be performed by adding an energy term which is the distance between these given points and the corresponding closest points on the curve.

As mentioned previously, image segmentation finally yields the totality of the energy minimization process that will place a regular contour at the edge of the object which we want to detect. Sometimes, we do not require the global optimum solution, since the initial contour can be provided interactively by the user to obtain a rough initial contour near the edge, even though a robust optimization scheme that converges to the minimum solution in an acceptable number of iterations is strongly desired. Some examples of image segmentation results using a Bspline snake which integrate gradient and region based energy are demonstrated in Fig. 4.5.

4.2.3 Geometric Deformable Models (Active Contours)

These models were introduced independently by Caselles [23] and Malladi [25] to propose an efficient solution addressing the primary limitations of the parametric deformable model using a geometric deformable model (level set). Advantages of the implicit contour formulation of the geometric deformable model over parametric formulation include: (1) no parameterization of the contour, (2) topological flexibility, (3) good numerical stability, and (4) straightforward extension of the 2D formulation to higher dimensions.

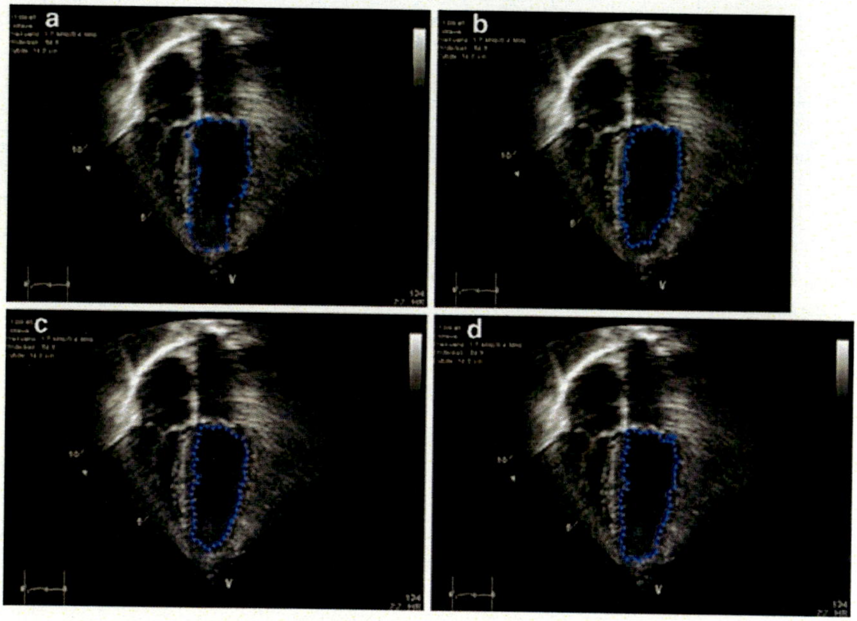

Fig. 4.5 B-Spline snake performance with respect to noisy image (i.e. echo cardiogram): (**a**) Initial contour by means of manual contouring; (**b**) Segmentation using only gradient-based energy; anatomical details of the upper part of the heart can be captured since this region has a bright area, but not in the lower part due to the noises; (**c**) Region-based energy, noisy problems can be solved but not anatomical details in the upper part; (**d**) Result using combination of 75% region-based and 25% edge-based energies; the advantage of both energies can be taken and limitations can be solved, thus enabling a better result

These models are based on curve evolution theory and the level set method [23, 26] proposed by Sethian and Osher [27] to track the surface interface and shape evolution in physical situations. Using this approach, curves and surfaces are evolved using only geometric measures, resulting in an evolution that is independent of parameterization. As in the other types of deformable models, the evolution is coupled with the image data in such a way that the process recovers object boundaries. The evolving curves and surfaces can be represented implicitly as a level set of a higher-dimensional function, so the evolution is independent of parameterization. As a result, topological changes can be handled automatically.

4.2.3.1 Curve Evolution

Let $C(t,p)$ be a kind of closed curves where t parameterizes the family and p the given curve, where $0 \leq p \leq 1$. As a closed curve, we assume that $C(0,t) = C(1,t)$ and similarly for the first derivatives for closed curves. Using this curve definition, the curve shortening flow, in the sense that the Euclidean curve length shrinks as quickly as possible when the curve evolves, can be obtained from the first variation of the length functional [28]:

$$\frac{\partial C}{\partial p} = \kappa \vec{N}, \tag{4.20}$$

where κ is the local mean curvature of the contour at a point, and $\vec{N}$ is the unit inward normal. For the intrinsic property of being closed, the curve under the evolution of the curve shortening flow will continue to shrink until it vanishes. By adding a constant ν, which we will refer to as the *"inflation term,"* the curve tends to grow and counteracts the effect of the curvature term when κ is negative [29]:

$$\frac{\partial C}{\partial p} = (\nu + \kappa)\vec{N}. \tag{4.21}$$

A stopping evolution term can be introduced into the above framework by changing the ordinary Euclidean arc-length function along the curve C to a geodesic arc length, by multiplying with a conformal factor g, where $g = g(x,y)$ is a positive differentiable function that is defined based on the given image $I(x,y)$. From the first variation of the geodesic curve length function, we obtain a new evolution equation by combining both the internal property of the contour and the external image force:

$$\frac{\partial C}{\partial p} = g(\nu + \kappa)\vec{N} - \nabla g. \tag{4.22}$$

Equation (4.22) gives us an elegant curve evolution definition for deformable model formulation applicable for curve evolution analysis. The main problem arises then in relation to how to represent the contour efficiently in terms of geometric and topologic stability as well as numerical implementation. One of the most common methods for representing the contour in using this scheme is the level set concept.

For a more detailed explanation of curve evolution in terms of image segmentation, interested readers are suggested to refer to the works of Casseles [23, 30], Kichenassamy [29], and Yezzi [31].

4.2.3.2 Level Set Concept

Curve evolution analysis using level set method views a curve as the zero level set of a higher-dimensional function $\phi(x,t)$. Generally, the level set function satisfies

$$\phi(x,t) \begin{cases} \phi(x,t) < 0 \text{ inside}\,\Omega(t) \\ \phi(x,t) = 0 \; C(t) \\ \phi(x,t) > 0 \text{ outside}\,\Omega(t) \end{cases},$$

where the artificial time t denotes the evolution process, $C(t)$ is the evolving curve, and $\Omega(t)$ represents the region enclosed by $C(t)$.

Figure 4.6 illustrates an important property of the level set method in handling the topological change in the object of interest. The first image shows a closed curve

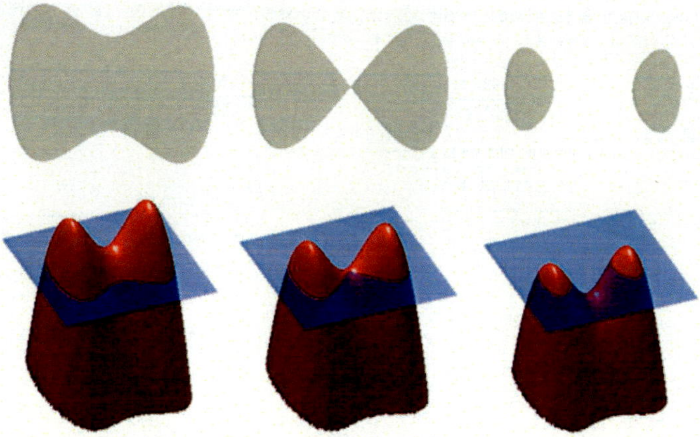

Fig. 4.6 Level set visualization

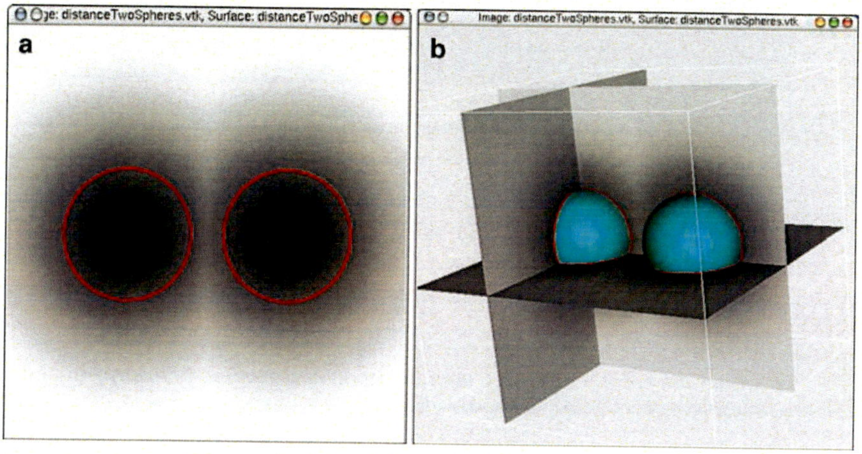

Fig. 4.7 Level set definition in three-dimensional objects: (**a**) An object represented implicitly using a 0th level set; (**b**) An example of a plane in the distance map representing the level set of the object

shape with a well-behaved boundary. The red surface below the object is a graph of a level set function ϕ determined by the shape, and the flat blue region represents the $x-y$ plane. The boundary of the shape is then the zero-th level set of ϕ, while the shape itself is the set of points in the plane for which ϕ has positive values (interior of the shape) or zero (at the boundary).

Extension of the level set method to higher dimensions is also possible and straightforward as shown in Fig. 4.7. Topological change in three dimensional space in this kind of deformable model can be observed in Fig. 4.8, where an initial surface defined as a sphere evolves to become a two-connected torus.

Mathematically speaking, instead of explicitly calculating the curve propagation directly on C, we can transform the contour implicitly using level set representation.

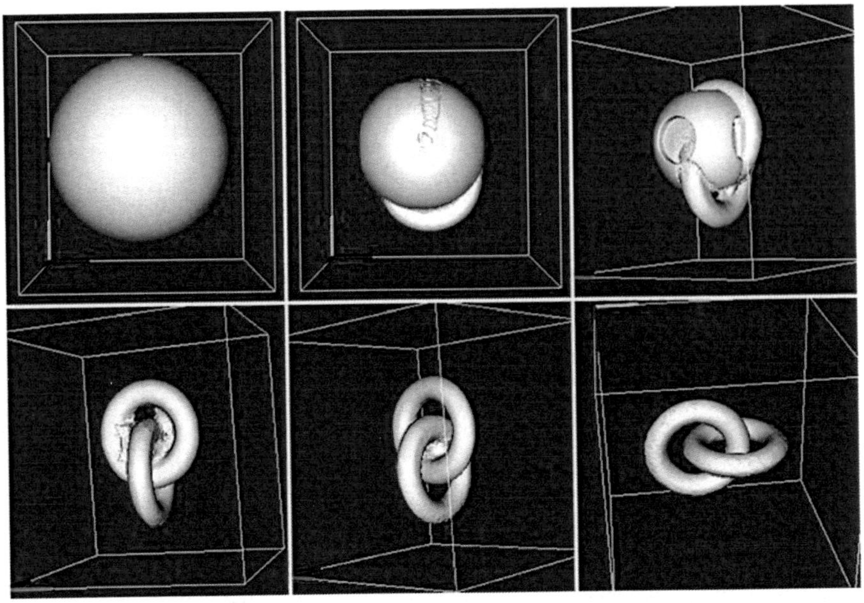

Fig. 4.8 Level set representation advantage in handling the topological change during segmentation. Initialized using a sphere the method can detect the connected chain as the final result

Thus, contour evolution, which is $\frac{\partial C}{\partial p} = F\vec{N}$, can be transformed in level set representation:

$$\frac{\partial \phi}{\partial t} = F|\nabla \phi|. \tag{4.23}$$

It would be very hard to elaborate upon this topology transformation using parameterized curve representation. One would need to develop an algorithm able to detect the moment the shape split (or merged), and then construct parameterizations for the newly obtained curves.

4.2.3.3 Geodesic Active Contour

Slightly different from the other models previously explained, this model does not impose any rigidity constraints (i.e. $w_2 = 0$); hence, the minimized energy is formulated as:

$$E(C) = \int_0^1 \underbrace{g(|\nabla I(C(s))|)\mathrm{d}s}_{\text{attraction term}} \quad \underbrace{\left|\frac{\partial C}{\partial p}\right|}_{\text{regularity term}} \mathrm{d}p. \tag{4.24}$$

Where g is a monotonically decreasing function, ds is the Euclidian arc-length element and L the Euclidian length of $C(t,p)$.

Using this formulation, we aim to detect an object in the image by finding the minimal-length geodesic curve that best takes into account the desired image characteristics. The stated energy in (4.24) can be minimized locally using the *steepest descent* optimization method, as demonstrated in [30]. It shows that in order to deform the contour towards the minimum local solution with respect to the geodesic curve length in Riemannian space, the curve evolves according to the following equation:

$$\frac{\partial C}{\partial t} = g\kappa\vec{N} - (\nabla g \cdot \vec{N})\vec{N}. \tag{4.25}$$

The segmentation result can be achieved in the equilibrium state where $\frac{\partial C}{\partial t} = 0$. Following the level set method for the curve evolution in (4.25), we obtain the curve evolution using the geodesic active contour in terms of the level set:

$$\frac{\partial \phi}{\partial t} = g\kappa|\nabla\phi| + \nabla g \nabla\phi. \tag{4.26}$$

Finally, in order to accelerate the convergence and place the model in the correct boundary, we integrate an elastic term in the curve evolution that will pull the model towards the desired object [30] and writing the curvature κ explicitly:

$$\frac{\partial \phi}{\partial t} = g \cdot \mathrm{div}\left(\frac{\nabla\phi}{|\nabla\phi|}\right)|\nabla\phi| + \underbrace{\nabla g \nabla\phi}_{\text{elastic term}} + v\nabla g|\nabla\phi|. \tag{4.27}$$

The role of this elastic term can be observed in Fig. 4.9 in which the upper-row active contour evolves without any elastic term, and hence the converged contour does not match the desired contour.

4.2.3.4 Chan–Vese Deformable Model

One limitation of the geodesic active contour lies in its dependence on image energy represented by the gradient. In order to stop the curve evolution, we need to define $g(|\nabla I|)$ which defines the edges of the object. In practice, discrete gradients are bounded, and so the stopping function is never zero on the edges, and the curve may pass through the boundary. Moreover, if the image is very noisy, the isotropic smoothing Gaussian has to be strong, which will smooth the edges as well.

To overcome these problems, Chan and Vese [32–34] proposed stopping process based on the general Mumford–Shah formulation of image segmentation [15], by minimizing the functional:

$$E^{MS}(f,C) = \mu.\mathrm{Length}(C) + \lambda \int_{\Omega} |f - f_0|^2 \mathrm{d}x\mathrm{d}y + \int_{\Omega\backslash C} |\nabla f|^2 \mathrm{d}x\mathrm{d}y, \tag{4.28}$$

where $\mathrm{f}_0 : \Omega \rightarrow \mathrm{R}$ is a given image to be segmented, and μ and λ are positive parameters. The solution image f is formed in smooth regions R_i and sharp

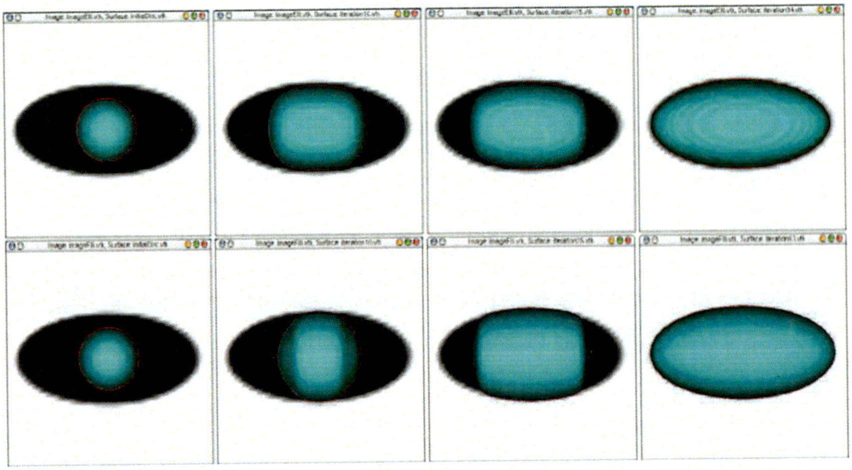

Fig. 4.9 Role of the elastic term in geodesic active contour: In the upper row, the model is evolved without the elastic term; in the lower row, an elastic term is used

boundaries C. A reduced form of this problem consists simply in the restriction of E^{MS} to piecewise constant functions and finding a partition of Ω such that f in Ω equals a constant.

To minimize the functional energy, Chan and Vese proposed a two-phase segmentation as follows:

$$\min_{C,C_o,C_b}\left\{\mu\int_{\Omega}\delta(\phi)|\nabla\phi|\mathrm{d}x\mathrm{d}y+\nu\int_{\Omega}H(\phi)\mathrm{d}x\mathrm{d}y+\lambda_o\int_{\mathrm{inside}(C)}|f-c_o|^2H(\phi)\mathrm{d}x\mathrm{d}y+\right\}$$

$$+\lambda_o\int_{\mathrm{outside}(C)}|f-c_o|^2(1-H(\phi))\mathrm{d}x\mathrm{d}y,$$

where ϕ is the level set function and $H(\phi)$ is the Heaviside function.

Generally, the above deformable models implemented by means of the level set method suffer from a slower speed of convergence than parametric deformable models due to their computational complexity. Application of the Chan-Vese deformable model to segmentation on a low signal, noisy image is shown in Fig. 4.10.

4.3 Comparison of Deformable Models

Whereas geometric deformable models often have been presented as an improvement (and even superior) to classical snakes [23, 25, 30], one might argue that they actually are complementary to each other. In a wide range of image segmentation

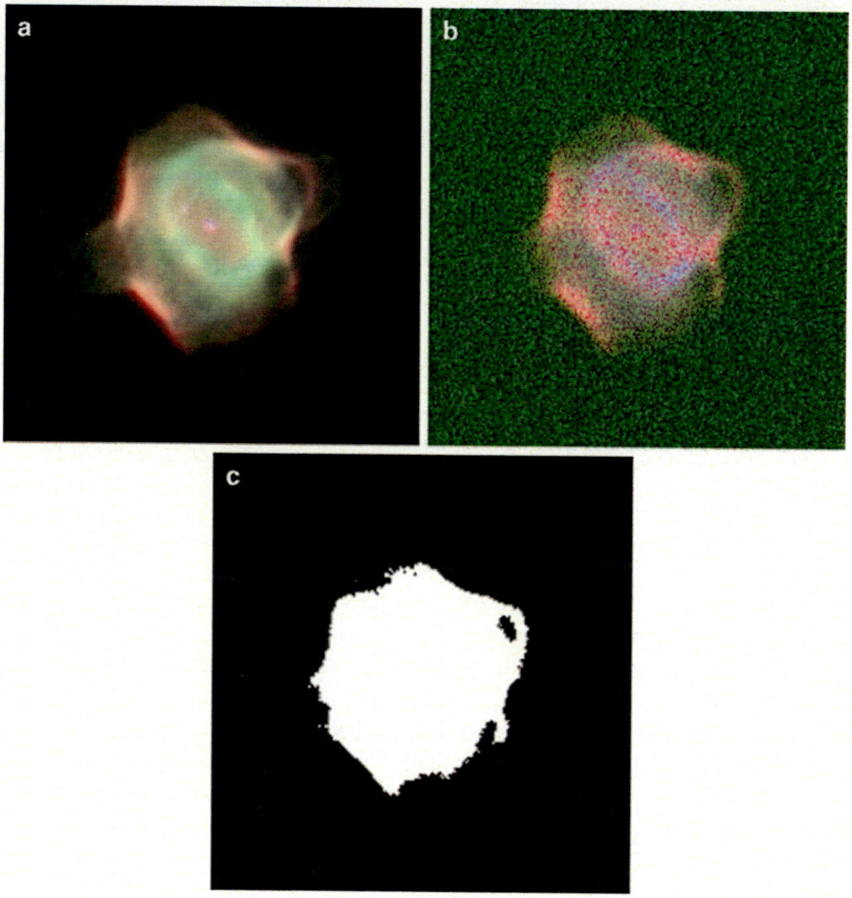

Fig. 4.10 Image segmentation using a Chan–Vese deformable model: (**a**) The original image; note that it has a very weak border; (**b**) The degraded image with added salt-and-paper noises; (**c**) A segmentation result represented as a binary mask (the white area is the detected object)

applications there is a trade-off between desirable properties. Instances of such important trade-offs are the level of automation versus control, and topological flexibility versus *a priori* constraints. Other important properties are the role of the initialization, existence of a unique solution, dependence of the result on parameter choice and the robustness of the method with respect to noise, and imperfect image data.

Table 4.1 summarizes the main properties of a number of deformable model types which have been described in the previous section. We categorize the deformable models into three approaches following the previous presentation.

Classical and explicit parametric snakes are well-known in performing well if the initialization is close to the desired objects. Moreover, since a parameterization is

Table 4.1 Comparison of different types of deformable models. We indicate whether the approach satisfies some properties (+) or not (−) or it is partially fulfilled ( ). These properties are topological flexibility (a); steady final state (b); possibility of segmenting multiple objects (c); support to a priori knowledge (d); independence from initialization (e); absence of ad-hoc parameters (f)

Deformable model type	a	b	c	d	e	f
Discrete Points						
– Classical model	−	−	−		−	
– Balloon	−	−	−		+	−
– Gradient vector Flow	−	−	−		+	+
Parametric Snake						
– B-Spline	−		−		+	−
Geometric						
– Geodesic active contour	+	+	+	−	+	
– Vesse–Chan model	+	+	+	−	+	

available, morphological properties can directly be derived from the representation. A drawback is that they are easily trapped by spurious edges representing local minimum solution when the initialization is poor. This problem has been tackled by the balloon approach, where an additional inflation force pushes the level sets over insignificant edges. This introduces an additional arbitrary parameter on which the result strongly depends.

Interestingly, the same difference is present between the classical parametric snake model and its geodesic version. In the latter approach, one parameter less is required, but as a consequence the result depends more strongly on the initialization.

4.4 Applications

4.4.1 Bone Surface Extraction from Ultrasound

In order to extract bone surfaces from ultrasound images by following the expert reasoning from clinicians, bones can be characterized by a strong intensity change from a bright pixel group to a global dark area below since the high absorption rate of bones generates an acoustic shadow behind them. Discontinuities are normally not present on the bone surface; hence the segmentation method should produce a smooth contour.

Considering all of these, an open active contour initialized below the edge in the acoustic shadow was proposed, going from one side of the image to another, and evolving vertically towards the top of the image until it meets the bone surface. A posterior treatment is then necessary to choose only real bone surface points in the image. A deformable model with an external balloon force is applied for this segmentation, but we only take into account the vertical component of the normal

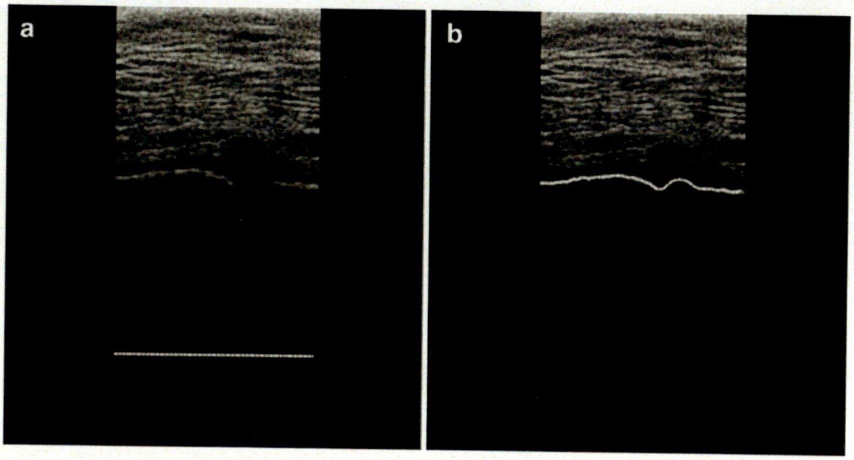

Fig. 4.11 Snake evolution, initialized with a simple contour at the bottom of image (**a**) and then evolved until it reaches convergence

to the curve and the horizontal point positions of the snake are fixed. This approach allows the snake to move in the vertical direction. We also require an open contour with all points able to freely move vertically, as illustrated in Fig. 4.11.

Then the discretization of the energy minimization scheme gives a stiffness matrix A similar to (4.15), except for the elements related to the two-model extremities. The elements of the matrix A become:

$$a_0 = a_N = 2\frac{w_1}{h^2} + 3\frac{w_2}{h^4} \quad b_0 = b_N = -\left(\frac{w_1}{h^2} + 3\frac{w_2}{h^4}\right).$$

A new local energy definition [35] is proposed to accommodate the image intensity changes and reduce the noise effect during the segmentation process. This regional energy can be defined as a difference between mean intensities in the regions above, below, and in the considered location. When this difference is negative, then a penalization value is applied, and the final energy term is defined as the product of this region-based measurement and the gradient term, and Fig. 4.12 illustrates the role of this energy in deformable model evolution.

Since the active contour initialization goes from one side to another in the image, and considering that the real bone contour does not always behave in a similar fashion, we need to perform a posterior point selection in which only points having a high enough intensity are retained.

A method has been proposed to improve its performance [36] by applying a Gradient Vector Flow in the narrow band around the evolved contour. Furthermore, addressing serial ultrasound images, snake initialization for the next slice can be obtained from the retrospective results of previous slices. This Gradient Vector Flow imparts bi-directional force to the evolved contour; in consequence, the model is capable of moving back when placed above the desired contour. To reduce

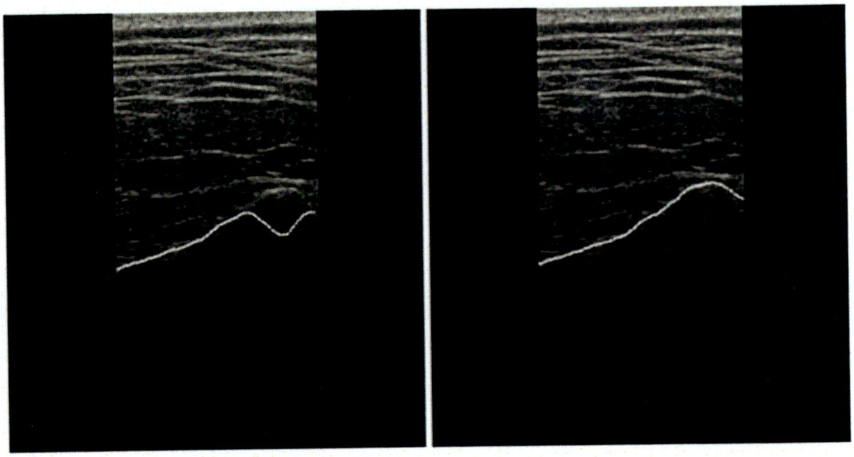

Fig. 4.12 Role of the regional energy term in bone surface segmentation. The model on the left evolves without a regional term, while such a term is applied on the right hand figure

computation time, Gradient Vector Flow and other image energy computations are confined within a thin narrow band around the evolved curve.

For the purpose of bone surface reconstruction from ultrasound imagery, other enhancements have been proposed [37] following the presented model. A set of bone contours is first extracted from a series of free-hand 2D B-Mode localized images, using an automatic segmentation method based on snakes with region-based energy as previously described. This data point-set is then post-processed to obtain a homogeneous re-sampling point-grid form. For each ultrasound slice, the algorithm first computes an approximation of the bone slice center. Using these central points, it approximates a line corresponding to the central axis of the bone. Then, for each slice, it: (a) updates the value of the central point as the intersection point of the line and the corresponding slice plane; (b) casts rays from the center towards the surface at regular intervals (spaced by a specific angle); (c) computes the new vertex as the intersection between rays and segments connecting the original points.

Three-dimensional B-Spline is applied to approximate the surface departing from these points. The method ensures a smooth surface and helps to overcome the problem of false positives in segmentation. Model reconstructions from localized ultrasound images using the proposed method are shown in Figure 4.13 and 4.14.

4.4.2 Spinal Cord Segmentation

4.4.2.1 Spinal Cord Measurements

This study has been conducted for quantitative assessment of disease progression in multiple sclerosis using MRI. For this study we use IR-FSPGR volume data as

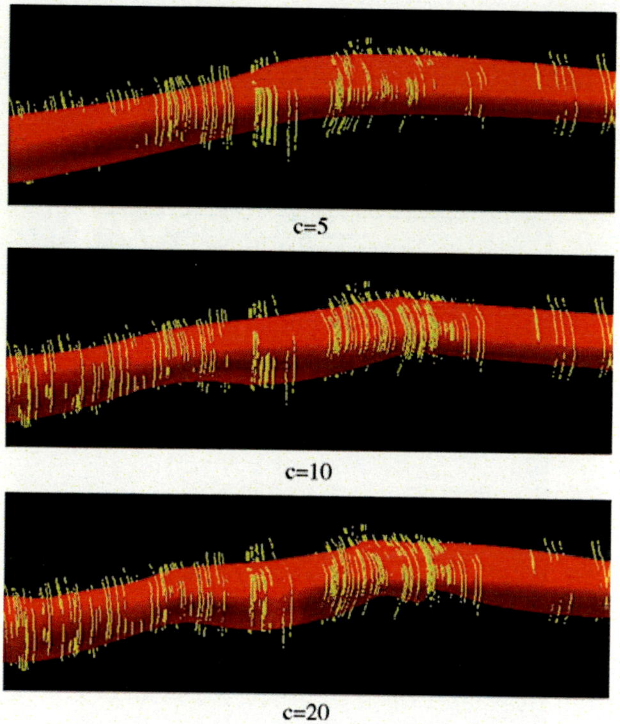

Fig. 4.13 Reconstructed surface of a real radius. Results for different numbers of control points c

Fig. 4.14 Reconstructed radius and ulna. Data scanned from a real subject

shown in Fig. 4.15, in which the spinal cord under analysis can be characterized by a bright structure against a dark background (representing CFS), normally with a cylindrical topology. Segmentation difficulties can arise due to artifacts, noise, and proximity to other structures such as vertebrae.

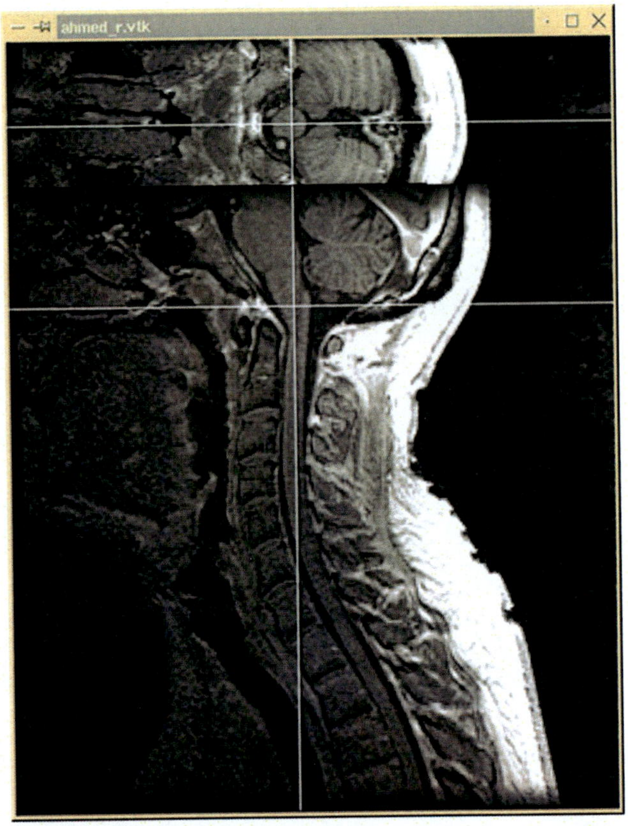

Fig. 4.15 MRI Image in which the spinal cords need to be segmented

Atrophy is generally assessed by measuring the cross-sectional areas at specific levels (typically C2–C5) along the cervical cord. This protocol introduces several uncertainties, including the choice of the level at which the measurements should be performed, the cord orientation, as well as the cord segmentation process itself. To provide non-biased area measurements, an MRI image, or part of the image, often needs to be re-formatted or acquired with slices perpendicular to the cord.

Moreover, the spinal cord cross-sectional area has often been measured either manually or using intensity-based 2D processing techniques. The limitations of such methods are various: measurements are restricted to a predefined level at which cords and slices are orthogonal; intensity-based segmentation is hindered by intensity variations caused by surface coils typically used during acquisition; 2D measurements are more prone to being biased by partial volume effects than 3D measurements; manual analysis is more time-consuming and more sensitive to intra- and inter-operator variability.

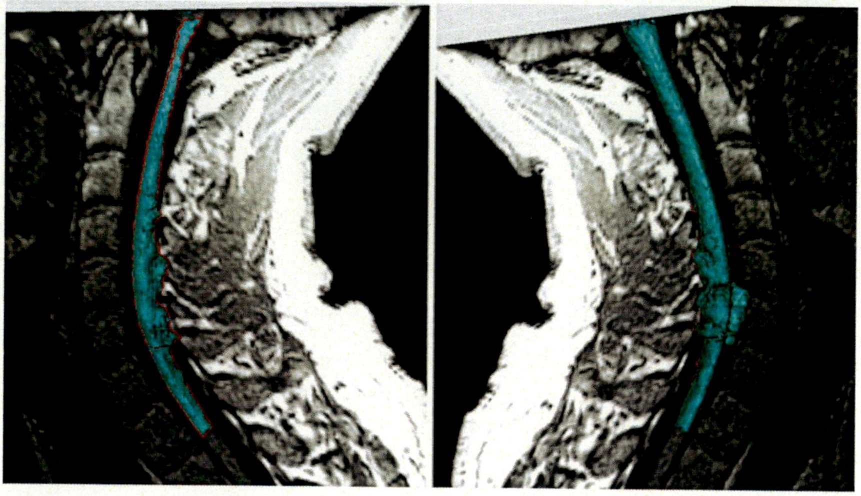

Fig. 4.16 Segmentation of an MRI Image using a naive geodesic active contour

4.4.2.2 Segmentation Using Geodesic Active Contour

We propose to apply a geometric deformable model to perform the segmentation without any image preprocessing of the data image. The model is initialized by placing a sphere in the spinal cord and letting this sphere evolve until the process reaches the convergence. Figure 4.16 shows that the segmentation method encounters difficulties in extracting the spinal cord at lower levels of vertebrae, due to the proximity of these organs in the image. Using a surface evolution-based segmentation method, such as geodesic active contour, the evolved surface passes over the vertebrae border.

We propose an intensity integration approach to solve this organ concurrence problem in a specific area, by applying a contrast-based selection to the surface of the organ in order to drive the curve evolution bi-directionally according to that contrast. Figure 4.17 illustrates the idea of contrast-based selection: suppose that the gray box in the middle is the spinal cord, the two black boxes represent the vertebrae and the red contour is the evolved curve. When the contour is placed in the spinal cord then ∇g and $\nabla \phi$ have different signs and the contour should be evolved towards the spinal cord border. But, when the contour placed in the vertebrae then ∇g and $\nabla \phi$ have the same sign and the contour should be evolved inversely.

Such an approach can be integrated directly into the geodesic active contour's evolution as:

$$\frac{\partial \phi}{\partial t} = \left[g \cdot \operatorname{div}\left(\frac{\nabla \phi}{|\nabla \phi|} \right) |\nabla \phi| + \operatorname{sign}(\nabla g \nabla \phi)(\nabla g \nabla \phi) + \nu \nabla g \right] \|\nabla \phi\|.$$

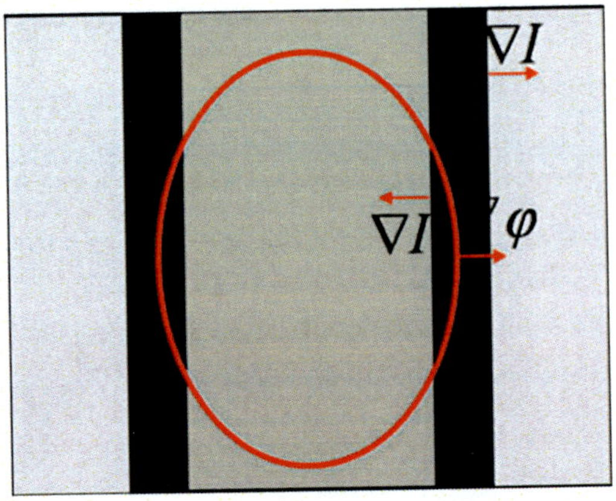

Fig. 4.17 Selective contrast

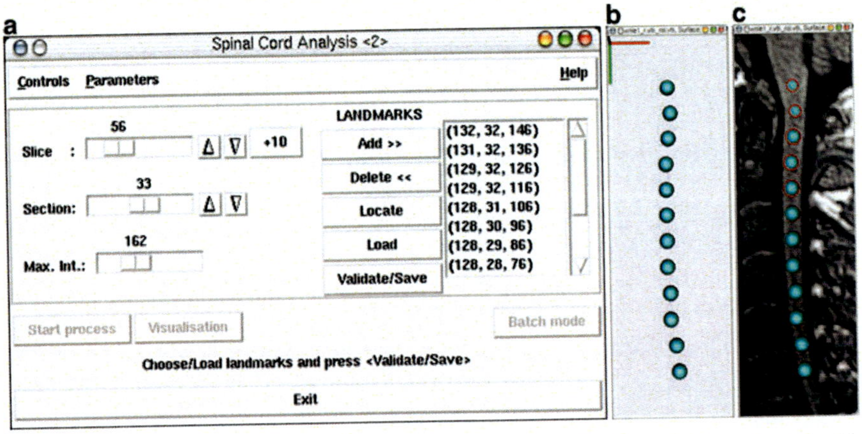

Fig. 4.18 User interface places the spheres in the spinal cord area (**a**) and the initialization begins (**b** and **c**)

We also enhanced the interactivity of the method, such that the user is allowed to initialize the method using various spheres in the spinal cord. These spheres can be provided interactively by means of a user interface in order to browse through axial slices and then locate the center of the sphere in that slice. It is not necessary to place the spheres in the middle of the spinal cord, but they should be situated entirely inside the spinal cord region in the MRI image.

The curve evolution process of the proposed deformable model for spinal cord segmentation purposes, deviating from the initial curve in Fig. 4.18, prior to reaching the convergence, is visualized in Fig. 4.19.

Applying this type of geometric deformable model, we find that the spinal cord can be obtained with better quality in a shorter computation time.

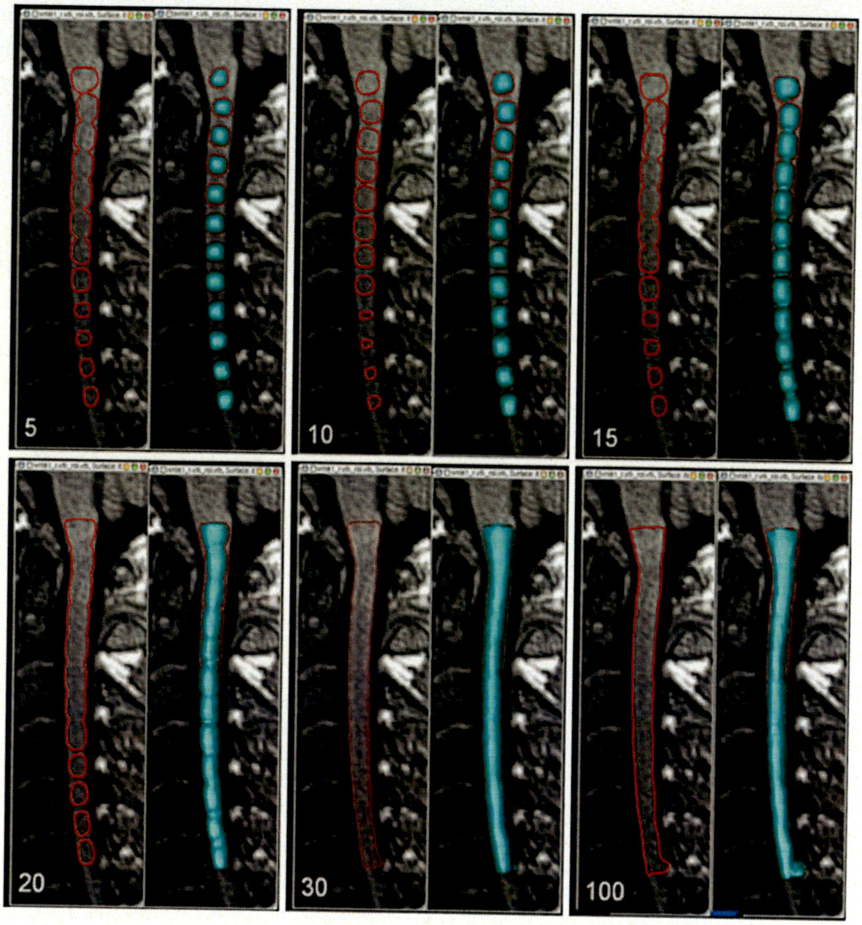

Fig. 4.19 Surface evolution during the segmentation process of spinal cord from the MRI image (the number in the *left corner* of each image represents the number of elapsed iterations)

4.5 Conclusion

This chapter describes some of the basic concepts of deformable models and their application in different cases of image segmentation. Image segmentation plays a critical role in almost all aspects of image analysis; it has opened a wide range of challenging problems oriented towards accurate featuring and geometric extraction of different types of images. The deformable model successfully overcomes the limitation of classical low-level image processing by providing an elegant and compact representation of shapes and objects in image analysis.

To gain the best performance of segmentation, the particular deformable model should be carefully chosen according to the application context. In general practice, the parametric deformable model runs faster than geometric ones but its typical shape representation is considerable lower.

References

1. Terzopoulos, D., McInerney, T.: Deformable models in medical image analysis: A survey. Med. Image Anal. **1**, 91–108 (1996)
2. Falcao, A., Udupa, J., Samarasekera, S., Sharma, S.: User steered image segmentation paradigm: Livewire and livelane. Graph. Models Image Process. **60**(4), 233–260 (1998)
3. Mortensen, E.N.: Interactive segmentation with intelligent scissors. Graph. Models Image Process. 60 (5), 349–384 (1998)
4. Falcao, A. et al.: A 3D generalization of user-steered live-wire segmentation. Med. Image Anal. 4(4), 389–402. (2000)
5. Schenk, A.M., Guido, P., Peitgen, H.-O.: Efficient semiautomatic segmentation of 3D objects in medical images. In: Medical Image Computing and Computer-Assisted Intervention, MICCAI 2000, vol. 1935, pp. 186–195. (2000)
6. Grady, L., et al. Random Walks for Interactive Organ Segmentation in Two and Three Dimensions: Implementation and Validation. , 773–780. 2005
7. Pham, D.L., Xu, C., Prince, J.L.: Current methods in medical image segmentation. Annu. Rev. Biomed. Eng. 2(1), 315–337 (2000)
8. McInerney, T., Terzopoulos, D.: Deformable models in medical image analysis: A survey. Med. Image Anal. **1**(2), 91–108 (1996)
9. Kirbas, C., Quek, F.: A review of vessel extraction techniques and algorithms. , ACM Comput. Surv. **36**(2), 81–121 (2004)
10. Suri, J.S., et al.: Shape recovery algorithms using level sets in 2-D/3-D medical imagery(part-II): a state-of-the-art review. Pattern Anal. Appl. **5**(1), 77–89 (2002)
11. Suri, J.S., et al.: A review on MR vascular image processing algorithms: Acquisition and prefiltering: part I. IEEE Trans. Inform. Technol. Biomed. 6(4), 324–337 (2002)
12. Kass, M., Witkin, A., Terzopoulos, D.: Snakes: Active contour models. Int. J. Comput. Vis. 321–322 (1998)
13. Fischler, M., Elschlager, R.: The representation and matching of pictorial images. ., IEEE Trans.– Comput. 22, 67–92 (1973)
14. Widrow, B.: The "rubber-mask" technique. Pattern Recogn. **5**, 175–211 (1973)
15. Williams, D., Shah, M.: A fast algorithm for active contours and curvature estimation. Comput.Vis. Graph. Image Process. Image Underst. **55**(1), 14–26 (1992)
16. Cohen, L.D.: On active contour models and balloons. , Comput. Vis. Graph. Image Process. Image Underst. 211–218 (1991)
17. Cohen, L.D., Cohen, I.: Finite-element methods for active contour models and balloons for 2D and 3D images. IEEE Trans. Pattern Anal. Mach. Intell. 1131–1147 (1993)
18. Xu, C., Prince, J.L.: Generalized gradient vector flow external forces for active contours. Signal Process. **71**(2), 131–139 (1998)
19. Xu, C., Prince, J.L.: Snakes, shapes, and gradient vector flow. IEEE Trans. Image Process. **7**(3), 359–369 (1998)
20. Amini, A.A., Weymouth, T.E., Jain, R.C.: Using dynamic programming for solving variational problems in vision. IEEE Trans. Pattern Anal. Mach. Intell. 12(9), 855 (1990)
21. Saad, Y.: Iterative methods for sparse linear systems, 2nd edn. s.l. SIAM Publisher, Philadelphia (2003)

22. Menet, S., Saint-Mark, P., Medioni, G.: B-snakes: implementation and application to stereo. Image Underst. Workshop. 720–726 (1997)
23. Caselles, V., Kimmel, R., Sapiro, G.: Geodesic active contours. Int. J. Comput. Vis. 22(1), 61–79 (1997)
24. Duncan, L.H., Staib, J.S.: Boundary fitting with parametrically deformable models. Trans. Pattern Recogn. Mach. Intell. **14**(11), 1061–1075 (1997)
25. Malladi, R., Sethian, J.A., Vemuri, B.C.: Shape modeling with front propagation: A level set approach. IEEE Trans. Pattern Anal. Mach. Intell. 17(2), 158–175 (1995)
26. Vemuri, B.C., Y., J. Yeand. Image registration via level-set motion: Applications to atlas-based segmentation. Medical Image Analysis, 7(1), 1–20 (2003)
27. Osher, S., Sethian, J.A.: Fronts propagating with curvature-dependent speed: algorithms based on Hamilton-Jacobi formulations. J. Comput. Phys. **79**, 12–49 (1988)
28. Tanenbaum, A.: Three snippets of curve evolution theory in computer vision. Math. Comput. Model. **24**, 103–119 (1996)
29. Kichenassamy, S., et al.: Conformal curvature flows: From phase transitions to active vision. , Archive of Rational Mechanics and Analysis **136**, 275–301 (1996)
30. Caselles, V., et al.: A geometric model for active contour. Numerische Mathematik **66**, 1–31 (1993)
31. Yezzi, A., et al.: A geometric snake model for segmentation of medical imagery. IEEE Trans. Med. Imaging, 16, 199–209 (1997)
32. Chan, T.F., Sandberg, B., Vese, L. Active contours without edges for vector valued images. J. Vis. Commun. Image Represent. **11**, 130–141 (2000)
33. Chan, T.F., Vese, L.A.: Active contours without edges. IEEE Trans. Image Process. 266–277 (2001)
34. Chan, T.F., Vese, L.A.: A Multiphase level set framework for image segmentation using the Mumford and Shah model. Int. J. Comput. Vis. 50(3), 271–293 (2002)
35. Alfiansyah, A., Streichenberger, R., Kilian, P., Bellemare, M.-E., Coulon, O.: Automatic segmentation of hip bone surface in ultrasound images using an active contour. In: CARS'2006. Computed Assisted Radiology and Surgery, in International Journal of Computer Assisted Radiology and Surgery, 1, supplement 1, 115–116, Osaka Japan, 2006.
36. Alfiansyah, A., Ng, K.H., Lamsudin, R.: Deformable model for serial ultrasound images segmentation: application to computer assisted hip athropasty. Singapore: s.n., In: International Conference on BioMedical Engineering (2008)
37. Lopez Perez, L., Le Maitre, J., Alfiansyah A., Bellemare, M.-E.: Bone Surface reconstruction using localized freehand ultrasound imaging. In: Vancouver: 30th Annual International IEEE EMBS Conference, pp. 2964–2967. (2008)
38. Montagnat, J., Delingette, H., Ayache, N.: A review of deformable surfaces: topology, geometry and deformation. Image Vis. Comput. **19**, 1023–104037 (2001)
39. Vray, D., et al.: {3D Quantification of Ultrasound Images: Application to Mouse Embryo Imaging In Vivo}. 2002
40. Liu, Y.J. et al.: Computerised prostate boundary estimation in ultrasound images using the radial bas-relief method. , Med. Biol. Eng. Comput. **35**, 4450–4454 (1997)
41. Tauber, P., Batatia, H., Ayache, A.: Robust B-Spline Snakes for Ultrasound Images Segmentation. s.l.: IEEE, 2004
42. Liu, F., et al.: Liver segmentation for CT images using GVF snake. Med. Phy. 32(12), 3699–3706 (2005)

Chapter 5
Fat Segmentation in Magnetic Resonance Images

David P. Costello and Patrick A. Kenny

5.1 Introduction

Over the past two decades, many authors have investigated the use of magnetic resonance imaging (MRI) for the analysis of body fat and body fat distribution. However, accurate isolation of fat in MR images is an arduous task when performed manually. In order to alleviate this burden, numerous automated and semi-automated segmentation algorithms have been developed for the quantification of fat in MR images. This chapter will discuss some of the techniques and models used in these algorithms, with a particular emphasis on their application and implementation.

When segmenting MR images the largest variable in the process is the image itself. What acquisition parameters give optimal image quality, in terms of signal to noise ratio (SNR), contrast, uniformity, and boundary definition? An appropriate MRI pulse sequence aims to generate adequate image contrast in the shortest imaging time. MRI can also introduce measurement errors and image artifacts as a result of the imaging process, which will complicate the segmentation process. The potential impact of artifacts such as intensity inhomogeneities and partial volume effect (PVE) on image segmentation will be discussed briefly.

Body fat volume and fat distribution provide key risk indicators for a number of diseases including non-insulin-dependent diabetes mellitus (NIDDM) and coronary heart disease (CHD) [1]. Traditionally, anthropometric measurements such as body mass index (BMI), waist to hip ratio, abdominal circumference, and caliper tests have been used to estimate total body fat and body fat distribution [2–4]. These methods indirectly infer an estimate of body fat based on a known correlation

---

D.P. Costello (✉)
Mater Misericordiae University Hospital and University College, Dublin, Ireland
e-mail: dcostello@mater.ie

G. Dougherty (ed.), *Medical Image Processing: Techniques and Applications*, Biological and Medical Physics, Biomedical Engineering, DOI 10.1007/978-1-4419-9779-1_5,

with underwater weighing [3]. Obesity is defined in terms of BMI [5], which is expressed as:

$$\mathrm{BMI} = \frac{\mathrm{Weight\,(kg)}}{\mathrm{Height}^2\,(\mathrm{m}^2)}. \tag{5.1}$$

However, BMI is a metric that fails to distinguish fat mass from lean mass. This renders BMI ineffective as a metric for assessing sample groups such as athletes, who because of their increased muscle mass can be categorized as overweight or obese even if their percentage body fat is below normal.

Body fat distribution can be analysed using volumetric images of the body. Both MRI and computed tomography (CT) are sophisticated volumetric imaging techniques that facilitate delineation of regional fat deposits in the body [6]. It is difficult to justify whole body CT for fat quantification because of its high radiation dose, especially when MRI exists as a non-ionising alternative [7]. Seidell et al. [8] show a strong correlation between fat volumes measured using both modalities. As a result of this, radiation dose remains the main reasons for selecting MRI over CT to analyse body fat.

5.2 Imaging Body Fat

Image acquisition is the most important step when quantifying fat with MRI. Good contrast, resolution, SNR and homogeneity are required for accurate image segmentation. Contrast in MR images depends on proton density, longitudinal relaxation time ($T1$) and transverse relaxation time ($T2$). As a result of the significant difference in T1 between fat and soft tissue, T1-weighted (T1w) MR sequences are used to enhance fat contrast. Numerous MRI pulse sequences are used to generate high contrast images for fat analysis, including: spin echo (SE), gradient echo (GE), Inversion recovery, three-Point Dixon method and water saturated (WS) balanced steady-state free precession (WS b-SSFP or TrueFISP) pulse sequences fat [6,8–17].

Both SE and GE sequences are used to image body fat [9–15]. GE sequences are faster than SE but provide decreased homogeneity, lower SNR and reduced T1 contrast. However, in some circumstances short scan times are necessary to reduce motion artifact. Fast-SE[1] (FSE) sequences reduce imaging time while maintaining homogeneity. Barnard et al. [9], compared fast and conventional spin echo pulse sequences and found good correlation between the two. However, when compared to GE sequences, FSE sequences are relatively slow for whole body imaging [18]. Brennan et al. [15], successfully used a T1w GE pulse sequence to demonstrate the relationship between whole body fat and BMI.

A number of authors have used advanced pulse sequences to improve contrast between fat and soft tissue. One variation on Gradient echo pulse sequence is a b-SSFP sequence. Both WS [19] and non-WS b-SSFP [6] pulse sequences have

[1]Fast spin echo techniques acquire between 2 and 16 lines of k-space during each TR.

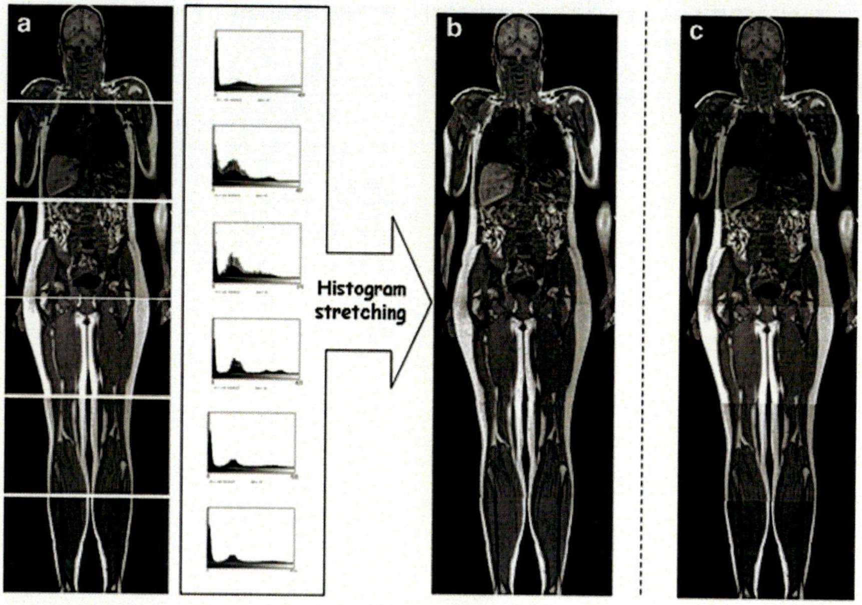

Fig. 5.1 (**a**) Whole body MR scan acquired in 6 stations; (**b**) the stitched image after post processing and (**c**) stitched image with no post processing

been used to image body fat. Peng et al. [19] proposed using a WS b-SSFP pulse sequence for fat quantification. WS b-SSFP [16, 19] allows for rapid acquisition of high contrast images containing fat and soft tissue. Peng et al. [19] found that this sequence performed better than a WS Fast-SE sequence and simplifies the segmentation process (see Sect. 5.4.2) [16].

Once a pulse sequence is selected, sequence parameters (e.g. repetition time (TR), echo time (TE), number of signal averages (NSA) and echo train length) should be set to optimize both acquisition time and image contrast. Acquisition time is a key factor when selecting a pulse sequence, as long scan times increases the likelihood of motion artifacts. The presence of motion artifact in the visceral cavity can make it very difficult to quantify fat. Scan time can be reduced by shortening acquisition time, reducing the NSA and increasing the echo train length. Changing these factors can also lead to a decrease in image contrast and SNR, which should be investigated prior to scanning large cohorts of patients.

There are a number of other imaging parameters that must also be considered. These include field of view (FOV), scan orientation, matrix size, slice thickness, and slice gap. FOV size affects the severity of inhomogeneities across the image. A larger FOV results in a less homogeneous B–field (magnetic–field) across the image. Homogeneity is particularly important when acquiring coronal whole body data sets. Consequently, it is normal practice to split a whole body scan into a number of smaller sections, known as stations, to reduce inhomogeneities caused by

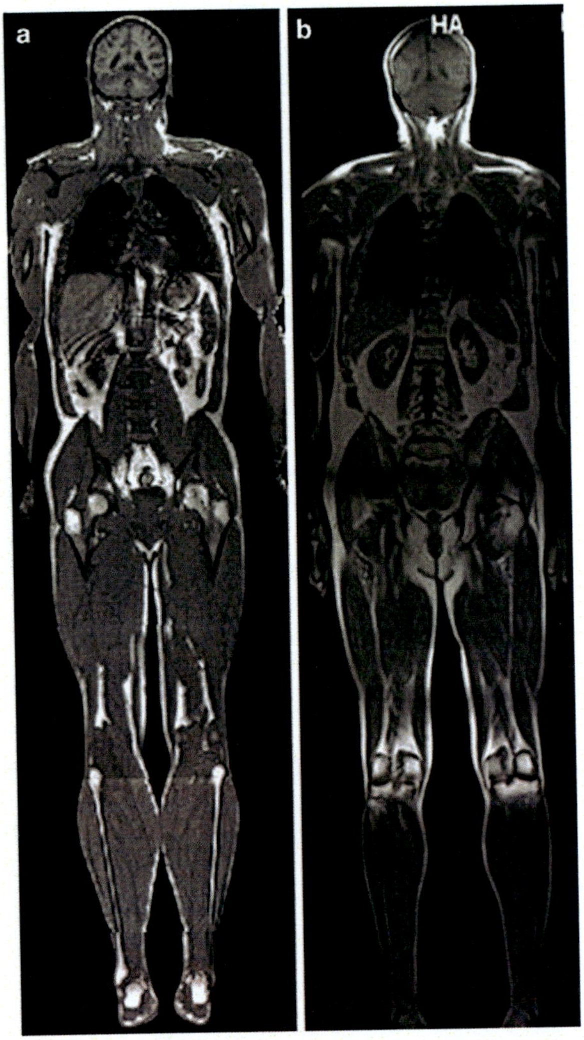

Fig. 5.2 (**a**) Intensity corrected whole body image and (**b**) whole body image with no intensity correction

variation in the B–field. The resultant sections are then stitched together to form a whole body data set as illustrated in Fig. 5.1 and 5.2. Figure 5.1c and 5.2b illustrate the intensity differences that can occur between stations in the same slice.

Intensity differences between stations for tissues of the same class can hamper image segmentation. Brennan et al. [15] describe a method of histogram matching to compensate for intensity differences between each station. The first step is to identify the soft tissue peak in the histogram of each station. After which, all soft tissue peaks are aligning to match the gray-scale distribution across all stations. Images are then stitched according to the coordinates in the DICOM header information, illustrated in Fig. 5.1b. Transverse MR images of the body are often used in the literature to image body fat. Opting to use transverse orientation over coronal has a number of advantages. Transverse images allow the use of a smaller

FOV, which reduces non-uniformities in the image. Smaller self-contained sections also remove the need for post processing such as image stitching and histogram matching. Transverse images of the abdomen make it possible to get an estimate of visceral fat volume from a small number of sample slices, which reduces scan time significantly [20].

Slice thickness and voxel size both influence SNR and PVE. Larger voxels increased SNR but increase the incidence of partial volume voxels (PVE is discussed in Sect. 5.3.1). Therefore, it is important to find a balance between SNR and PVE to optimize the segmentation process. Increasing the gap between slices can reduce acquisition time without compromising image quality. A slice gap of 100% will half acquisition time. Fat volume in the gap can be estimated using interpolation.

5.3 Image Artifacts and Their Impact on Segmentation

5.3.1 Partial Volume Effect

PVE occurs when a single voxel contains a mixture of two or more tissue types, (e.g. at the interface of fat and soft tissue or fat and air), resulting in blurring at boundaries. Figure 5.3 illustrates the effect of the PVE on subcutaneous fat. MR images have limited resolution which increases the probability of the PVE occurring [21]. Voxels affected by the PVE have an intermediate gray level intensity, which is determined by the proportion of each tissue type contained within that voxel [22]. This effect is observed at interfaces between gray and white matter in the brain and fat and soft tissue throughout the body. The PVE can cause fuzziness (or uncertainty) at the boundaries of two tissues classes [8]. This inhibits the use of edge detection methods due to ambiguity at the interface of tissue classes leading to a lack of clear edges [8, 23].

Edge-based segmentation methods aim to find borders between regions by locating edges (or boundaries) within images. MR images have a relatively low resolution and SNR when compared to other modalities such as CT. PVE and low

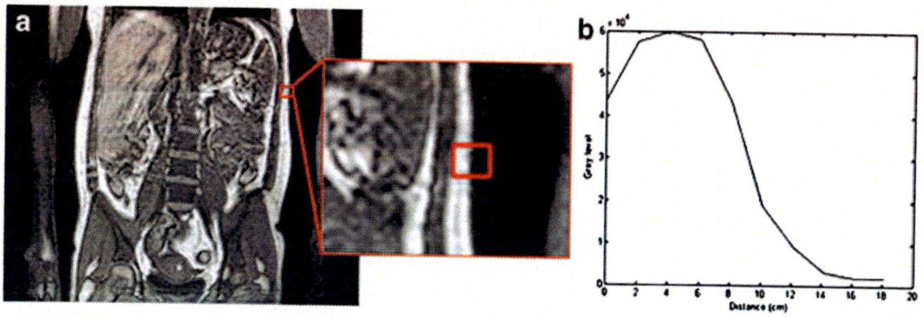

Fig. 5.3 (**a**) T1w GE image containing fat, soft tissue, and background (**b**) is a profile through plot of the region outlined in *red* in image (**a**)

SNR can result in uncertainty at the boundary of objects in MR images which limits the use of edge detection algorithms [8]. Figure 5.3b illustrates the problem encountered by edge detectors when faced with the PVE. Where does fat stop and background start? The use of edge detection algorithms can lead to incomplete segmentation of boundaries in MR images. PVE is a three-dimensional phenomenon and can affect large volumes of tissue in a local area. This can affect the performance of global segmentation techniques, discussed in Sect. 5.4.

5.3.2 *Intensity Inhomogeneities*

Intensity inhomogeneities can have a significant impact on segmentation and quantitative analysis of MR images. They are caused by non-uniformity in the RF field (B_1), irregularities in the main magnetic field (B_0), susceptibility effects of normal tissue and receiver coil sensitivity profile [24]. This artifact can cause the appearance of skewed or bimodal fat peaks in the image histogram [25]. As a result, clinical MR images require some processing before segmentation and analysis can take place. Inhomogeneities in MR images can be modeled as a multiplicative bias field [23]. The bias field is characterized as a gradual change in intensity within segmentation classes across the entire image which cannot be attributed to random noise [23]. It can degrade the performance of the intensity–based segmentation algorithm, as the model assumes spatial invariance between tissues of the same class across the entire image. An expression for the biased image is given in (5.1).

$$f_{\text{biased}} = f_{\text{original}}(x,y)\,\beta(x,y) + n(x,y), \tag{5.2}$$

where f_{biased} is the degraded image, f_{orignal} is the image without degradation or noise and $n(x,y)$ is random noise. A number of approaches are investigated in the literature for the correction of the bias field (26–30). The impact of intensity inhomogeneities on thresholding is illustrated in Fig. 5.4.

Siyal et al. [26] and Rajapakse et al. [23] reviewed a number of approaches to reduce the appearance of the intensity inhomogeneities. These approaches can be split into two categories, retrospective and prospective modeling. Prospective modeling uses prior knowledge of the bias field, which can be obtained by imaging a homogeneous phantom. Homogeneous phantoms only provide a good estimate of the bias field for objects of similar size to the phantom. When imaging patients, the dimensions of the scanned volume can vary from patient to patient and also between sections of the same patient (e.g. the legs and the torso). The volume of the area being imaged changes the loading on the receiver coils' in the MRI scanner, which in turn alters the coils sensitivity profile [27]. To account for this Murakami et al. [28] performed a calibration scan directly on the patient to estimate the bias field. Prospective modeling of the bias field can be impractical for studies containing large numbers of patients, as imaging time increases due to the need for additional phantom/patient scans [27].

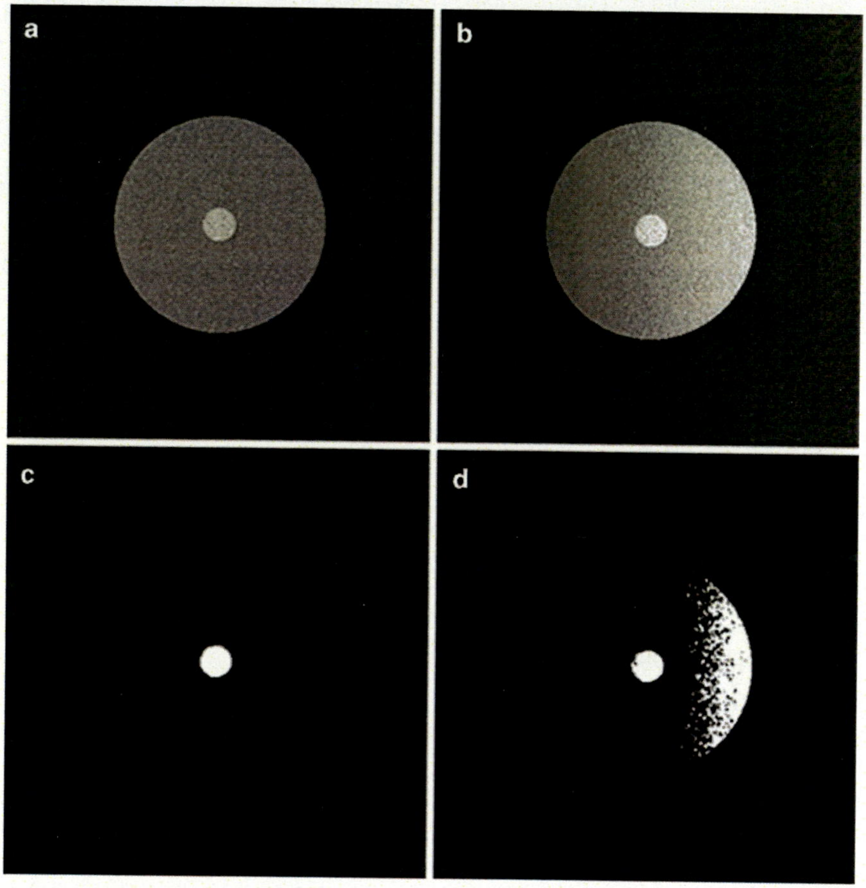

Fig. 5.4 (**a**) Ideal unbiased image (**b**) Product of the ideal image and bias field, (**c**) and (**d**) are the corresponding thresholded images

Retrospective methods are more practical to implement, as they are based on image processing and do not require any additional scan time. A number of methods to retrospectively correct the bias field were used in the literature [27, 29–32]. In an MR image, the bias field is considered to have a low spatial frequency, while anatomical structures are likely to consist of higher spatial frequencies. Therefore, it is possible to model the bias field based on the low frequency components of the original image [33]. This model can then be used with (5.2) to correct the inhomogeneity. Guillemaud et al. [34] proposed a method that changes the multiplicative bias field into an additive one, by applying homomorphic filtering to the logarithmic transform of the image data. An extension to this correction method which combines homomorphic filtering and normalized convolution was also proposed by Guillemaud et al. [32]. Both of these methods can affect high frequency components in the image. However, the second approach uses normalized convolution to compensate for this.

Yang et al. [35] proposed the use of overlapping mosaics to segment fat in MR images affected by intensity inhomogeneities. This segmentation technique is an example of adaptive thresholding and is discussed further in Sect. 5.4.6.

5.4 Overview of Segmentation Techniques Used to Isolate Fat

Most of the segmentation algorithms discussed in this section are 'hard segmentation algorithms', i.e. a definite label is assigned to every voxel in the image (e.g. fat or non-fat). Some consideration will be given to soft segmentation algorithms and their usefulness in dealing with the PVE. These algorithms take into consideration the proportion of each tissue type in every voxel.

Once an image has been segmented, the volume of fat (V_F) contained within an image is calculated using:

$$V_{Fat} = N_{Fat\_Voxels} \times V_{voxel}. \tag{5.3}$$

where N_{Fat_Voxels} is the number of voxels classified as fat in the image and V_{voxel} is the volume of a single voxel. The total fat in kilograms can be calculated by multiplying this value by the density of fat [15].

5.4.1 *Thresholding*

Thresholding is the simplest forms of image segmentation. It is a real-time segmentation technique that is both fast and computationally inexpensive. Thresholding transforms a gray-scale image $f(i,j)$, into a binary image $g(i,j)$, based on a threshold value, T. The process is summarized as:

$$\begin{aligned} g(i,j) &= 1 \quad \text{for } f(i,j) > T, \\ g(i,j) &= 0 \quad \text{for } f(i,j) \leq T. \end{aligned} \tag{5.4}$$

In its simplest form thresholding is a manual process in which the user interactively selects a threshold value (T), based on the distribution of gray-levels in the image histogram, to create a binary image similar to those shown in Fig. 5.5.

Manual thresholding, like all subjective processes, is open to inter and intra-operator variability. Figure 5.5c, d are examples of alternative segmentation results that were obtained using alternative threshold values.

At the outset, some authors used manual thresholding to quantify fat in MR image. However, in an effort to reduce subjectivity, Chan et al. [11] set a strict protocol for threshold selection. The threshold was selected as the minima between the soft tissue and fat peaks in the image histogram. Chan's method shows good correlation with BMI for a sample group of patients [11]. One drawback of this

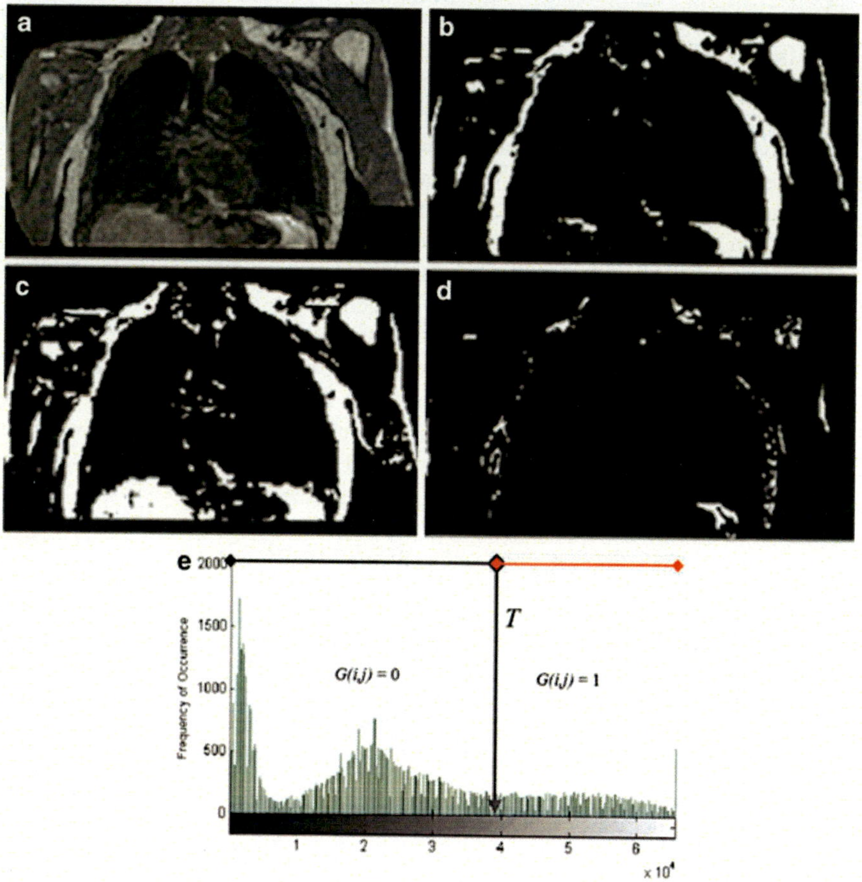

Fig. 5.5 (**a**) T_1-Weighted GE image (**b**) manually thresholded image (**c**) over-thresholded (**d**) under-thresholded (**e**) image histogram

approach is that MR image histograms can have multiple minima between tissue peaks as a result of random noise and inhomogeneities. This can cause ambiguity when manually selecting a threshold value. Another approach used in the literature is to preset a threshold for all subjects based on the manual analysis of a group of healthy controls [13, 36]. This system of segmentation is very rigid and can require user interaction to reclassify mislabelled pixels [13]. One way to avoid variability is to automate the process of thresholding to select an optimum threshold value.

5.4.2 *Selecting the Optimum Threshold*

Subjectively choosing an image threshold is a relatively simple task. However, the objective selection of an optimum threshold can be much more complex. Many algorithms have been developed for the automated selection of optimum thresholds (see, e.g., Zhang et al. [37], Sezgin et al. [38]). Six categories of automated thresholding, including histogram shape information, clustering and entropy methods have been proposed [38]. Histogram–shape–based methods threshold an image based on the peaks, valleys, or curvature of the smoothed images histogram. Clustering–based methods group the elements of an image histogram into two or more tissue classes based on a predefined model. A variety of techniques have been proposed in the literature for automatic threshold selection in gray-scale images. These methods include shape-based algorithms including peak and valley thresholding [39,40] and clustering methods such as the Otsu method [41].

The Otsu method is one of the most referenced thresholding methods in the literature for finding an optimal threshold [41,42]. This method is a non-parametric, unsupervised clustering algorithm used for the automatic selection of an optimal threshold [41]. Optimal thresholds are calculated by minimizing the weighted sum of within-class variance of the foreground and background pixels. The weighted sum of within-class variance, σ_w^2, can be expressed as:

$$\sigma_w^2 = W_b\sigma_b^2 + W_f\sigma_f^2, \tag{5.5}$$

where W_b and W_f are the number of voxels in the background and foreground, respectively, and σ_b^2 and σ_f^2 are the variance in the background and foreground.

Otsu's thresholding is an iterative algorithm which calculates all possible threshold values for the image and the corresponding variance on each side of the threshold. The threshold is then set as the value which gives the maximum value for σ_w^2.

Otsu Algorithm

- Compute histogram
- Set up initial threshold value
- Step through all possible thresholds
  - Compute σ_w^2 for each one
- The optimum threshold corresponds to the maximum σ_w^2

Thresholding using this method gives satisfactory results when the number of voxels in each class is similar. MR images used for the analysis of body fat usually contain at least three tissue classes, soft tissue, fat and background. An extension of the Otsu method known as Multilevel Otsu thresholding can be used to segment images with more than two tissue classes. The Otsu method was used in Fig. 5.6 to segment fat, soft tissue and background.

Using Multilevel Otsu thresholding complete segmentation is not always possible as illustrated in Fig. 5.6b. To compensate, a morphological hole–filling operation was carried out resulting in Fig. 5.6c. Lee and Park [43] found that when foreground area in an image is small relative to the background, segmentation errors will occur. The Otsu method also breaks down in images with a low SNR.

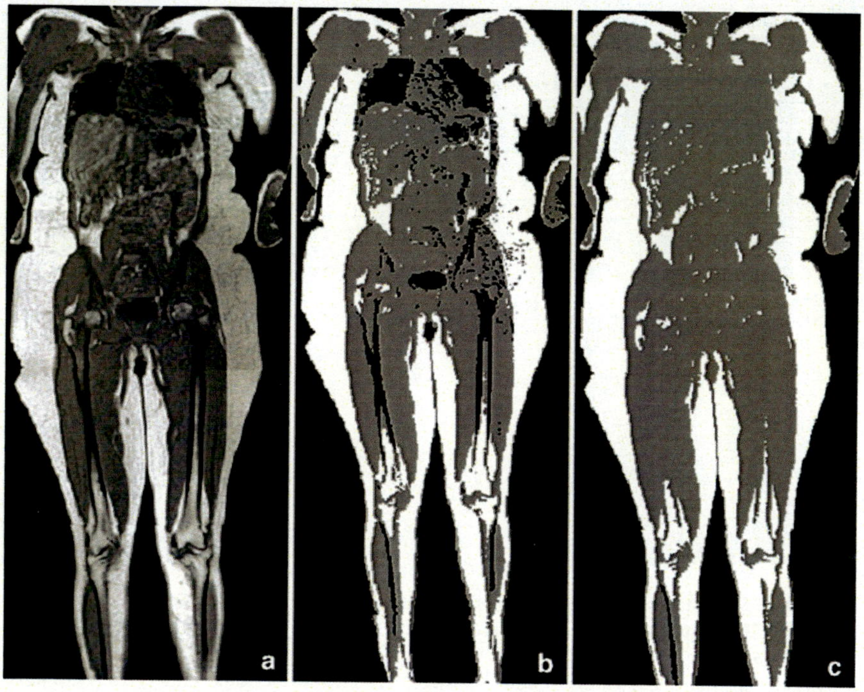

Fig. 5.6 (**a**) T1w GE image containing fat and soft tissue, (**b**) image segmentation using a Multi-Otsu method and (**c**) segmented image corrected using morphological operator

In MRI, the water signal can sometimes obscure the fat peak in the image histogram and make it difficult to use histogram–based global–segmentation techniques to locate the optimum threshold. WS sequences such as b-SSFP (or FISP) and T1w FSE can be used to simplify the image segmentation process [19]. Peng et al. [19] compared Water-suppressed T1w TSE and WS b-SSFP and found that SNR and contrast were superior in WS b-SSFP. In later work, Peng et al. [44] introduced a simple automated method to quantify fat in water saturated MR images. This technique is based on an ideal model of the image histogram and global thresholding. Figure 5.7 illustrates the effect of water saturation on the image histogram.

Peng's segmentation model assumes that all voxels beyond the peak fat value ($S_{\max}$) in Fig. 5.7e are fat and all voxels between 0 and $S_{\max}$ are partial volume fat voxels. On average, partial volume fat voxels are 50% fat [16]. Therefore, the threshold value, S_{th}, is set to $S_{\max}/2$. Once a threshold value is calculated classification of subcutaneous and visceral fat is completed manually. Using water–saturated MR images removes the obstacle of overlapping peaks from the image histogram, which facilitates simple thresholding. Segmentation results shown in Fig. 5.7e, f are very different because of the improved contrast in (d), demonstrating that an optimal imaging protocol can greatly simplify the segmentation process.

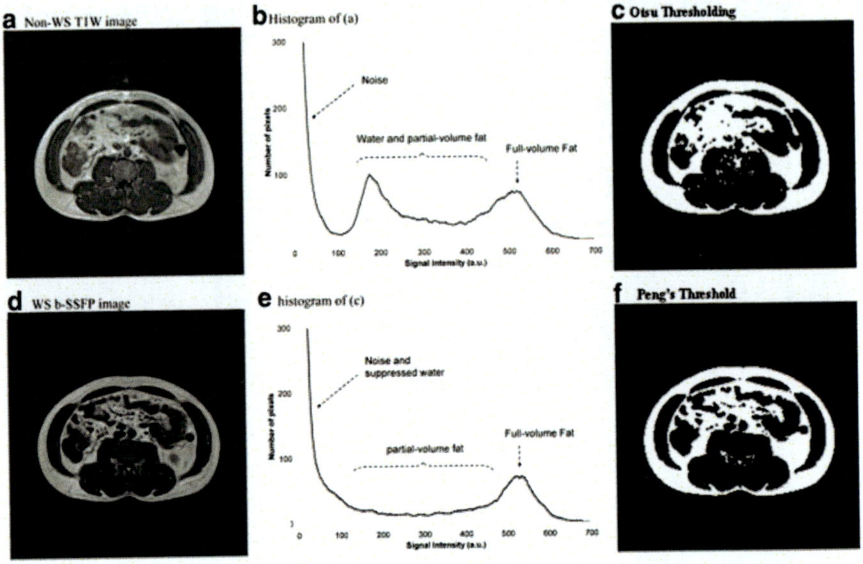

Fig. 5.7 Images and their corresponding intensity histograms obtained using T1W TSE and WS b-SSFP sequences. Both images are acquired from the same anatomic slice of the same subject with breath hold. The T1W TSE image (**a**) is shown to have lower contrast between fat and nonfat. Water and partial-volume fat signal are also in the same gray level range as shown in (**b**), making automated fat quantification difficult. The WS b-SSFP image (**d**), however, shows negligible water signal, leading to fat-only images. The corresponding histogram (**e**) shows that signal from suppressed water, partial-volume fat, and full-volume fat are delineated. This makes it possible to perform automated, yet accurate fat quantification. This material is reproduced with permission of John Wiley & Sons, Inc. [44] with the addition of images (**c**) and (**f**). Image (**c**) is the result of OTSU thresholding applied to (**a**) and (**f**) is the result of the segmentation technique described by Peng et al. [44]

5.4.3 Gaussian Mixture Model

When overlapping peaks cannot be avoided the Gaussian mixture model (GMM) can be used to model complex probability density functions (PDF), such as image histograms, $f(x)$, as k overlapping Gaussians. This is illustrated in Fig. 5.8 and can be expressed as:

$$f(x) = \sum_{i=1}^{k} p_i N\left(x|\mu_i, \sigma_i^2\right), \tag{5.6}$$

where μ_i and σ_i are the mean and standard deviation of the ith Gaussian, respectively. p_i is the mixing proportion of the constituent Gaussians, N, used to model the PDF of the image. It satisfies the conditions:

$$p_i > 0, \quad \sum_{i=1}^{k} p_i = 1. \tag{5.7}$$

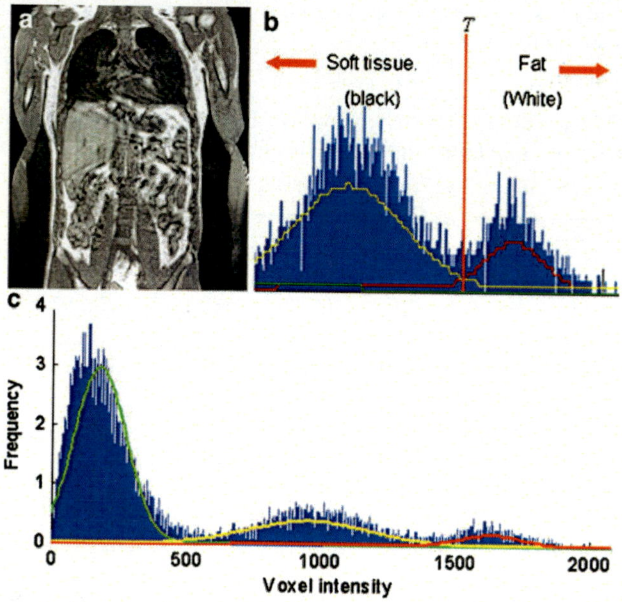

Fig. 5.8 (**a**) T1-weighted GE image containing fat and soft tissue, (**b**) shows the threshold value *T* illustrates the threshold location using a *red line*. Voxels with gray level intensities higher than *T* are classified as fat. (**c**) Image histogram of (**a**) and its constituent Gaussians estimated using the GMM

Each Gaussian cluster is then modeled using the product of (5.8), the general equation for a Gaussian distribution and its probability (p_i).

$$N(\mu_i, \sigma_i) = \frac{1}{\sigma_i\sqrt{2\pi}} \exp\left(-\frac{(x-\mu_i)^2}{2\sigma_i^2}\right). \tag{5.8}$$

Figure 5.8c, illustrates an image histogram containing 3 tissue classes and its constituent Gaussians calculated using a GMM. One of the most common algorithms used to calculate p_i, μ_i and σ_i is the expectation maximization (EM) algorithm [45].

The GMM assumes that the PDF of all constituent tissue types in an image histogram are Gaussian and that tissue of the same class is uniform in intensity throughout the image or region of image to be segmented. When segmenting fat in MR images, k is usually estimated using *a priori* knowledge as 3 (fat, soft tissue and background).

The probability of an event (pixel/voxel) belonging to the ith distribution is given by

$$P(x|\Theta) = \sum_{i=1}^{k} p_i N(x|\mu_i, \sigma_i), \tag{5.9}$$

where

$$\Theta = \{\mu_{i=1},\ldots,\mu_{i=k},\sigma_{i=1},\ldots,\sigma_{i=k},p_{i=1},\ldots,p_{i=k}\}. \qquad (5.10)$$

The EM algorithm is a two-step iterative algorithm consisting of an expectation step (E-step) and a maximization step (M-step). The algorithm can be initialized using k-means clustering or by equal partitioning of the data into k regions or by automated seed initialization [46]. The expected log likelihood function for the complete data is calculated using the E-step and is defined by $Q(\Theta,\widehat{\Theta}(t))$ using the estimated parameters $\widehat{\Theta}(t)$.

$$Q(\Theta,\widehat{\Theta}(t)) \equiv E[\log N\,(X,Y|\Theta)|X,\widehat{\Theta}(t)]. \qquad (5.11)$$

The function $Q(\Theta,\widehat{\Theta}(t))$ contains two arguments Θ denotes the parameters that will be optimized in order to maximize the likelihood and $\widehat{\Theta}(t)$ corresponds to the estimated values. X is the observed data and remains constant while Y is the missing data, which is controlled by the underlying distributions.

The second step in the algorithm is the M-step. This step uses the maximized values from (5.11) above to generate a new set of parameters and is given by:

$$\widehat{\Theta}(t+1) = \arg\max_{\Theta} Q(\Theta,\widehat{\Theta}(t)). \qquad (5.12)$$

If we have an estimate of the means (μ_i) and standard deviation (σ_i) of the constituent Gaussians we can compute the probability (p_i) of a point (gray-level) in the histogram belonging to the kth Gaussian. The maximization step updates the Gaussian parameters using (5.13), (5.14) and (5.15).

$$p_i^{\text{new}} = \frac{1}{k}\sum_{i=1}^{k} N(i|x,\widehat{\Theta}(t)). \qquad (5.13)$$

$$\mu_i^{\text{new}} = \frac{\sum_{i=1}^{k} xN(i|x,\widehat{\Theta}(t))}{\sum_{i=1}^{k} N(i|x,\widehat{\Theta}(t))}. \qquad (5.14)$$

$$\sigma_i^{\text{new}} = \frac{\sum_{i=1}^{k} N(i|x,\widehat{\Theta}(t))(x_i-\mu_i^{\text{new}})(x_i-\mu_i^{\text{new}})^T}{\sum_{i=1}^{k} N(i|x,\widehat{\Theta}(t))}. \qquad (5.15)$$

The algorithm iterates between (5.11) and (5.12) until convergence is reached. Based on the Gaussian estimated using the EM algorithm an optimized threshold is calculated that minimizes misclassification error. The threshold for fat is set to the closest gray-level corresponding to the intersection of the fat and soft tissue Gaussians, as illustrated in Fig. 5.8b. Figure 5.13 in Sect. 5.5 is an example of an image which has been segmented successfully using the GMM. Lynch et al. [46]

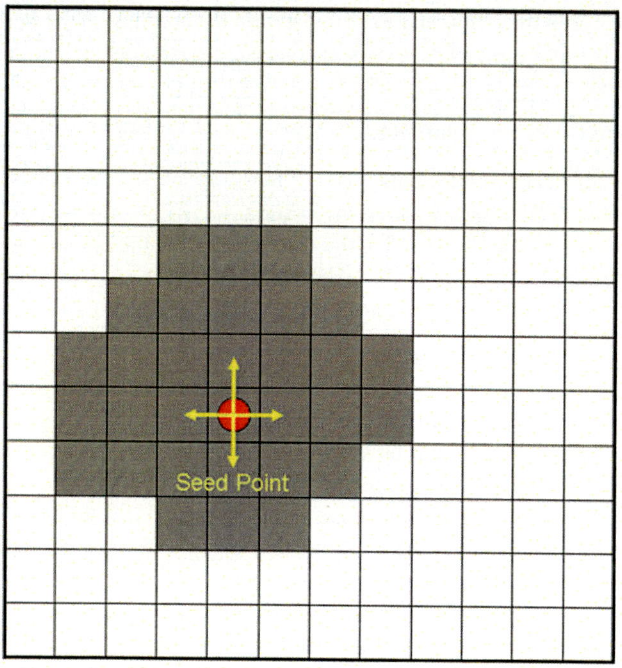

Fig. 5.9 Seed point selection voxels in a region of interest. Each of the nearest neighbor voxels are illustrated using *yellow arrows*

proposed a novel approach to initialize cluster centers based on histogram analysis. It used the GMM to obtain a more robust segmentation of fat in MR images. No objective assessment of this method was carried out.

5.4.4 Region Growing

Region growing is an image segmentation technique that creates regions based some predefined homogeneity criteria such as texture, gray-level or color. Gray-level is the characteristic most commonly used when segmenting fat in MR images. Region growing aims to merge voxels or small regions in an image into larger ones based on a homogeneity criteria. The first step in any region-growing algorithm is the selection of a seed point, as illustrated in Fig. 5.9. This can be a single voxel or a small region of voxels and can be selected manually or automatically. Next, a homogeneity criterion is set. For example, if the gray-level of the neighboring voxel is between two threshold values, merge the voxel (region) with the seed point. The seed voxels nearest neighbors are illustrated using the yellow arrows in Fig. 5.9. Each new voxel in the region becomes a growth point and is compared to its nearest neighbors using the same homogeneity criteria. Growth of the region ceases when no new growth points can be created within the confines of the homogeneity criteria. Growth or merging of regions will continue iteratively until the stopping criterion is reached.

Region Growing Algorithm

- Select a seed point/points
- Define a homogeneity/merging criteria
- Join all voxels connected to the seed that follow the homogeneity criteria to form a region
- Stop the algorithm when no adjacent voxels agree with the homogeneity criteria

Figure 5.10 shows an image containing a number of manually selected seed points and the resultant segmentation using region growing. Unlike thresholding, region growing can differentiate objects based on their spatial location, which enables the classification of different fat categories. In Fig. 5.10b, note also that bone marrow adipose tissue and the liver are not classified as fat. Fat classification as distinct from segmentation is an important issue and is discussed in detail in Sect. 5.5.

Region growing can be implemented using a number of algorithms, including region merging, region splitting and split and merge algorithms [39, 47]. Split and merge algorithms can be used to automate region growing, removing the need for subjective seed point selection [48]. Automated seed selection does not always guarantee complete segmentation due to the unpredictability of seed point selection and may therefore require manual addition of extra seed points after initial segmentation. In noisy images, it is possible for holes to appear in the segmented regions, post processing may be used to reduce these.

Siegel et al. [6] used a region growing technique in its simplest form to segment fat in transverse MR images of the abdomen. However, region growing is sometimes combined with edge detection to reduce the occurrence of over- and under-segmentation. In a study by Yan et al. [49], a three-step segmentation process to isolate skeletal structures in CT images was employed. The first step was the application of a three-dimensional region-growing algorithm with adaptive local thresholds, which was followed by a series of boundary correction steps. This approach is both accurate and reliable for high contrast, low noise CT images. Subsequently, a similar technique was applied to the segmentation of fat in MR images by Brennan et al. [15].

Brennan et al. [15], used a four-step automated algorithm for the quantification of fat in MR images. Their algorithm was based on initialization of fat regions using conservative thresholding followed by steps of boundary enhancement, region growing and region refining (post processing). A weak correlation was found between total body fat quantified using MRI and BMI. The authors attribute this to the flawed nature of BMI measurements. A more appropriate measure of accuracy would have been comparison with manual delineation of fat by an expert radiologist (the gold standard). Combination of both region growing and edge–based segmentation algorithms contradicts work by Rajapakse and Kruggel [23], who state that the use of region and edge detection schemes are unsuitable in MR images due

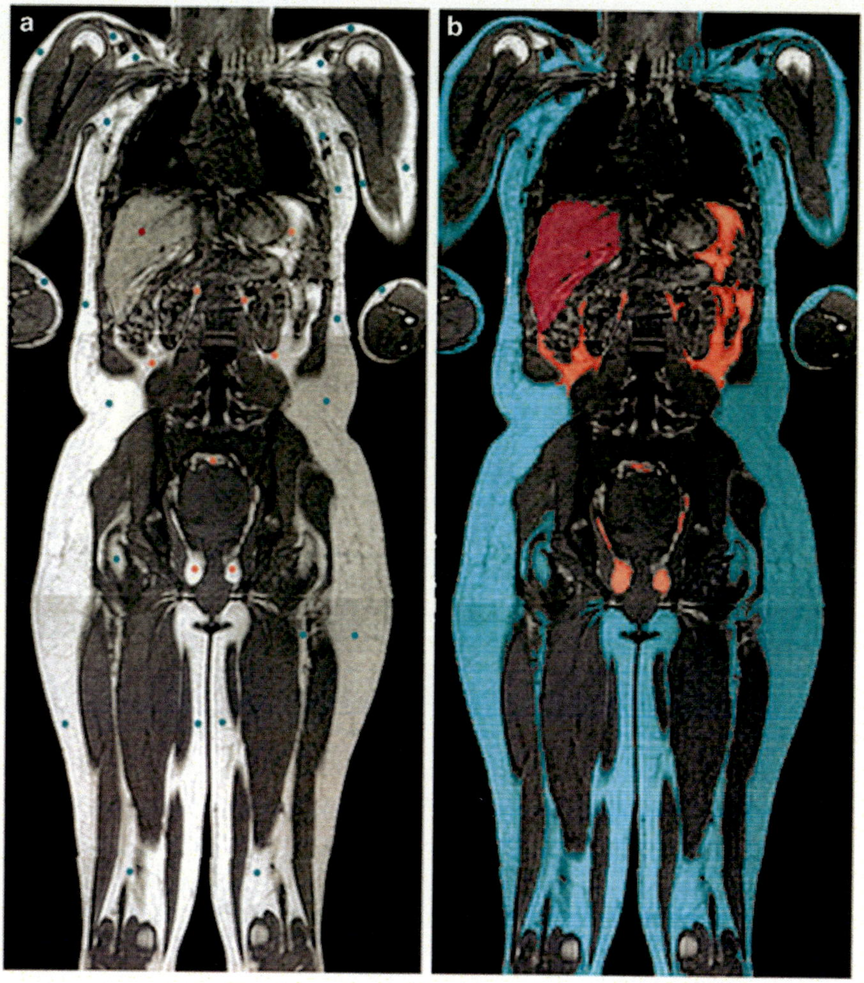

Fig. 5.10 (**a**) A T1w GE image and the seed points used for region growing and (**b**) is the resultant segmentation following region growing, visceral fat is labeled with red, other fat is labeled *blue* and the liver *purple*. Each labeled group is quantified separately

to the lack of clearly defined edges. Despite this, the data presented by Brennan et al. were well segmented. Brennan's method did not classify and label body fat. Further steps are required to develop classification algorithm.

5.4.5 *Adaptive Thresholding*

Due to intensity inhomogeneities in MR images a single threshold may not be sufficient for the entire image. Segmentation using adaptive thresholding can

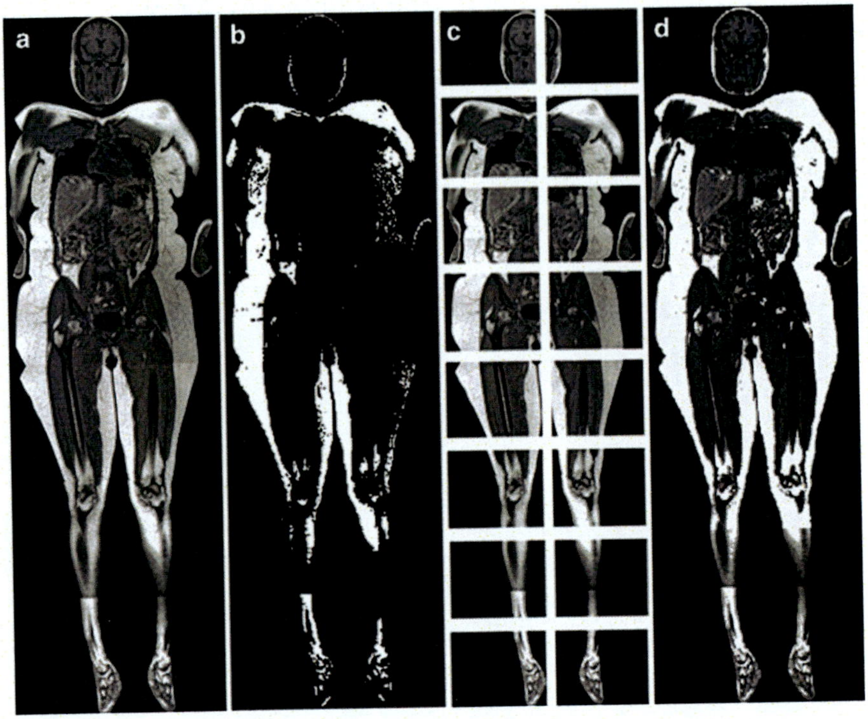

Fig. 5.11 (**a**) Whole body T_1-weighted GE image affected by intensity inhomogeneities; (**b**) Result of global segmentation using the GMM on (**a**); (**c**) Sub-images used for adaptive thresholding; and (**d**) is the result of adaptive thresholding

compensate for intensity inhomogeneities [39]. Adaptive segmentation can be achieved by dividing an image into a number sub-images as shown in Fig. 5.11c [50]. Each sub-image is then segmented using one of the segmentation algorithms discussed in Sect. 5.4.2.

Two factors must be considered when selecting the size of the sub-images:

(1) They must be small enough so the impact of the intensity inhomogeneity is minimal across each of their areas
(2) They must contain enough voxels to maintain a workable SNR

Figure 5.11a is an example of an image that is affected by intensity inhomogeneities and (b) is the result of global segmentation using the GMM algorithm. Using adaptive segmentation a significant improvement can be seen in Fig. 5.11d. If the sub-images cannot be made small enough to reduce the impact of the intensity inhomogeneities, a technique which uses overlapping mosaics may be used [35], this is discussed in Sect. 5.4.6.

Local adaptive thresholding (using a sliding window) can be an effective segmentation algorithm in the presence of inhomogeneities [51]. This technique thresholds individual voxels using the mean or median value of their surrounding $n \times n$ neighborhood. MR images acquired for fat analysis can contain large monotone regions consisting of a single tissue type. In order to achieve meaningful segmentation the size of the neighborhood must be large enough to contain more than one tissue class at any point tin the image. This should be considered when selecting the neighborhood size.

5.4.6 Segmentation Using Overlapping Mosaics

Yang et al. [35] developed a method to segment fat in MR images using overlapping mosaics. The segmentation technique consists of 3 steps:

(1) Mosaic bias field estimation
(2) Adipose tissue segmentation
(3) Consistency propagation

Following smoothening (low pass filtering) to remove noise the expression for the biased image in (5.2) becomes:

$$f'_{\text{biased}}(x,y) = f_{\text{original}}(x,y)\beta(x,y), \tag{5.16}$$

where $f'_{\text{biased}}(x,y)$ is the image after filtering. Assuming the bias field varies gradually across the entire image, log $(\beta(x,y))$ can be approximated by a piecewise linear function. Yang divides $f'_{\text{biased}}(x,y)$ into a array of overlapping mosaics or sub-images (T_{ij}). Within each of the sub-images Log $(\beta(x,y))$ is assumed to be first order linear, therefore, $\forall(x,y) \in T_{ij}$:

$$\log f'_{\text{biased}}(x,y) = \log(f_{\text{original}}(x,y)) + a_{ij}\left(x - x_{ij}^{(0)}\right) + b_{ij}\left(y - y_{ij}^{(0)}\right) + c_{ij}, \tag{5.17}$$

where $(x_{ij}^{(0)}, y_{ij}^{(0)})$ is the upper left voxel in each sub image. Optimal values for a and b are estimated by maximizing the function:

$$P = \sum_{\xi}\left(\sum_{(x,y)\in T_{ij}} \delta(\log(f'_{\text{biased}}(x,y)))\left(-a_{ij}\left(x - x_{ij}^{(0)}\right) - b_{ij}\left(y - y_{ij}^{(0)}\right) - \xi\right)\right)^2, \tag{5.18}$$

where $\delta(x) = 1$ when $x = 0$, and ξ is the gray-scale intensity index of the image histogram. C_{ij} is not calculated because it affects voxels in the sub–image uniformly, causing a change in position of gray–levels in the image histogram but not the shape. Once the image is corrected, the skewed and bimodal peaks discussed in Sect. 5.3.2 appear more distinctive.

When intensity inhomogeneities are corrected, image segmentation is carried out using a multi-level thresholding technique on each sub-image. Segmentation

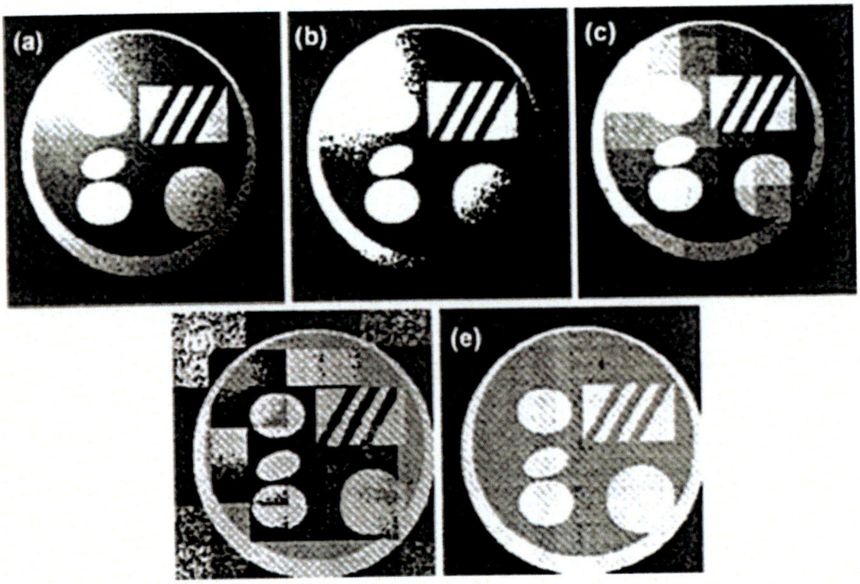

Fig. 5.12 The intermediate processing results of the proposed algorithm applied to a synthetic phantom. (**a**) The original image with an SNR of 20 dB; (**b**) the optimum manually segmented result by using intensity thresholding; (**c**) the bias corrected image by using overlapping mosaics; (**d**) the result of initial segmentation; (**e**) the final fat distribution after applying inter-mosaic consistency propagation. This material is reproduced with kind permission from Springer Science+Business Media B.V [35]

is based on an automated thresholding technique that minimises the total variance within the three classes, fat soft tissue and background, giving 2 threshold values ξ_1 and ξ_2 [16].

A measure of confidence, λ_{ij}, of the segmentation result is calculated, as not all sub images will contain all three tissue classes.

$$\lambda_{ij} = \frac{\underset{\xi \geq \xi_2}{meanH_{ij}(\xi)}}{\underset{\xi < \xi_1}{meanH_{ij}(\xi)}}, \tag{5.19}$$

$H_{ij}(\xi)$ is the log transform of the image histogram. λ_{ij} is likely to be large when all three tissue classes are present in the sub-image. However, when only one or two tissue classes are present λ_{ij} will be much smaller indicating misclassification. The mosaic tile with the highest value of λ_{ij} is used as a seed for consistency propagation. The regions of overlap between the seed tile and its nearest neighbors are compared. If any conflicting segmentation results are present, then the value for ξ_2 in neighboring tile is changed to that of the seed. This process is propagated to all tiles within the image until segmentation result like those shown in Fig. 5.12 are achieved. Peng et al. [16], compared this technique to the gold standard, manual segmentation, and found that the mean percentage between the two was 1.5%.

5.5 Classification of Fat

To appreciate the complexities associated with quantifying fat in medical images, it is important to know what exactly needs to be measured. The most common conflict in the literature is in the terminology used, (i.e. fat or adipose tissue) [52]. The difference between fat and adipose tissue is important when quantifying fat in MR images. Bone marrow in a typical T1w imaging sequence has the same graylevel as body fat. However, bone marrow adipose tissue is not classified as fat because it is connected to haematopoietic activity[2] and not to obesity [53, 54]. Classification is further complicated by the subdivision of fat into three main categories: total body fat [15], visceral fat and subcutaneous fat [6]. Whole body fat includes the measurement of all adipose tissue except bone marrow and adipose tissue contained in the head, hands and feet [52]. A summary of the proposed classification of adipose tissue within the body is given by Shen et al. [52] and is presented in Table 5.1. Examination of body fat distribution involves the analysis of two or more of the fat categories outlined in Table 5.1.

Global segmentation algorithms such as thresholding require extra steps to classify fat. This can be achieved manually by drawing a region of interest around areas such as the viscera. Figure 5.13 shows the result of manual classification

Table 5.1 Proposed classification of total body adipose tissue as given by Shen et al. [52]

Adipose tissue compartment	Definition
Total adipose tissue	Sum of adipose tissue, usually excluding bone marrow and adipose tissue in the head, hands, and feet
Subcutaneous adipose tissue	The layer found between the dermis and the aponeuroses and fasciae of the muscles. Includes mammary adipose tissue
Superficial subcutaneous adipose tissue	The layer found between the skin and a fascial plane in the lower trunk and gluteal-thigh area
Deep subcutaneous adipose tissue	The layer found between the muscle fascia and a fascial plane in the lower trunk and gluteal-thigh areas
Internal adipose tissue	Total adipose tissue minus subcutaneous adipose tissue
Visceral adipose tissue	Adipose tissue within the chest, abdomen, and pelvis
Non-visceral internal adipose tissue	Internal adipose tissue minus visceral adipose tissue
Intramuscular adipose tissue	Adipose tissue within a muscle (between fascicles)
Perimuscular adipose tissue	Adipose tissue inside the muscle fascia (deep fascia), excluding intramuscular adipose tissue
Intermuscular adipose tissue	Adipose tissue between muscles
Paraosseal adipose tissue	Adipose tissue in the interface between muscle and bone (e.g., paravertebral)
Other non-visceral adipose tissue	Orbital adipose tissue; aberrant adipose tissue associated with pathological conditions (e.g., lipoma)

[2] Hematopoietic activity: pertaining to the formation of blood or blood cells.

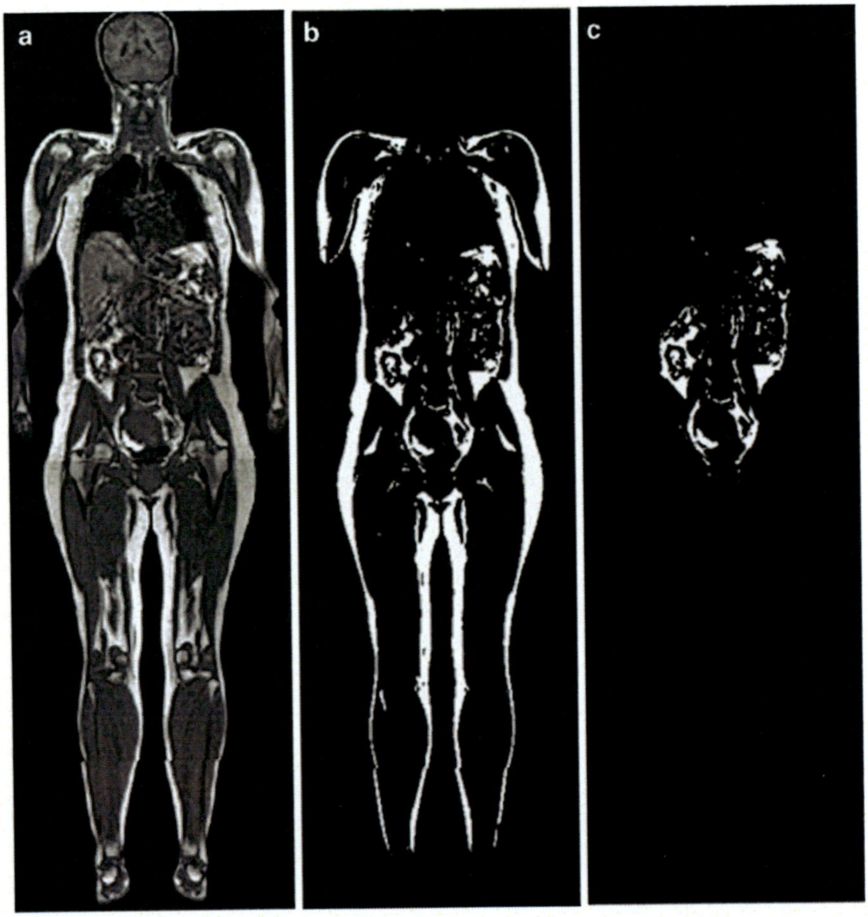

Fig. 5.13 (**a**) T1w GE image (**b**) global segmentation using the GMM and (**c**) classification of visceral fat

after global segmentation using the GMM. Region growing allows for classification based on spatial location. This is illustrated in Fig. 5.10 where both the liver and visceral fat are differentiated from fat in the image. Further work is required to fully automate classification.

5.6 Conclusions

This chapter reviewed the challenges associated with the quantification of fat in MR images. Accurate segmentation and analysis of fat using MRI is not a trivial matter. Ideally, MR images acquired for fat analysis should contain three distinct tissue classes: fat, soft tissue and air. However, as a result of the many artifacts inherent to MRI this ideal image model is rarely attained. Therefore, consideration must be given to both the PVE and intensity inhomogeneities when segmenting fat in MR images. It is crucial that fat segmentation is approached with a thorough understanding of these artifacts and the limitations they present. A number of techniques used to segment fat in the presence of inhomogeneities were outlined in this chapter.

Image acquisition is the most important step in the quantification and analysis of fat using MRI. Before scanning patients, the imaging sequence should be optimized to achieve a balance between image contrast and acquisition time. Selection of an appropriate imaging sequence, such as the WS–bSSFP can significantly reduce the complexity of the segmentation algorithm required.

Fat classification currently requires manual intervention and will remain a significant challenge in the future. Future work in this field will investigate the prospect of fully automating the classification process and the use of soft segmentation algorithms involving fuzzy sets to overcome the PVE.

References

1. British Nutrition Foundation Obesity Task Force: Obesity: the report of the British Nutrition Foundation Task Force, John Wiley & Sons (1999)
2. Sarría, A., Moreno, L.A., et al.: Body mass index, triceps skinfold and waist circumference in screening for adiposity in male children and adolescents. Acta Pædiatrica **90**(4), 387–392 (2001)
3. Peters, D., et al.: Estimation of body fat and body fat distribution in 11-year-old children using magnetic resonance imaging and hydrostatic weighing, skinfolds, and anthropometry. Am. J. Hum. Biol. **6**(2), 237–243 (1994)
4. Rush, E.C., et al.: Prediction of percentage body fat from anthropometric measurements: comparison of New Zealand European and Polynesian young women. Am. J. Clin. Nutr. **66**(1), 2–7 (1997)
5. (WHO), W.H.O. [cited; Available from:www.who.int/dietphysicalactivity/publication/facts/obesity/en/. (2008)
6. Siegel, M.J., et al.: Total and intraabdominal fat distribution in preadolescents and adolescents: measurement with MR imaging. Radiology **242**(3), 846–56 (2007)
7. Kullberg, J., et al.: Whole-body adipose tissue analysis: comparison of MRI, CT and dual energy X-ray absorptiometry. Br. J. Radiol. **82**(974), 123–130 (2009)
8. Seidell, J.C., Bakker, C.J., van der Kooy, K.: Imaging techniques for measuring adipose-tissue distribution–a comparison between computed tomography and 1.5-T magnetic resonance. Am. J. Clin. Nutr. **51**(6), 953–957 (1990)
9. Barnard, M.L., et al.: Development of a rapid and efficient magnetic resonance imaging technique for analysis of body fat distribution. NMR Biomed. **9**(4), 156–64 (1996)
10. Thomas, E.L., et al.: Magnetic resonance imaging of total body fat. J. Appl. Physiol. **85**(5), 1778–85 (1998)
11. Chan, Y.L., et al.: Body fat estimation in children by magnetic resonance imaging, bioelectrical impedance, skinfold and body mass index: a pilot study. J Paediatr. Child Health **34**(1), 22–28 (1998)
12. Kamel, E.G., McNeill, G., Van Wijk, M.C.: Change in intra-abdominal adipose tissue volume during weight loss in obese men and women: correlation between magnetic resonance imaging and anthropometric measurements. Int. J. Obes. Relat. Metab. Disord. **24**(5), 607–613 (2000)

13. Ross, R., et al.: Influence of diet and exercise on skeletal muscle and visceral adipose tissue in men. J. Appl. Physiol. **81**(6), 2445–2455 (1996)
14. Terry, J.G., et al.: Evaluation of magnetic resonance imaging for quantification of intraabdominal fat in human beings by spin-echo and inversion-recovery protocols. Am. J. Clin. Nutr. **62**(2), 297–301 (1995)
15. Brennan, D.D., et al.: Rapid automated measurement of body fat distribution from whole-body MRI. AJR Am. J. Roentgenol. **185**(2), 418–23 (2005)
16. Peng, Q., et al.: Automated method for accurate abdominal fat quantification on water-saturated magnetic resonance images. J. Magn. Reson. Imaging. **26**(3), 738–46 (2007)
17. Kovanlikaya, A., et al.: Fat quantification using three-point dixon technique: in vitro validation. Acad. Radiol. **12**(5), 636–639 (2005)
18. Goyen, M.: In: Goyen, M. (ed.) Real Whole Body MRI Requirements, Indications, Perspectives, 1 edn., vol. 1, p. 184. Berlin, Mc Graw Hill (2007)
19. Peng, Q., et al.: Water-saturated three-dimensional balanced steady-state free precession for fast abdominal fat quantification. J. Magn. Reson. Imaging . **21**(3), 263–271 (2005)
20. Warren, M., Schreiner, P.J., Terry, J.G.: The relation between visceral fat measurement and torso level–is one level better than another? The Atherosclerosis Risk in Communities Study, 1990–1992. Am. J. Epidemiol. **163**(4), 352–358 (2006)
21. Dugas-Phocion, G., et al.: Improved EM-Based tissue segmentation and partial volume effect quantification in multi-sequence brain MRI. In: Lecture Notes in Computer Science, Medical Image Computing and Computer-Assisted Intervention – MICCAI 2004, vol. 3216, p. 7. (2004)
22. González Ballester, M.Á., Zisserman, A.P. Brady, M.: Estimation of the partial volume effect in MRI. Med. Image Anal. **6**(4), 389–405 (2002)
23. Rajapakse, J.C., Kruggel, F.: Segmentation of MR images with intensity inhomogeneities. Image Vis. Comput. **16**(3), 165–180 (1998)
24. Li, X., et al.: Partial volume segmentation of brain magnetic resonance images based on maximum a posteriori probability. Med. Phys. **32**(7), 2337–2345 (2005)
25. Horsfield, M.A., et al.: Incorporating domain knowledge into the fuzzy connectedness framework: Application to brain lesion volume estimation in multiple sclerosis. Med. Imaging IEEE Trans. **26**(12), 1670–1680 (2007)
26. Siyal, M.Y., Yu, L.: An intelligent modified fuzzy c-means based algorithm for bias estimation and segmentation of brain MRI. Pattern Recognit. Lett. **26**(13), 2052–2062 (2005)
27. Yun, S., Kyriakos, W.E., et al.: Projection-based estimation and nonuniformity correction of sensitivity profiles in phased-array surface coils. J. Magn. Reson. Imaging. **25**(3), 588–597 (2007)
28. Murakami, J.W., Hayes, C.E. Weinberger, E. Intensity correction of phased-array surface coil images. Magn. Reson. Med. **35**(4), 585–590 (1996)
29. Pham, D.L., Prince, J.L.: An adaptive fuzzy C-means algorithm for image segmentation in the presence of intensity inhomogeneities. Pattern Recognit. Lett. **20**(1), 57–68 (1999)
30. Nie, S., Zhang, Y., Li, W., Chen, Z.: A novel segmentation method of MR brain images based on genetic algorithm. IEEE International Conference on Bioinformatics and Biomed. Eng. 729–732 (2007)
31. Wells, W.M., et al.: Adaptive segmentation of MRI data. Med. Imaging IEEE Trans. **15**(4), 429–442 (1996)
32. Guillemaud, R.: Uniformity correction with homomorphic filtering on region of interest. In: Image Processing, 1998. ICIP 98. Proceedings. 1998 International Conference on (1998)
33. Behrenbruch, C.P., et al.: Image filtering techniques for medical image post-processing: an overview. Br. J. Radiol. **77**(suppl_2), S126–132 (2004)
34. Guillemaud, R., Brady, M.: Estimating the bias field of MR images. Med. Imaging IEEE Trans. **16**(3), 238–251 (1997)
35. Yang, G.Z., et al.: Automatic MRI adipose tissue mapping using overlapping mosaics. MAGMA **14**(1), 39–44 (2002)
36. Leroy-Willig, A., et al.: Body composition determined with MR in patients with Duchenne muscular dystrophy, spinal muscular atrophy, and normal subjects. Magn. Reson. Imaging **15**(7), 737–44 (1997)

37. Zhang, Y.J., Gerbrands, J.J.: Comparison of thresholding techniques using synthetic images and ultimate measurement accuracy. In: Pattern Recognition, 1992. Vol.III. Conference C: Image, Speech and Signal Analysis, Proceedings., 11th IAPR International Conference on (1992)
38. Mehmet, S., Bulent, S.: Survey over image thresholding techniques and quantitative performance evaluation. J. Electron. Imaging **13**(1), 146–168 (2004)
39. Sonka, M., Hlavac, V., Boyle, R.: In: Hilda, G. (ed.) Image Processing, Analysis, and Machine Vision, International Student Edition, 3rd edn., vol. 1. Thomson, Toronto, p. 829. (2008)
40. Nualsawat, H., et al.: FASU: A full automatic segmenting system for ultrasound images. In: Proceedings of the Sixth IEEE Workshop on Applications of Computer Vision. 2002, IEEE Computer Society.
41. Otsu, N.: A threshold selection method from gray-level histograms. Syst. Man Cybern. IEEE Trans. **9**(1), 62–66 (1979)
42. Sezgin, M., Sankur, B.: Survey over image thresholding techniques and quantitative performance evaluation. J. Electron. Imaging **13**(1), 146–168 (2004)
43. Lee, H., Park, R.H.: Comments on 'An optimal multiple threshold scheme for image segmentation'. Syst. Man Cybern. IEEE Trans. **20**(3), 741–742 (1990)
44. Peng, Q., et al.: Automated method for accurate abdominal fat quantification on water-saturated magnetic resonance images. J. Magn. Reson. Imaging **26**(3), 738–746 (2007)
45. Dempster, A.P., Laird, N.M., R.D. B.: Maximum likelihood from incomplete data via the EM algorithm. J. R. Stat. Soc. Ser. B (Methodological) **39**(1), 1–38 (1977)
46. Lynch, M., et al.: Automatic seed initialization for the expectation-maximization algorithm and its application in 3D medical imaging. J. Med. Eng. Technol. **31**(5), 332–340 (2007)
47. Liang, Z.: Tissue classification and segmentation of MR images. Eng. Med. Biol. Mag. IEEE. **12**(1), 81–85 (1993)
48. Pham, D.L., Xu, C., Prince, J.L.: Current methods in medical image segmentation. Annu. Rev. Biomed. Eng. **2**, 315–37 (2000)
49. Yan, K., Engelke, K., Kalender, W.A.: A new accurate and precise 3-D segmentation method for skeletal structures in volumetric CT data. Med. Imaging IEEE Trans. **22**(5), 586–598 (2003)
50. Wee-Chung Liew, A., Yan, H., Yang, M.: Robust adaptive spot segmentation of DNA microarray images. Pattern Recogn. **36**(5), 1251–1254 (2003)
51. Dougherty, G.: Digital Image Processing for Medical Applications, 1 edn., vol. 1, p. 447. Cambridge University Press, New York (2009)
52. Shen, W., et al.: Adipose tissue quantification by imaging methods: A proposed classification. Obes. Res. **11**(1), 5–16 (2003)
53. Laharrague, P., Casteilla, L.: Bone Marow Adipose Tissue, 1 edn. Nutrition and health. Humana Press, New Jersey (2007)
54. Mantatzis, M., Prassopoulos P.: Total body fat, visceral fat, subcutaneous fat, bone marrow fat? What is important to measure? AJR Am. J. Roentgenol. **189**(6), W386 (2007); author reply W385

Chapter 6
Angiographic Image Analysis

Olena Tankyevych, Hugues Talbot, Nicolas Passat, Mariano Musacchio, and Michel Lagneau

6.1 Introduction

The important rise of medical imaging during the twentieth century, mainly induced by physics breakthroughs related to nuclear magnetic resonance and X-rays has led to the development of imaging modalities devoted to visualize vascular structures. The analysis of such *angiographic* images is of great interest for several clinical applications. Initially designed to generate 2D data, these imaging modalities progressively led to the acquisition of 3D images, enabling the visualization of vascular volumes.

However, such 3D data are generally huge, being composed of several millions of voxels, while the useful –vascular– information generally represents less than 5% of the whole volume. In addition to this sparseness, the frequent low signal-to-noise ratio and the potential presence of artifacts make the analysis of such images a challenging task. In order to assist radiologists and clinicians, it is therefore necessary to design software tools enabling them to extract as well as possible the relevant information embedded in 3D angiographic data.

One of the main ways to perform such a task is to develop *segmentation* methods, i.e., tools which (automatically or interactively) extract the vessels as 3D volumes from the angiographic images. A survey of such segmentation methods is proposed in Sect. 6.3. In particular, it sheds light on recent advances devoted to merge different image processing methodologies to improve the segmentation accuracy.

Another way to consider computer-aided analysis of 3D angiographic images is to provide human experts with a base of high-level anatomical knowledge which can possibly be involved in more specific analysis procedures such as vessel labelling.

O. Tankyevych (✉)
Université Paris-Est, Laboratoire d'Informatique Gaspard-Monge – UMR CNRS 8049, Paris, France
e-mail: tankyevo@esiee.fr

G. Dougherty (ed.), *Medical Image Processing: Techniques and Applications*, Biological and Medical Physics, Biomedical Engineering, DOI 10.1007/978-1-4419-9779-1_6,

Such knowledge can in particular be embedded in *vascular atlases* which are devoted to model qualitative and/or quantitative information related to vessels. A survey of different existing vascular atlases, and ways they can be created is presented in Sect. 6.4.

The purpose of this chapter is to provide some general background notions on 3D angiographic image analysis. Due to space limitations, it is impossible to propose an exhaustive overview on vessel segmentation and vascular knowledge modeling. Consequently, Sects. 6.3 and 6.4 propose partial, but hopefully relevant, states of the art on these topics. They present some of the most classical and/or recent related works, and some pointers on more complete surveys linked to the main topics of this chapter (or to connected research fields, for the sake of completeness). They also present some recent contributions of some of the authors, especially related to vessel segmentation.

6.2 Clinical Context

Vascular pathologies are one of the main causes of morbidity and mortality in the Western world, and thus constitute an important issue in public health. The causes are manifold, from traumatic lesions (due to accidents) to genetic vascular diseases (such as some arteriovenous malformations), *via* those linked to obesity and stress (such as atheromatosis and diabetes).

An anomaly affecting vessels can provoke perturbations in organ circulation as well as in tissues supplied by the involved vascular network. If the lumen of the arteries is shrunk, such as in an atheromatosis disease, blood flow will be affected and the associated organ will be insufficiently supplied, leading, in the worst cases, to ischemia, and then tissue death. The breaking of a vessel, normal (as in a trauma) or pathological (as in an aneurysm rupture), can cause hemorrhages.

The various angiographic imaging techniques need to determine the nature of potential and actual vascular problems, and to accurately identify the affected vessels, in order to select the most effective treatment. Magnetic resonance angiography (MRA) was developed during the last decades, and has the advantage of being non-invasive. X-ray angiography, and particularly computed tomography angiography (CTA), is invasive and irradiating, but remains effective in terms of image accuracy.

6.3 Vessel Segmentation

The segmentation of vascular structures from 3D images is a particularly challenging task. Here, the notion of segmentation is considered in a broad sense. From an image processing point of view, segmentation consists of partitioning an image into an object, i.e., a structure of interest, and a background, i.e., the remainder of the image volume. In the context of angiographic imaging, we consider that vessel segmentation includes (*a*) methods that detect either whole vessels (i.e., their lumen and/or walls) or their medial axes and/or (*b*) methods that perform low-level processing or high-level knowledge extraction (e.g., vein/artery discrimination [100, 103] or vessel labelling [14, 42]). We also consider some methods which could be classified as filtering, since their purpose is to perform vessel enhancement, which consists mainly of denoising, but also of vessel reconnection (e.g., in the case of stenosis, or of signal loss [27, 76]).

As discussed above, the difficulty in performing vessel segmentation is due to the sparseness of data, and the possible presence of irrelevant signal (other tissues, artifacts or noise). Moreover, anatomical properties of vessels are highly variable in size, appearance, geometry and topology, even more so in pathological cases such as aneurysms, stenoses, calcifications or arteriovenous malformations.

There exist several kinds of angiographic data, generally well-fitted for visualizing specific vascular structures, and consequently for dealing with specific clinical issues. The choice of a segmentation method is often linked to the type of images under consideration, the vessel(s) being studied and the clinical purpose. The next section discusses the various methodological segmentation strategies.

6.3.1 Survey of Vessel Segmentation Methods

6.3.1.1 General Overview

Several surveys devoted to 3D vascular segmentation have been proposed during the last decade. The survey proposed in [94] focuses on vessel segmentation from MRA images,[1] and divides them into skeleton methods (with an interest in medial axes) and non-skeleton ones (that aim at detecting whole vascular volumes). Another (globally similar) classification is proposed in [51], which deals more generally with vessel segmentation from any kind of data independently of their dimension or acquisition technique. The most recent survey [56] mainly refers to 3D vessel segmentation from MRA and CTA, and divides its description into (*a*) the *a priori* information which can be used for segmentation, (*b*) the basic tools using this information for detecting vessels, and (*c*) the methodological frameworks involving these tools, as well as a discussion on pre- and post-processing considerations.

In the next section, we introduce the segmentation methods divided into eight main categories corresponding to the main image processing strategies on which

[1]Part I of this survey [93] also describes MRA acquisition techniques.

they rely: region-growing, differential analysis, model-based filtering, deformable models, path finding, vessel tracking, statistical approaches, and mathematical morphology.[2]

6.3.1.2 Region-Growing Methods

Region-growing has been one of the first strategies considered for image segmentation [117], and in particular medical/angiographic ones.Basically, region-growing relies on two elements: one (or several) seed(s) [1] assumed to belong to the structure of interest to be segmented, and a propagation criterion, enabling the segmentation of the object from the seed, by iterative addition of adjacent voxels.

In the case of vessel segmentation, seeds are generally defined interactively inside vessels. The seeds can also be detected automatically, especially in the case where they constitute the root of a vascular tree [69]. The possible definition of several seeds can straightforwardly lead to an application of region-growing to vessel separation, and in particular, to vein/artery discrimination. In such a case, a set of seeds is defined for arteries and veins, respectively. A competitive region-growing is then performed, based on *ad hoc* propagation criteria (e.g., a measure of gray-scale connectedness in [100]).[3]

The propagation criterion is commonly based on intensity properties, related to the high-intensity vascular signal. However, more sophisticated properties can also be embedded in this segmentation strategy. In particular, it has been proposed to consider *a priori* knowledge related to the shape and size of the vessels to be segmented [68], or to their topology [78]. The correctness of the orientation of the vessels during the segmentation process has also been considered by proposing "wave propagation" strategies [115], which aim to constrain the segmentation front to remain normal to the vessel axis. It may be noticed that this kind of approach has been further used for vessel tracking methods (discussed later in the section). The concept of wave propagation has also led to the development of methods related to both deformable models (level-sets) and path-finding approaches, namely, fast-marching methods [61].

Region-growing methods rely on a simple algorithmic framework, which makes their development and use quite easy and induces a low (generally linear) computational cost. In addition, they guarantee termination which is not systematically available for other non-monotonic strategies. However, the connectivity hypothesis intrinsically associated with this strategy constitutes a weakness, since the method may fail in segmenting vessels in case of vascular signal loss (due to partial volume

[2]Due to limited space we heave omitted those methods which have resulted led to fewer publications, such as neural network-based methods [52].

[3]Note that, by duality, region-growing also provides solutions to segmenting vessels by skeletonization. In such a case, the growing process starts from a seed being a subset of the background (which can then be automatically defined), and generally includes topological constraints in the propagation criterion [27, 76].

effect, or flow artifacts, for instance). *A contrario*, the use of a criterion being too permissive may lead to leakage phenomena, and a final over-segmentation of vessels [66]. In this context, region-growing methods have often been preferentially devoted to the segmentation of large and/or well-contrasted vessels (for which intensity and connectivity hypotheses are generally reliable).

6.3.1.3 Differential Analysis

Vessels are generally bright structures within a dark background. If an image is viewed as the discrete analog of a function from $\mathbb{R}^3$ to $\mathbb{R}$, vessels then appear as the maxima of this function. Consequently, it may be possible to detect them by analyzing the differential properties of the image.

In order to deal with the discrete/continuous issue involved by this strategy, the (discrete) image is convolved with a series of Gaussian derivatives of different standard deviations and in different directions, and the responses obtained are combined into a matrix.

In the case of first derivatives analysis, this matrix, which is the covariance matrix of gradient vectors [2, 8], is called the *structure tensor*. Except for vessel segmentation, the first derivatives have also been involved in *diffusion filtering*, which consists of the propagation of information in the orientations suggested by these derivatives [60].

In the case of second derivatives analysis, the resulting information is gathered in the *Hessian matrix*. The main idea behind eigen analysis of the Hessian matrix is to extract one or more principal directions of the local structure of the image. This gives the direction of the minimal curvature, the principal direction in the tubular structure and a high curvature in the vessel cross-section plane, which makes the filter more efficient than line filters.

Compared with the image gradient, the Hessian matrix can capture the shape characteristics of objects, such as tubes, planes, blob surfaces or noise. In particular, the eigenvalues of the Hessian matrix can be combined into a *vesselness* function in order to describe plate-, blob-like and tubular objects [34, 53, 84].

These methods can be performed in multi-scale frameworks in order to detect objects of different sizes. It has to be noticed that the choice and number of the considered scales is particularly important in such methods. If performed at a unique scale, they do not detect vessels of different sizes, especially those out of the range of the considered scale. Conversely, if performed at numerous scales, they can potentially detect all the vessels but they become computationally quite expensive.

In addition, the robustness of such methods to noise is strongly related to the considered scale. For large scales, the blurring effect of Gaussian filtering tends to remove noise effects and, unfortunately, smaller objects. *A contrario*, for small scales, the noise is hardly corrected by this filtering, and the method may bias the derivative evaluation accuracy, thus requiring the incorporation of assumptions related to noise in the method [113].

Despite some weaknesses, which require specific care, derivative-based methods provide efficient solutions for detecting vessels, especially in a multi-scale framework, and have therefore often been considered for the design of segmentation methods based on model filtering (see next section) or for the guidance of deformable models, for instance.

6.3.1.4 Model-Based Filtering

In general, vessel appearance can be used as a prior for segmentation. In this case, such a prior can describe vessel specific characteristics: photometric (usually being brighter than the background) and/or geometric (curvilinear). The most simple are *intensity* and *geometry-based models*, which are often combined in deformable model methodologies (see next section). We will describe such models in the order of increasing complexity.

Intensity Models

Intensity models, which are among the simplest ones, strongly depend on the imaging modality. They can integrate brightness, contrast and gradient priors, but also imaging properties, like intensity ranges or intensity variation based on location, or even noise distribution [2] (see also Sect. 6.3.1.6 for a discussion of noise modeling).

In [111], a cylindrical parametric intensity model is directly fit to the image intensities through an incremental process based on a Kalman filter for estimating the radii of the vessels. While in [79], local neighborhood intensities are considered in a spherical polar coordinate system in order to capture the common properties for the different types of vascular points. A natural integration into this kind of models is a background description [85, 102].

While simple, intensity models are highly dependent on the nature of the images. Therefore, they have to be tuned for all kinds of circumstances, such as artifacts or other image distortions, as well as to compensate for image variability.

Geometry Models

The assumption that vessels are elongated thin objects, globally similar to tubes has been used for the design of several geometric models, such as *generalized cylinders*, *superellipsoids*, *Gaussian lines*, or *bar-like* profiles [9, 53, 102].

Based on second-order derivatives (see previous section), several models incorporating geometrical properties have been developed. In [34], an ideal cylinder is proposed in order to enhance vessels within a measure called *vesselness*, while in [84] a more general model incorporates elliptical shapes.

The bifurcation issue has also been considered, for instance in [3] where a bifurcation models is proposed and optimized based on vessel centerline information.

Geometry models are powerful tools for describing vessels and aiding their further extraction within tracking schemes or by deformation. However, these methods assume image regularities that are present in high-quality images, but not necessarily in noisier ones, nor in pathological cases. Furthermore, they often require careful parameter tuning, which may change from one data set to the next. They can be used together with the intensity models, often combined in probabilistic and/or statistical approaches contributing to decision-making whether pixel belong to a vascular structure or not.

6.3.1.5 Deformable Models

Deformable models aim at fitting a geometric hypersurface (e.g., a 2D surface in a 3D image), by moving it and modifying its shape from an initial model, under the guidance of several (generally antagonist) forces: *external* ("data-driven") ones, related to the image content, and *internal* ("model-driven") ones, devoted to preserve correct geometric properties (e.g., regularity). Such models have been intensively used in the field of image analysis due to the following advantages: arbitrary shape representation, topological adaptivity, sub-pixel precision, etc.

Among the most classical methods, *snakes* (often used in 2D in order to segment vessel cross-sections), have been considered, e.g., in [64], or in [46], where two (1D and 2D) snakes are used for both segmentation and stenosis quantification.

Level-sets constitute another classical type of deformable model, and rely on an Eulerian version of contour evolution with partial derivative equations. The contour is integrated as the zero-level of a higher dimension function (level-set). In [59], an original level-set based scheme deformed an initial boundary estimate toward the vascular structures in the image using a codimension-two regularization force, based on the vessel centerlines instead of the vessel surface (see Fig. 6.1a). Another level-set based method [62] estimated the background and vessel intensity distributions based on the intensity histogram, to more efficiently steer the level-set onto the vessel boundaries.

Several efforts have been conducted to improve deformable models in the quite specific case of elongated structures. In this context, [104] used flux maximization as an alternative curvature-based regularization to make surface normals evolve according to the gradient vector field. The key idea was to evolve a curve or a surface under constraints by incorporating not only the magnitude but also the direction of an appropriate vector field.

In [54], local variances are measured with first-order derivatives and are propagated according to their strengths and directions, with an optimally oriented flux reporting more accurate and stable responses and higher robustness to disturbances from adjacent structures in comparison with Hessian-based measures.

The major advantage of deformable model methods is that they are sensitive to weak edges and robust to noisy structures. However, the intensity variation inside

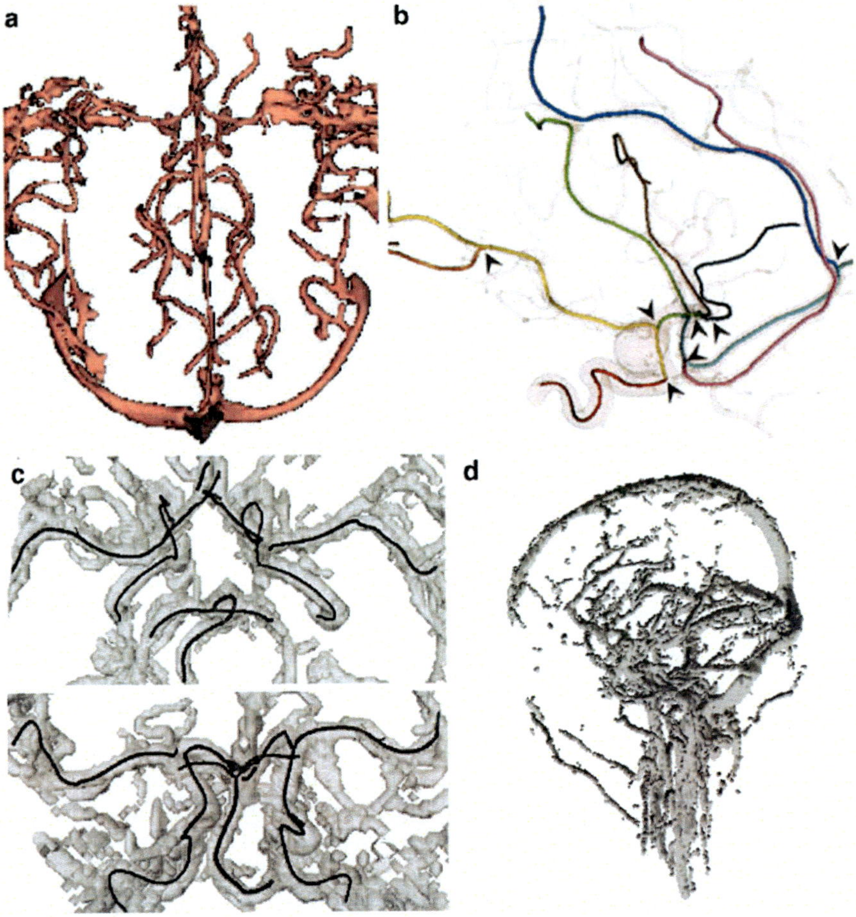

Fig. 6.1 Vessel segmentation examples. (**a**) Brain vessels segmentation based on deformable models. (**b**) Brain arteries segmentation based on path-finding and statistical approaches. (**c**) Brain arteries segmentation based on vessel tracking. (**d**) Brain vessels segmentation based on gray-level hit-or-miss transform. Illustrations from (**a**) [59], (**b**) [110], (**c**) [31], (**d**) [68]

vascular structures can generate significant intensity gradient with this undesired discontinuity stopping the contour evolution at these regions. Due to this local minima, the initial forces should be described with such precision that the final object borders are not far from the initial ones. Nonetheless, the evolution of the deformation can be a costly process. However, integrating vessel features and forces in powerful optimization schemes helps overcome these problems.

6.3.1.6 Statistical Approaches

Vessel segmentation based on statistical approaches generally relies on specific assumptions related to the intensity distribution of the vascular/non-vascular signals in MRA data (only very few statistical methods have been devoted to CTA, see, e.g., [32], which proposes a particle-filtering strategy for the segmentation of coronary arteries), and especially physical models of blood flow. If the number and the nature of these distributions is known correctly, it is possible to determine their respective parameters (and in particular the mean intensity characterizing the associated structures), via a standard Expectation-Maximization (EM) technique [23].

In MRA, two or three distributions are generally considered, for the blood, and the other anatomical structures and the background, respectively. They led, in particular to the definition of Gaussian-Gaussian-uniform [107] and normal-Rayleigh–2×normal [80] mixtures for time-of-flight (TOF) MRA, and Maxwell-Gaussian [19], Maxwell-Gaussian-uniform [17] mixtures for phase-contrast (PC) MRA. In [18], a hybrid model, enables one to choose between these two kinds of mixtures. Alternatively to these "constrained" mixture choices, [29] has proposed a linear combination of discrete Gaussians with alternate signs, involved in a modified EM, which adaptively deals with both laminar and turbulent (pathological) blood flow [28].

In the primarily considered strategies, the determination of the vascular intensity led to a straightforward segmentation by thresholding of the image (sometimes enriched by a hierarchical analysis of the image by octree decomposition [107]). From an algorithmic point of view, segmentation improvements were also performed by considering spatial information (i.e., statistical dependence) between neighbor voxels, by integrating Markov random fields (MRF) [38] in a post-classification correction step [80]. In other works, speed and phase information provided by PC-MRA were fused and involved in a maximum *a posteriori*-MRF framework to enhance vessel segmentation [17, 18].

Statistical methods globally inherit the strengths and weaknesses of the EM algorithm. First, they generally require one to establish hypotheses on the signal distribution. Moreover, they involve several parameters, for instance, weight, mean and standard deviation, of the distributions. The initialization of the segmentation process then requires special attention. Indeed, the convergence may depend on the quality of the initial distribution settings (sometimes automatically determined based on heuristic rules [17, 107]). As for any optimization strategy, the termination also requires one to decide whether the process has correctly converged or not (which is sometimes empirically determined, for instance by a maximal number of iterations [107]). Finally, since the segmentation process is strongly based on photometric properties (the results often consist of global or local thresholdings), higher-level knowledge such as geometric assumptions are hardly considered, and require post-processing steps based on a statistical framework [80], or, more efficiently the collaboration of alternative image processing techniques (see examples in Sect. 6.3.1).

6.3.1.7 Path Finding

Based on extremal intensity and connectedness criteria, the detection of a vessel segment (or more precisely its medial axis) can be expressed as the determination of a minimal cost path in a weighted graph modeling voxels, their neighborhood relations and their intensity.

Vessel segmentation based on such strategies can rely on standard minimal path finding techniques [25] (i.e., on "global" minimization strategies, while methods categorized in the next *Vessel tracking* section will rely on "local" (step-by-step) minimization strategies). This is, for instance, the case in [75].

Alternatively to classic path-finding methods, fast-marching strategies [101] have been considered. They are both related to the level-sets (see Sect. 6.3.1.5) and minimal path-finding methodologies (they remain, in particular, consistent with the continuous formulation of the minimal-path research). In contrast to fully discrete path-finding, they enable the determination of paths with a sub-voxel accuracy [4].

The methods based on path-finding are globally well-suited to the detection of vessel medial axes, especially in the case of small vessels which justifies their frequent use in coronary detection. (For larger vessels, the optimal path may diverge from the medial axis, leading to eccentric results [57].) However, efforts have also been conducted in developing segmentation methods that extract both vessel axes and vessel walls [7, 57], expressing the whole vascular volume segmentation as the minimization of a path in a space enriched with a supplementary "scale" dimension corresponding to the vessel radius.

Despite attempts to segment whole vascular trees [114], such methods generally remain devoted to the segmentation of vessel segments, thus requiring one to interactively provide at least an initial and final point [75, 108]. In this case, they may be robust to noise, and signal decrease (or short signal loss) along the vessel, especially in the case of stenoses. Since these methods are based on monotonic and/or finite algorithmic processes, their termination is guaranteed and their theoretical algorithmic cost is generally low. However, in practice, the computational cost may be high, and the provision of initial and final points can potentially enable its reduction by computing paths from both points simultaneously [75].

6.3.1.8 Tracking Methods

By opposition to path-finding methods, tracking methods consist of finding a vessel locally, by progressively determining successive segments composing it. Such an approach requires one to interactively propose a seed, namely the starting point of the tracking process, located in the vessel, and (in most cases) the direction in which the vessel has to be tracked.

This strategy has to be applied step by step, a small segment of the vessel being detected at each step. The principal issues to consider in such methods are the determination of a correct geometry of the detected segment (namely the determination of its cross section), the determination of the vessel axis, and the evaluation of the direction of the next segment to be found (i.e., the trajectory modification), or equivalently, the next point in the vessel. Less frequently, vessel tracking methods, such as the one proposed in [63], directly perform a more global iterative vascular volume detection, corrected, at each step, by the analysis of the induced vessel axis, which can in particular be constrained by *ad hoc* topological hypotheses.

The determination of the vessel cross-section at the current point (which enables in particular its correct repositioning on the vessel axis) can be performed according to several strategies. The use of a gradient-based measure is considered in [109] (a centerline measure based on the vessel profile then enables one to approximate the vessel centerline, even in case of non-circularity). The explicit determination of vessel cross-sections to estimate the vessel axis may however be avoided. In particular, it can be done by considering that the medial axis is necessarily located on a ridge point [5], which may be detected by second-derivatives criteria. Such an approach requires *a minima* the use of cross section information related to the size of the vessel (in order to determine the correct scale factor) and circularity hypotheses. It can also be performed based on a local optimisation of 3D models [102, 112], which may also lead to the determination of the vessel axis orientation. More classically, the next tracking point may be determined according to the best fit of a sphere modeling the vessel into the image [12, 47, 69].

Despite a few attempts to deal with the case of bifurcations, which can enable the recursive processing of a whole vascular tree [9, 13, 31], vessel tracking is especially well-suited to the segmentation of single vessels (see Fig. 6.1c). In this case, the termination has to be considered. Some methods require, in particular, the provision of both a start and an end point [109].

It should be noted that, similarly to the other local approaches (which aim at detecting a part of the vessels, and/or are guided by providing a seed), such methods present a generally low algorithmic/computational cost. However, they present some drawbacks related to the determination of multiple parameters and to possible error propagation (which characterize such local methods), potentially leading to incorrect segmentation if vessel orientation and/or axis is miscalculated at a given step, for instance due to a bifurcation, non circular section, or a strong axis curvature.

6.3.1.9 Mathematical Morphology Methods

Mathematical morphology (MM) is a well-established theory of non-linear, order-based image analysis. Fundamental texts on morphology include the books by Serra [86, 87], but more recent and more synthetic texts are also available, including the works by Soille [89] and by Najman and Talbot [72].

Filtering thin objects with morphology can be achieved using appropriate structuring elements. Typically, thin structuring elements include segments and paths, combined over families. To account for arbitrary orientation, one can use families of oriented segments and compute a supremum of openings or an infimum of closings as described in [90].

To account for noise or disconnection, families of incomplete segments can be used instead, yielding so-called *rank-max openings*, which are just as efficient and also described in the same reference.

Paths are elongated structuring elements, but they are not necessarily locally straight. Even though the size of families of paths grow exponentially with their length, there exists a recursive decomposition that makes the use of such families tractable [43]. As with segments, it is useful to account for some discontinuities using so-called *incomplete paths*. As with segments, there exists an efficient implementation [95]. *In fine*, path and segment operations are comparable in speed.

In [96], it is shown that path and segment morphological operators significantly outperform linear and steerable filters for the segmentation of thin (2D) structures, even in the presence of heavy noise. Paths operators have been extended to 3D in [44], and shown to outperform all other morphological filters for thin object segmentation in 3D, both for efficiency and performance.

Connected operators are also the supremum of openings or infimum of closings, but using families of structuring elements that are so large makes little sense. Instead, the concept of connectivity is used [83, 105]. The simplest of those is the area opening or closing. Informally, the area opening suppresses objects that are smaller in area than a given size λ. It extends readily to arbitrary lattices, and corresponds to a supremum of openings with a very large family of structuring elements: all the connected sets that have an area smaller than λ. In the continuum, this family is not countable, but in the discrete case it is still very large. Fortunately it is not implemented in this way. A very efficient way to implement this operator is via the *component tree* [65, 71, 82].

In [106], a scale-independent elongation criterion was introduced to find vascular structures, while in [11], component tree was mixed with classification strategies to segment 3D vessels in an automated fashion.

Other useful connected operators are thinnings rather than openings, as they make it possible to use more complex criteria for object selection, for instance, using elongation measures, that are not necessarily increasing.

Hit-or-miss transforms repeatedly use pairs of structuring elements (SEs) to select objects of interest, rather than single SEs. In [10, 68], authors used such operators for 3D vessel segmentation, including brain, liver and heart vessels (see Fig. 6.1d).

6.3.1.10 Hybrid Methods

Despite the huge amount of methodological contributions dedicated to 3D vessel segmentation, proposed during the last 20 years, the results provided by such segmentation methods generally remain less than perfect.

The handling of under-segmentation (especially in the case of small vessels, whose size is close to the image resolution, of signal decrease, or of the partial volume effect) and over-segmentation (especially in the case of neighboring anatomical structures, or of high intensity artifacts), the robustness to image degradations (low signal-to-noise ratio), the ergonomy (automation or easy interaction), the low computational cost, the guarantee of termination and convergence, and the accuracy of the result (for instance, the ability to provide results at a higher resolution than the image one) are desirable properties for such methods. Unfortunately, none is generally exempt from drawbacks, even in the frequent (and justified) case where the method is devoted to a quite specific task, vascular structure, and/or image modality.

As nearly all the main strategies of image processing have been –not fully satisfactorily– investigated to propose solutions to this issue, a reasonable trend during the last years has consisted of designing hybrid segmentation methods obtained by crossing methodologies. An alternative way to overcome this issue is to inject more guiding knowledge in the segmentation processes, which justifies – among other reasons– the generation of anatomical vascular models, as discussed in Sect. 6.4. These strategies aim, in particular, at taking advantage of (distinct and complementary) advantages of different segmentation techniques.

A synthetic overview of such hybrid methods is now presented.

Principal Strategies

Hybrid vessel segmentation methods present a range of possible solutions for overcoming certain weaknesses of each method and combining their advantages.

One of the most popular hybrid methods is a combination of multi-scale differential analysis within vessel detection schemes as in [34, 84] with deformable models, such as level-sets [15], B-spline snakes [33], and maximum geometric flow [24, 103].

The deformable method with energy minimizing functionals has also been combined with statistical region-based information in a multi-scale feature space for automatic cerebral vessel segmentation [45]. The tracking strategies were reinforced by gradient flux of circular cross-sections as in [55], while in [35] a multiple hypothesis tracking was used with Gaussian vessel profile and a statistical model fitting.

In [110], a probabilistic method for axis finding was used within a tracking with minimal path finding strategy together with a possible used guidance (see Fig. 6.1b). This method is especially well-fitted for pathological cases.

Multi-scale morphology has been used with Gabor wavelets (providing vessel size and direction) filters in [92]. The advantage of the Gabor wavelet is that it is capable of tuning to specific frequencies, allowing estimation of the vessel dimension, while the morphological top-hat filter enhances the contrast between vessel structures and background.

6.4 Vessel Modeling

6.4.1 Motivation

6.4.1.1 Context

The availability of accurate knowledge related to anatomical structures is of precious use in nearly all the fields related to medical image analysis. Knowing where an organ is located, its shape, dimensions, functions, cellular or chemical composition, and its spatial relations or collaborations with other organs constitutes the basics of anatomy and medicine.

In the case of vessels, and more generally of vascular trees,[4] anatomical knowledge can be classified into three categories:

- *Morphological* properties: what is the shape of a vessel (cross-section, trajectory), its size (diameter, cross-section area), its orientation, etc.?
- *Structural* properties: what is the topology of a vascular network (number of branches, bifurcations, presence of cycles/anastomoses), its position, its spatial relations with other organs, etc.?
- *Functional* properties: what are the vascular territories of an organ (i.e., what parts of an organ are supplied by a given branch of a vascular network)?

In the field of angiographic image analysis, the question of functional properties, and more specifically the partition of an organ into vascular territories has not been intensively considered. In the case of coronaries, the vascular territories are generally implicitly provided by the different branches of the coronary tree (the knowledge of such regions is of actual importance for determining the parts of the heart being affected by vessel stenoses, and possible subsequent heart attack). Computational modeling of these branches has been carefully studied for several years. In the case of cerebral vasculature, the different areas of the brain supplied by the main branches originated from the Willis polygon have been described long ago

---

[4]From a structural (and more especially from a topological) point of view, the terminology of "tree" is generally incorrect for most of vascular networks, despite its frequent use in the literature devoted to angiographic imaging. In the sequel of the chapter we will generally distinguish vascular *trees* from vascular *networks*. This distinction will be clarified in the next sections.

in the medical literature (see, e.g., [70, 97, 98] for recent contributions),[5] but these areas have not yet played a crucial role in angiographic/medical image analysis despite their potential helpfulness. Similar considerations can be made for the liver vascular networks, and in particular the portal network, the branches of which define the main hepatic anatomical segments [21].

6.4.1.2 Usefulness

The other two kinds of anatomical knowledge, namely, morphological and structural, have been the subject of a several publications related to medical image analysis dating back to the end of the 1980s. In particular, the coronary tree and the (arterial and venous) networks of the brain have been considered.

The first studies, related to the heart, have essentially been devoted to gather and model structural information related to the coronary arteries in order to assist the radiologists in their analysis of vessels from CT data, especially for the diagnosis and follow-up of stenoses and their consequences on heart blood supply. The globally simple structure of the coronary tree and its (relative) invariance has led to the design of the first vascular models. Such vascular models will be referred to as *atlases*[6] in the sequel of the chapter. A survey of this first family of (deterministic) atlases is proposed in Sect. 6.4.2.

More recent studies, essentially devoted to the cerebrovasculature, have intended to gather and model morphological information related to potentially complex vascular networks. It should be noted that, in contrast to vascular structures such as coronary arteries or hepatic vessels, vascular networks such as the cerebral ones are not actually tree structures. A *vascular tree* originates from a single vessel, which progressively divides itself (by bifurcations) into branches, leading to an arborescent hierarchy (which, in particular, does not present any cycle). Vascular networks, such as the cerebral ones, present a more complex organisation. They can originate from several vessels, which divide themselves to refine into smaller branches, but which can also join together to give birth to new vessels, or present cyclic structures, as anastomoses. To the complexity induced by the structure of such vascular networks, one must add the complexity induced by the nature of the vessels visualized in the considered images (both veins and arteries, large and small vessels), from the image modalities (generally non-injected data, especially in MRA, in the case of brain vessels), but also from the anatomical variability of the vessels (from both morphological and structural points of view). In this context, the design of no longer deterministic, but *statistical* atlases had to be considered. A survey of this second

[5]The notion of "vascular areas", which actually does not match the anatomical notion of vascular territories, was introduced in [78] in order to propose a partition of the cerebral volume to facilitate the segmentation of vessels from PC-MRA data.

[6]The notion of an atlas has been the subject of a quite intensive research activity during the last 15 years, especially in the field of brain imaging. The emergence of *computational anatomy* [39, 50, 99] is, in particular, a direct expression of such research activities.

family of atlases is proposed in Sect. 6.4.3. One of the main uses of such statistical atlases is the guidance of automated vessel segmentation procedures [68], which is a crucial step for several medical image analysis applications.

6.4.2 Deterministic Atlases

The first works on vascular atlases have consisted in developing *deterministic* models of the vessels. By deterministic, we mean that a model is a (representative) example of what can be considered a vascular network. Although being a good (and actually useful) representation of the anatomical truth, such a deterministic atlas is however not necessarily able to take in consideration in an accurate way the interindividual variability. Broadly speaking, these atlases can be seen as a direct transcription of the models described (both textually and visually) in the anatomy literature. The pioneering works related to this topic were actually based on this approach.

6.4.2.1 Pioneering Works

To the best of our knowledge, the first vascular atlas generated from angiographic data was developed for the modeling of coronary arteries [26]. This "hand-made" atlas consists of a (piecewise linear) skeleton modeling the main coronary artery segments and branches, and providing information on topology (e.g., position of the bifurcations), position and trajectory of vessels. Starting from 2D arteriographies of 37 patients, vessels were manually segmented from two orthogonal views, from the origin of the coronaries to the most distal visible point on each considered branch. A total set of approximately 100 points was then regularly sampled on each segmented tree, leading to a 3D mean positioning of each point. An interactive choice of the structures to be visualized, and the visualization angle allowed the generation of 2D projections of the atlas. In the same period, a second approach was proposed in [36], relying on a model composed of two orthogonal planes embedding a structural and spatial representation of each one of the left and right coronary trees. Based on this pseudo-3D reference, a symbolic description of the arteries was proposed, providing in particular information on branch names and hierarchy, position, (qualitative) orientation, or vascular territories. This description was made by use of declarative programming with each predicate formalizing a given information related to a vessel, while some more general rules modeled heuristic information, such as continuity or angular limits at bifurcations.

In contrast to methods such as [36], which rely on bases of semantic knowledge, those which took advantage of the emerging technologies offered by computer graphics at the end of the 1980s (such as [26]), gave rise to related strategies, essentially based on graph modeling and geometric information.

6.4.2.2 Graph-Based and Geometric Atlases

Among the methods aiming at generating deterministic atlases, one can distinguish those based on graphs, and those based on geometry. The first ones essentially focus on a symbolic description of the vascular structures (independently from their embedding in the 3D space, i.e., from their anatomical reality), while the second ones especially aim at defining such models as objects which "match at best" a spatial reality.

One of the main uses of such atlases, is the labeling of coronary branches, i.e., the automatic naming of vessels, in order to assist radiological analysis. The extraction of reliable vascular information from cardiovascular data (generally 2D or 3D CT angiography) is of precious use for coronary disease assessment. In this context, it is not only required to segment these vessels (which is a non-trivial task, subject to strong research efforts by the medical image analysis community [67]), but also to be able to name each branch of the coronary tree, in order to facilitate the radiological analysis. Such a highly semantic task can not, of course, be carried out without using high-level *a priori* anatomical knowledge. Based on these considerations, several vascular atlases have been involved in –and sometimes specifically designed for– this labeling task [14, 30, 36, 41, 42].

Graph-Based Atlases

The extraction of a graph modeling the structure of a vascular network (i.e., assigning an edge to each vessel branch, and a node to each junction/bifurcation) has been a purpose frequently considered by the first vessel segmentation methods devoted to 3D angiographic data [37, 115]. Note that the main weakness of these first approaches was propagation of segmentation errors in the obtained model.

A solution proposed in [30] relies on the data collected, and validated, in [26]. It proposed to define both a symbolic graph-based atlas, which models the tree structure of the coronary arteries, and to couple it with a geometric 3D atlas which models spatial and geometric relationships. Unclassically, the nodes of the graph represent vessel segments while the edges model their bifurcations. Each node of the graph is then associated with a vessel name, a width, but also a list of points located on the vessel medial axis. This information then intrinsically provides a geometric model of the vessels.

In order to automatically build a graph-model of a vascular tree without depending on possible errors inherited from the segmentation process of real images, an alternative consists of generating such a graph from a realistic anatomical phantom. This is the approach proposed in [14] for generating a graph-based atlas of the coronary arteries. The use of a phantom enables to easily obtain a segmentation (which can be validated *a posteriori*) and to derive, by a topological post-processing, curvilinear structures enabling to define a graph structure. In [14], such an atlas can be achieved by storing at each edge/vessel segment information attributes such as its name, length, orientation and diameter.

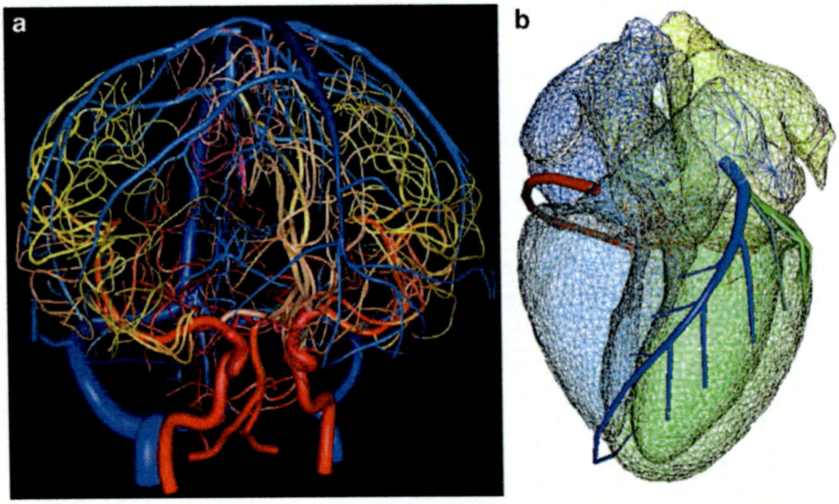

Fig. 6.2 Geometry-based atlases. (**a**) Atlas of the cerebral vascular (arterial and venous) networks. (**b**) Atlas of the whole heart and of the coronary arteries. Illustrations from (**a**) [74] and (**b**) [58]

Geometry-Based Atlases

The works described above have focused on vascular structures presenting simple properties, namely the coronary arteries, out of their anatomical neighboring context. In recent works, efforts have been conducted to design vascular atlases related to more complex structures. These contributions rely, in particular, on the use of geometric models, and specifically surfacic meshes.

In [74], a geometric atlas of the whole cerebral vascular network was proposed (see Fig. 6.2a). This network was quite complex, being composed of veins and arteries of varying sizes (at the resolutions available in 3D CT and MR angiographic data, namely 0.5 mm), organized in a non-arborescent fashion. The generation process, based on the TOF MRA of a healthy patient, was composed of several iterative steps, the most crucial of which was segmentation (performed manually, for the sake of correctness), medial axes determination and topology correction (also performed interactively), vessel surface generation, quantitative knowledge extraction and vessel labeling. It led to a quite accurate vascular atlas providing information on the type of vessels (arteries or veins), their position in the intracranial volume, their name, size and topology. Such an atlas, essentially designed with high-level image processing tools, but in a basically manual fashion, however, remains strongly related to the only patient involved in the image acquisition process.

In [58], a geometric atlas of the whole heart, made of surfacic meshes corresponding to different anatomical structures, was proposed (see Fig. 6.2b). In addition to the coronary arteries, it also modeled several anatomical structures such as the heart chambers and the trunks of the connected vasculature (the model generation of which is beyond the scope of this chapter). The information used for generating this vascular atlas consisted of measurements from [26], which helped to create a first vascular model. In order to correctly fit this model on its neighboring cardiac structures, a registration step was carried out.[7] The registration process was driven by the medial axes of the main artery segments, interactively delineated from 27 3D CT data (which had previously been involved in the mesh generation of the other anatomical structures). It was based on an incremental relaxation of the authorized degrees of freedom, first accepting rigid (translation, rotation) transformation, then scaling, and finally, affine transformation. In contrast to the vascular atlas proposed in [74] for the cerebrovasculature, the one presented here, despite the relative simplicity of the modeled vascular tree, was sufficiently specific to model spatial relationships with neighboring –non vascular– structures. This is the first (and to our knowledge, the only) vascular atlas offering such a property. In contrast to the previous atlas, this one was created (at least partially) thanks to the vascular information provided by 3D CT data of several patients. As stated in the synthetic description of the generation protocol, this required one to be able to process heterogeneous anatomical knowledge, possibly presenting variability. In particular, this implies the consideration of tools enabling one to fuse the information related to several patients in a unified result. In the present case, this was done by considering registration. In Sect. 6.4.3, it will be shown that based on similar registration-based strategies, it is possible to obtain results which are no longer deterministic, but statistical, enabling in particular the modeling of the interindividual variability.

6.4.3 *Statistical Atlases*

In the above section, we have considered the vascular atlases which can be qualified as *deterministic*, in the sense that they present a model of vasculature which could be seen as the vascular network of a representative patient among the population. In this section, we now focus on non deterministic vascular atlases, and more especially on *statistical* ones, which are intuitively less similar to a hard anatomical model, but which aim at gathering and modeling more completely and efficiently the characteristics of a whole population of patients.

6.4.3.1 Anatomical Variability Handling

When designing an anatomical model (in the present case, a vascular atlas), two questions have to be considered carefully:

1. How to model the invariant information, i.e., the set of characteristics shared by the whole population?

---

[7]The reader interested in registration – which is an issue strongly linked to (vascular) atlas generation – may complete the study of this chapter by reading the following surveys [48, 116].

2. How to model the interindividual variability, i.e., the set of varying characteristics among this population, in a unified framework?

The deterministic atlases described in Sect. 6.4.2 actually provide efficient answers to the first question. However, since they are based on a hard (graph or geometric) model, their ability to handle interindividual variability is not obvious.[8]

Most of the contributions devoted to deterministic vascular atlas generation propose various (partial) solutions to this issue. In the preliminary works on coronary modeling [26], it is mentioned that there exist three variants in the coronary trees structure: right dominant (10%), balanced anatomic distribution (80%), and left dominant (10%)[9]. In [36], such variations are considered by exhaustively modeling each induced branch distribution (in this case by integrating them into the symbolic base of knowledge).

In the case of cerebral vessels, there also exists a strong interindividual variability from both topological and geometrical points of view [81]. In order to cope with this issue in the case of the vascular atlas proposed in [74] (and obtained from a single patient), a straightforward solution is an exhaustive list of each topological variation described in the anatomy literature [73]. In [41], a more unified solution is proposed for the modeling and storing of such interindividual topological variations. This is done by initially considering a classical graph-based modeling (see Sect. 6.4.2.2) enriched by fusing of several anatomical models/graphs into a "vascular catalog", composed of a graph of all variations, and a discrimination matrix. This matrix helps to extract these variations as graphs similar to those of [14, 30]. (See also [40] for a more theoretical/methodological contribution related to the same concepts.)

The handling of variability proposed in these contributions is essentially based on characteristics related to the structure of the vascular networks. This is a straightforward consequence of the modeling strategies which are primarily based on topological data structures. In particular, the quantitative variations are generally omitted from these atlases, and the answer of these methods to the second question actually remains partial.

Some recent contributions try to propose complementary answers to this second question. They are specifically devoted to coping with the issue of modeling the variability of anatomical characteristics which can be quantified, for instance the size, orientation, position, or even the shape of the vessels. To this end, they propose to generate non-deterministic (namely *statistical*) atlases.

---

[8]However, this fact does not represent a crippling drawback, since the relevance of interindividual variability handling is essentially modulated by the applications requiring the designed atlas. In particular, deterministic atlases must not be considered as less (or more) relevant than statistical ones.

[9]Note that the information on coronary arteries gathered in [26] corresponds to a sample of patients with "normal-sized hearts". This example illustrates the general necessity to constraint some anatomical hypotheses if we hope to finally obtain a useful model from a finite (and generally restricted) set of patients. Such a consideration remains valid when considering statistical models: a classical example is the restriction to either healthy or non-healthy people in the considered pool of patients.

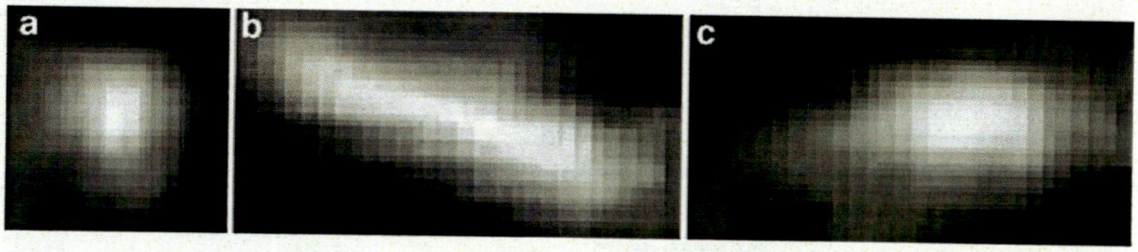

Fig. 6.3 Atlas for the portal vein entry, in the liver. (**a**) Sagittal, (**b**) coronal, and (**c**) axial slices. Illustration from [68]

6.4.3.2 Recent Works

The methods described hereafter are mainly devoted to design vascular atlases of vessels or vascular trees/networks from a set of patients/images presenting possible anatomical variations. In all these contributions, the input data consist of vascular volumes extracted from angiographic images. Similarly to most of the methods for deterministic atlas generation, those for statistical atlas generation then strongly rely on vessel segmentation. The following methods have been classified according to the degree of complexity of the modeled anatomical information.

Shape model

When the vascular structures of interest are sufficiently simple, for instance when only a vessel, or a vessel segment has to be modeled, a first (and straightforward) approach for generating vascular atlases can consist in creating shape models. Such models can be defined by computing the mean image of data obtained from the segmentation of a (learning) image database. This mean image of binary functions can be seen as a fuzzy function with values in the interval $[0,1]$.

In [69], such a model (which can in particular be involved in subsequent segmentation procedures [68]) was proposed for a vessel segment, namely the entrance of the portal vein, in the liver. The atlas, built from a database of 15 segmented images of the portal vein entry, is illustrated in Fig. 6.3.

Density Atlas

When the vascular structures become more complex, in particular in the case where a whole vascular tree/network is considered, a straightforward mean image gathering each patient vascular information is no longer sufficient to accurately generate a satisfactory vascular atlas. In this more difficult context, it becomes necessary to develop adequate strategies for fusing several vascular images onto a coherent anatomical reference. Such strategies thus require the use of a registration procedure.

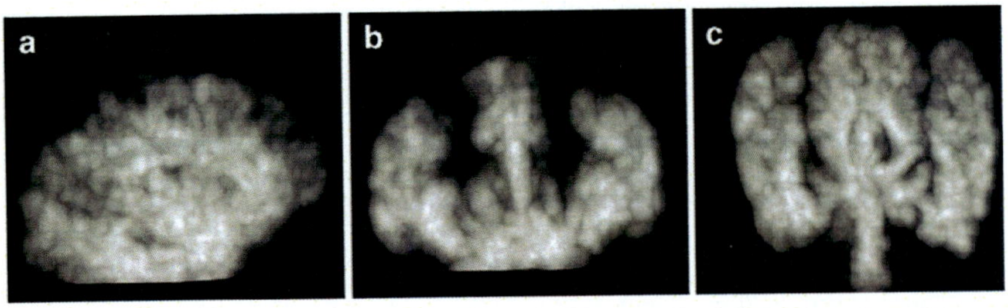

Fig. 6.4 Atlas of the cerebral arteries. (**a**) Sagittal, (**b**) coronal, and (**c**) axial maximum intensity projections. Illustration from [20] (with kind permission from Springer Science+Business Media: MICCAI 2003, Tissue-based affine registration of brain images to form a vascular density atlas, volume 2879 of LNCS, 2003, p. 12, D. Cool et al., Fig. 1)

Intuitively, a first and natural way to proceed consists in attempting to register all the (segmented) vascular networks onto a chosen one, considered as the reference. Such an approach has been developed in [16], where the reference network is first skeletonized and then processed to provide a distance map (providing the distance to the closest vessel). The other segmented vascular networks are then registered (by affine transformation) on this template. The mean and variance images obtained from the distance maps of all the registered images finally provide a kind of probabilistic vascular atlas.

If such an approach enables the discrimination between healthy and non-healthy patients (especially in the case of arterio-venous malformations), it is actually not sufficient to accurately model the vessels. This is due, in particular, to the lack of a morphological reference. In order to correct this drawback, it is possible to perform registration no longer to angiographic data, but to associated morphological images. This alternative approach was proposed in [20], where each angiographic data (in this case, cerebral MRA) was associated with a T2-weighted MRI of the same patient. An affine registration procedure was then applied between these T2 data, leading to morphology-based deformation fields which were then applied on distance maps similar to the ones previously described, leading to an atlas consisting of a vascular density map. A result for cerebral arteries, obtained from 9 patients is depicted in Fig. 6.4. Nonetheless, for creating vascular atlases as density fields, a recent strategy was proposed in [88] for cardiovascular CT data. In contrast to the case of cerebral MRA data, which requires the simultaneous use of MRI data, the considered CTA images contained both morphological (cardiac) structures and angiographic ones. It was then possible to directly perform registration on such data. After (non-rigid) registration of the main vessel centerlines of each image with the chosen reference image and estimation of the closest centerline for each point of the image, a mean-shift clustering was performed in order to assign each point to one of the three main vessel clusters. An artery-specific density at each point was then computed from a covariance analysis. The result obtained from 85 CTA is depicted in Fig. 6.5.

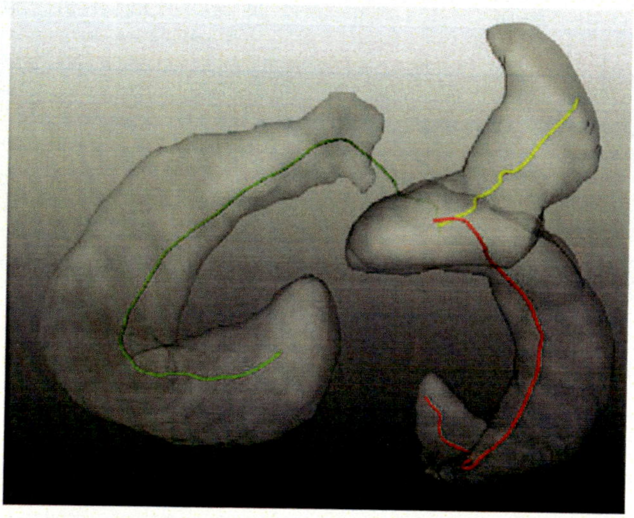

Fig. 6.5 3D visualisation of the vascular atlas (here, thresholded density field) for the three main coronary arteries (centerlines of which, for a given CTA, are depicted in green, red and yellow). Illustration from [88], (©2010 IEEE)

Enriched atlas

The statistical atlas generation protocols presented above are essentially devoted to a density field generation. This density field models information related to a "vascular presence probability" and possibly a shape, when the interindividual variability is sufficiently low.

It may, however, be useful to be able to model more accurate information, related for instance to size and orientation. An approach described in [77] proposed such a method. It requires as input more information than a simple segmentation, namely a segmented volume (as in [69]), the associated medial axes (as in [16, 20, 88]), but also information on the vessel orientation (it should be noted that all these information elements may be obtained from a segmented volume, by using adequate methods[10]). It also requires correct deformation fields in order to register these different data with an anatomical reference. In order to do so, each angiographic image (in the current case, cerebral PC-MRA phase image) was associated with a morphological image (namely the associate PC-MRA magnitude image), following a strategy similar to the one proposed in [20]. Non-rigid registration of these morphological images then provided the deformation fields in order to match the segmentation, medial axes and orientation maps associated with each image onto

[10]See, e.g., [22] for a robust medial axis computation method.

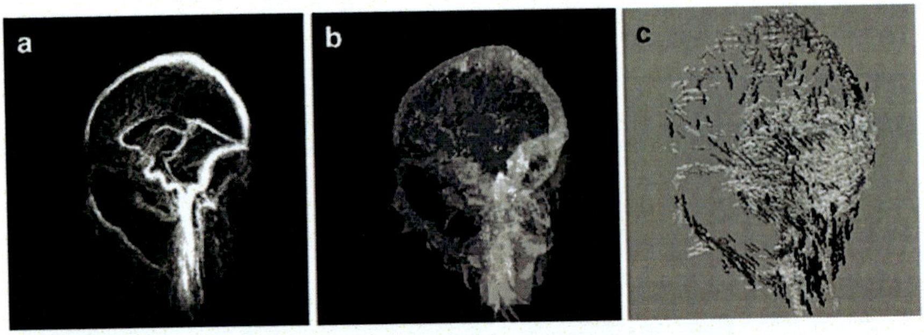

Fig. 6.6 Atlas of the cerebral vascular network. (**a**) Vascular density, visualized as a maximum intensity projection (sagittal view). (**b**) Mean vessel diameters, visualized as a maximum intensity projection (sagittal view). (**c**) 3D visualisation of a part of the orientation image. Illustration from [77]

the anatomical reference. The mean and variance values for each one of these scalar and vectorial attributes finally led to a vascular density field (as in [16, 20, 88]), but also to size and orientation intervals at each vascular point of the atlas. Such an atlas, modeling both cerebral veins and arteries, built from 16 patients is partially illustrated in Fig. 6.6.

Remaining Challenges

The vascular atlas generation protocols discussed in the previous section provide, in contrast to most of the ones devoted to deterministic atlases, the way to fuse information from a potentially large set of data in a globally automated fashion. Such automation requirements however induce several conditions related to vessel segmentation which can be performed automatically as already discussed in Sect. 6.3.1, but which (at least in the case of vessels) still does not offer perfect results. However, it should be noted that in the context of atlas generation, a sufficient condition for a correct use of automatically segmented data would be the guarantee that they do not present any false positives (the presence of false negatives being possibly compensated by the possibly high number of segmented data).

Another crucial issue related to non-deterministic atlas generation is the availability of efficient registration methods. The most recent methods [77, 88] are based on non-rigid registration techniques, the accuracy of which (in contrast to rigid or even affine registration) is probably a *sine qua non* condition to obtain satisfactory results. Since such non-rigid registration algorithms have probably reached a sufficient degree of efficiency for the processing of dense images (such as morphological cerebral data, for instance), the development of efficient registration procedures in the case of sparse – and more especially of angiographic – data seems to remain, despite a few recent works [6, 16, 49, 91], a globally open question.

References

1. Adams, R., Bischof, L.: Seeded region growing. IEEE Trans. Pattern Anal. Mach. Intell. **16**(6), 641–647 (1994)
2. Agam, G., Armato, S. III, Wu, C.: Vessel tree reconstruction in thoracic CT scans with application to nodule detection. IEEE Trans. Med. Imaging **24**(4), 486–499 (2005)
3. Antiga, L., Steinman, D.A.: Robust and objective decomposition and mapping of bifurcating vessels. IEEE Trans. Med. Imaging **23**(6), 704–713 (2004)
4. Avants, B.B., Williams, J.P.: An adaptive minimal path generation technique for vessel tracking in CTA/CE-MRA volume images. In: Medical Image Computing and Computer-Assisted Intervention – MICCAI 2000. Lecture Notes in Computer Science, vol. 1935, pp. 707–716. Springer, Berlin (2000)
5. Aylward, S.R., Bullitt, E.: Initialization, noise, singularities, and scale in height ridge traversal for tubular object centerline extraction. IEEE Trans. Med. Imaging **21**(2), 61–75 (2002)
6. Aylward, S.R., Jomier, J., Weeks, S., Bullitt, E.: Registration and analysis of vascular images. Int. J. Comput. Vis. **55**(2–3), 123–138 (2003)
7. Benmansour, F., Cohen, L.D.: A new interactive method for coronary arteries segmentation based on tubular anisotropy. In: International Symposium on Biomedical Imaging – ISBI 2009, pp. 41–44. IEEE (2009)
8. Bigun, J., Bigun, T., Nilsson, K.: Recognition by symmetry derivatives and the generalized structure tensor. IEEE Trans. Pattern Anal. Mach. Intell. **26**(12), 1590–1605 (2004)
9. Boldak, C., Rolland, Y., Toumoulin, C.: An improved model-based vessel tracking algorithm with application to computed tomography angiography. Biocybern. Biomed. Eng. **23**(1), 41–64 (2003)
10. Bouraoui, B., Ronse, C., Baruthio, J., Passat, N., Germain, P.: 3D segmentation of coronary arteries based on advanced Mathematical Morphology techniques. Comput. Med. Imaging Graph. **34**(5), 377–387 (2010)
11. Caldairou, C., Passat, N., Naegel, B.: Attribute-filtering and knowledge extraction for vessel segmentation. In: International Symposium on Visual Computing – ISVC 2010, Lecture Notes in Computer Science, vol. 6453, pp. 13–22. Springer, Berlin (2010)
12. Carrillo, J.F., Hernández Hoyos, M., Davila-Serrano, E.E., Orkisz, M.: Recursive tracking of vascular tree axes in 3D medical images. Int. J. Comput. Assist. Radiol. Surg. **1**(6), 331–339 (2007)
13. Carrillo, J.F., Orkisz, M., Hernández Hoyos, M.: Extraction of 3D vascular tree skeletons based on the analysis of connected components evolution. In: Computer Analysis of Images and Patterns – CAIP 2005, Lecture Notes in Computer Science, vol. 3691, pp. 604–611. Springer, Berlin (2005)
14. Chalopin, C., Finet, G., Magnin, I.E.: Modeling the 3D coronary tree for labeling purposes. Med. Image Anal. **5**(4), 301–315 (2001)
15. Chen, J., Amini, A.A.: Quantifying 3-D vascular structures in MRA images using hybrid PDE and geometric deformable models. IEEE Trans. Med. Imaging **23**(10), 1251–1262 (2004)
16. Chillet, D., Jomier, J., Cool, D., Aylward, S.: Vascular atlas formation using a vessel-to-image affine registration method. In: Medical Image Computing and Computer-Assisted Intervention – MICCAI 2003. Lecture Notes in Computer Science, vol. 2878, pp. 335–342. Springer, Berlin (2003)
17. Chung, A.C.S., Noble, J.A., Summers, P.: Fusing speed and phase information for vascular segmentation of phase contrast MR angiograms. Med. Image Anal. **6**(2), 109–128 (2002)
18. Chung, A.C.S., Noble, J.A., Summers, P.E.: Vascular segmentation of phase contrast magnetic resonance angiograms based on statistical mixture modeling and local phase coherence. IEEE Trans. Med. Imaging **23**(12), 1490–1507 (2004)
19. Chung, A.C.S., Noble, J.A., Summers, P.E., Brady, M.: 3D vascular segmentation using MRA statistics and velocity field information in PC-MRA. In: Information Processing in Medical Imaging – IPMI 2001. Lecture Notes in Computer Science, vol. 2082, pp. 461–467. Springer, Berlin (2001)

20. Cool, D., Chillet, D., Guyon, J.P., Foskey, M., Aylward, S.: Tissue-based affine registration of brain images to form a vascular density atlas. In: Medical Image Computing and Computer-Assisted Intervention – MICCAI 2003. Lecture Notes in Computer Science, vol. 2879, pp. 9–15. Springer, Berlin (2003)
21. Couinaud, C.: Le foie, études anatomiques et chirurgicales. Masson, Paris (1957)
22. Couprie, M., Coeurjolly, D., Zrour, R.: Discrete bisector function and Euclidean skeleton in 2D and 3D. Image Vis. Comput. **25**(10), 1519–1698 (2007)
23. Dempster, A.P., Laird, N.M., Rubin, D.B.: Maximum likelihood from incomplete data via the EM algorithm. J. R. Stat. Soc. Ser. B (Methodological) **39**(1), 1–38 (1977)
24. Descoteaux, M., Collins, D.L., Siddiqi, K.: A geometric flow for segmenting vasculature in proton-density weighted MRI. Med. Image Anal. **12**(4), 497–513 (2008)
25. Dijkstra, E.W.: A note on two problems in connection with graphs. Numerische Mathematik **1**, 269–271 (1959)
26. Dodge, J.T. Jr., Brown, B.G., Bolson, E.L., Dodge, H.T.: Intrathoracic spatial location of specified coronary segments on the normal human heart. Circulation **78**(5), 1167–1180 (1988)
27. Dokládal, P., Lohou, C., Perroton, L., Bertrand, G.: Liver blood vessels extraction by a 3-D topological approach. In: Medical Image Computing and Computer-Assisted Intervention – MICCAI 1999. Lecture Notes in Computer Science, vol. 1679, pp. 98–105. Springer, Berlin (1999)
28. El-Baz, A., Farag, A.A., Gimel'farb, G.L., El-Ghar, M.A., Eldiasty, T.: A new adaptive probabilistic model of blood vessels for segmenting MRA images. In: Medical Image Computing and Computer-Assisted Intervention – MICCAI 2006. Lecture Notes in Computer Science, vol. 4191, pp. 799–806. Springer, Berlin (2006)
29. El-Baz, A., Farag, A.A., Gimel'farb, G.L., Hushek, S.G.: Automatic cerebrovascular segmentation by accurate probabilistic modeling of TOF-MRA images. In: Medical Image Computing and Computer-Assisted Intervention – MICCAI 2005. Lecture Notes in Computer Science, vol. 3749, pp. 34–42. Springer, Berlin (2005)
30. Ezquerra, N., Capell, S., Klein, L., Duijves, P.: Model-guided labeling of coronary structure. IEEE Trans. Med. Imaging **17**(3), 429–441 (1998)
31. Flasque, N., Desvignes, M., Constans, J.M., Revenu, M.: Acquisition, segmentation and tracking of the cerebral vascular tree on 3D magnetic resonance angiography images. Med. Image Anal. **5**(3), 173–183 (2001)
32. Florin, C., Paragios, N., Williams, J.: Particle filters, a quasi-Monte-Carlo-solution for segmentation of coronaries. In: Medical Image Computing and Computer-Assisted Intervention – MICCAI 2005. Lecture Notes in Computer Science, vol. 3749, pp. 246–253. Springer, Berlin (2005)
33. Frangi, A.F., Niessen, W.J., Hoogeveen, R.M., van Walsum, T., Viergever, M.A.: Model-based quantitation of 3-D magnetic resonance angiographic images. IEEE Trans. Med. Imaging **18**(10), 946–956 (1999)
34. Frangi, A.F., Niessen, W.J., Vincken, K.L., Viergever, M.A.: Multiscale vessel enhancement filtering. In: Medical Image Computing and Computer-Assisted Intervention – MICCAI 1998. Lecture Notes in Computer Science, vol. 1496, pp. 130–137. Springer, Berlin (1998)
35. Friman, O., Hindennach, M., Kühnel, C., Peitgen, H.O.: Multiple hypothesis template tracking of small 3D vessel structures. Med. Image Anal. **14**(2), 160–171 (2009)
36. Garreau, M., Coatrieux, J.L., Collorec, R., Chardenon, C.: A knowledge-based approach for 3-D reconstruction and labeling of vascular networks from biplane angiographic projections. IEEE Trans. Med. Imaging **10**(2), 122–131 (1991)
37. Gerig, G., Koller, T., Székely, G., Brechbühler, C., Kübler, O.: Symbolic description of 3-D structures applied to cerebral vessel tree obtained from MR angiography volume data. In: Information Processing in Medical Imaging – IPMI 1993. Lecture Notes in Computer Science, vol. 687, pp. 94–111. Springer, Berlin (1993)
38. Graffigne, C., Heitz, F., Pérez, P., Prêteux, F., Sigelle, M., Zerubia, J.: Hierarchical Markov random field models applied to image analysis: A review. In: Neural Morphological and

Stochastic Methods in Image and Signal Processing, 1995, SPIE Proceedings, vol. 2568, pp. 2–17. SPIE (1995)
39. Guimond, A., Meunier, J., Thirion, J.P.: Average brain models: A convergence study. Comput. Vis. Image Underst. **77**(2), 192–210 (2000)
40. Hall, P.: On the addition and comparison of graphs labeled with stochastic variables: Learnable anatomical catalogs. J. Comb. Optim. **5**(1), 43–58 (2004)
41. Hall, P., Ngan, M., Andreae, P.: Reconstruction of vascular networks using three-dimensional models. IEEE Trans. Med. Imaging **16**(6), 919–930 (1997)
42. Haris, K., Efstratiadis, S.N., Maglaveras, M., Papas, C., Gourassas, J., Louridas, G.: Model-based morphological segmentation and labeling of coronary angiograms. IEEE Trans. Med. Imaging **18**(10), 1003–1015 (1999)
43. Heijmans, H., Buckley, M., Talbot, H.: Path openings and closings. J. Math. Imaging Vis. **22**, 107–119 (2005)
44. Hendriks, C.L.L.: Constrained and dimensionality-independent path openings. IEEE Trans. Image Process. **19**(6), 1587–1595 (2010)
45. Hernandez, M., Frangi, A.F.: Non-parametric geodesic active regions: Method and evaluation for cerebral aneurysms segmentation in 3DRA and CTA. Med. Image Anal. **11**(3), 224–241 (2007)
46. Hernandez, M., Frangi, A.F., Sapiro, G.: Three-dimensional segmentation of brain aneurysms in CTA using non-parametric region-based information and implicit deformable models: Method and evaluation. In: Medical Image Computing and Computer-Assisted Intervention – MICCAI 2003. Lecture Notes in Computer Science, vol. 2879, pp. 594–602. Springer, Berlin (2003)
47. Hernandez Hoyos, M., Orłowski, P., Piatkowska-Janko, E., Bogorodzki, P., Orkisz, M.: Vascular centerline extraction in 3D MR angiograms for phase contrast MRI blood flow measurement. Int. J. Comput. Assist. Radiol. Surg. **1**(1), 51–61 (2006)
48. Holden, M.: A review of geometric transformations for nonrigid body registration. IEEE Trans. Med. Imaging **27**(1), 111–128 (2008)
49. Jomier, J., Aylward, S.R.: Rigid and deformable vasculature-to-image registration: A hierarchical approach. In: Medical Image Computing and Computer-Assisted Intervention – MICCAI 2004. Lecture Notes in Computer Science, vol. 3216, pp. 829–836. Springer, Berlin (2004)
50. Joshi, S., Davis, B., Jomier, M., Gerig, G.: Unbiased diffeomorphic atlas construction for computational anatomy. NeuroImage **23**(Supplement 1), S151–S160 (2004)
51. Kirbas, C., Quek, F.: A review of vessel extraction techniques and algorithms. ACM Comput. Surv. **36**(2), 81–121 (2004)
52. Kobashi, S., Kamiura, N., Hata, Y., Miyawaki, F.: Volume-quantization-based neural network approach to 3D MR angiography image segmentation. Image Vis. Comput. **19**(4), 185–193 (2001)
53. Krissian, K., Malandain, G., Ayache, N., Vaillant, R., Trousset, Y.: Model-based detection of tubular structures in 3D images. Comput. Vis. Image Underst. **80**(2), 130–171 (2000)
54. Law, M.W.K. sand Chung, A.C.S.: Weighted local variance-based edge detection and its application to vascular segmentation in magnetic resonance angiography. IEEE Trans. Med. Imaging **26**(9), 1224–1241 (2007)
55. Lesage, D., Angelini, E.D., Bloch, I., Funka-Lea, G.: Design and study of flux-based features for 3D vascular tracking. In: International Symposium on Biomedical Imaging – ISBI 2009, pp. 286–289. IEEE (2009)
56. Lesage, D., Angelini, E.D., Bloch, I., Funka-Lea, G.: A review of 3D vessel lumen segmentation techniques: Models, features and extraction schemes. Med. Image Anal. **13**(6), 819–845 (2009)
57. Li, H., Yezzi, A.J.: Vessels as 4-D curves: Global minimal 4-D paths to extract 3-D tubular surfaces and centerlines. IEEE Trans. Med. Imaging **26**(9), 1213–1223 (2007)
58. Lorenz, C., von Berg, J.: A comprehensive shape model of the heart. Med. Image Anal. **10**(4), 657–670 (2006)

59. Lorigo, L.M., Faugeras, O.D., Grimson, W.E.L., Keriven, R., Kikinis, R., Nabavi, A., Westin, C.F.: CURVES: Curve evolution for vessel segmentation. Med. Image Anal. **5**(3), 195–206 (2001)
60. Mahadevan, V., Narasimha-Iyer, H., Roysam, B., Tanenbaum, H.L.: Robust model-based vasculature detection in noisy biomedical images. IEEE Trans. Inform. Technol. Biomed. **8**(3), 360–376 (2004)
61. Manniesing, R., Niessen, W.J.: Local speed functions in level set based vessel segmentation. In: Medical Image Computing and Computer-Assisted Intervention – MICCAI 2004. Lecture Notes in Computer Science, vol. 3216, pp. 475–482. Springer, Berlin (2004)
62. Manniesing, R., Velthuis, B.K., van Leeuwen, M.S., van der Schaaf, I.C., van Laar, P.J., Niessen, W.J.: Level set based cerebral vasculature segmentation and diameter quantification in CT angiography. Med. Image Anal. **10**(2), 200–214 (2006)
63. Manniesing, R., Viergever, M.A., Niessen, W.J.: Vessel axis tracking using topology constrained surface evolution. IEEE Trans. Med. Imaging **26**(3), 309–316 (2007)
64. McInerney, T., Terzopoulos, D.: Medical image segmentation using topologically adaptable surfaces. In: Computer Vision, Virtual Reality and Robotics in Medicine and Medical Robotics and Computer-Assisted Surgery - CVRMed-MRCAS 1997. Lecture Notes in Computer Science, vol. 1205, pp. 23–32. Springer, Berlin (1997)
65. Meijster, A., Wilkinson, H.: A comparison of algorithms for connected set openings and closings. IEEE Trans. Pattern Anal. Mach. Intell. **24**(4), 484–494 (2002)
66. Metz, C., Schaap, M., van der Giessen, A., van Walsum, T., Niessen, W.: Semi-automatic coronary artery centerline extraction in computed tomography angiography data. In: International Symposium on Biomedical Imaging – ISBI 2007, pp. 856–859. IEEE (2007)
67. Metz, C., Schaap, M., van Walsum, T., van der Giessen, A., Weustink, A., Mollet, N., Krestin, G., Niessen, W.: Editorial: 3D segmentation in the clinic: A grand challenge II- Coronary artery tracking. In: Medical Image Computing and Computer-Assisted Intervention – MICCAI 2008 – Workshop: Grand Challenge Coronary Artery Tracking (2008). http://hdl.handle.net/10380/1399
68. Naegel, B., Passat, N., Ronse, C.: Grey-level hit-or-miss transforms – Part II: Application to angiographic image processing. Pattern Recognit. **40**(2), 648–658 (2007)
69. Naegel, B., Ronse, C., Soler, L.: Using grey-scale hit-or-miss transform for segmenting the portal network of the liver. In: International Symposium on Mathematical Morphology – ISMM 2005. Computational Imaging and Vision, vol. 30, pp. 429–440. Springer SBM (2005)
70. Naidich, T.P., Brightbill, T.C.: Vascular territories and watersheds: A zonal frequency analysis of the gyral and sulcal extent of cerebral infarcts. Part I: the anatomic template. Neuroradiology **45**(8), 536–540 (2003)
71. Najman, L., Couprie, M.: Building the component tree in quasi-linear time. IEEE Trans. Image Process. **15**(11), 3531–3539 (2006)
72. Najman, L., Talbot, H. (eds.): Mathematical Morphology: From Theory to Applications. Wiley, London (2010)
73. Nowinski, W., Thirunavuukarasuu, A., Volkau, I., Marchenko, Y., Aminah, B., Puspitasari, F., Runge, V.: A three-dimensional interactive atlas cerebral arterial variants. NeuroInformatics **7**(4), 255–264 (2009)
74. Nowinski, W., Volkau, I., Marchenko, Y., Thirunavuukarasuu, A., Ng, T., Runge, V.: A 3D model of human cerebrovasculature derived from 3T magnetic resonance angiography. NeuroInformatics **7**(1), 23–36 (2009)
75. Olabarriaga, S.D., Breeuwer, M., Niessen, W.J.: Minimum cost path algorithm for coronary artery central axis tracking in CT images. In: Medical Image Computing and Computer-Assisted Intervention – MICCAI 2003. Lecture Notes in Computer Science, vol. 2879, pp. 687–694. Springer, Berlin (2003)
76. Passat, N., Ronse, C., Baruthio, J., Armspach, J.P., Bosc, M., Foucher, J.: Using multimodal MR data for segmentation and topology recovery of the cerebral superficial venous tree. In: International Symposium on Visual Computing – ISVC 2005. Lecture Notes in Computer Science, vol. 3804, pp. 60–67. Springer, Berlin (2005)

77. Passat, N., Ronse, C., Baruthio, J., Armspach, J.P., Maillot, C.: Magnetic resonance angiography: From anatomical knowledge modeling to vessel segmentation. Med. Image Anal. **10**(2), 259–274 (2006)
78. Passat, N., Ronse, C., Baruthio, J., Armspach, J.P., Maillot, C., Jahn, C.: Region-growing segmentation of brain vessels: An atlas-based automatic approach. J. Magn. Reson. Imaging **21**(6), 715–725 (2005)
79. Qian, X., Brennan, M.P., Dione, D.P., Dobrucki, W.L., Jackowski, M.P., Breuer, C.K., Sinusas, A.J., Papademetris, X.: A non-parametric vessel detection method for complex vascular structures. Med. Image Anal. **13**(1), 49–61 (2009)
80. Sabry Hassouna, M., Farag, A.A., Hushek, S., Moriarty, T.: Cerebrovascular segmentation from TOF using stochastic models. Med. Image Anal. **10**(1), 2–18 (2006)
81. Salamon, G., Huang, Y.P.: A Radiological Anatomy of the Brain. Springer, Berlin (1976)
82. Salembier, P., Garrido, L.: Binary partition tree as an efficient representation for image processing, segmentation and information retrieval. IEEE Trans. Image Process. **9**(4), 561–576 (2000)
83. Salembier, P., Serra, J.: Flat zone filtering, connected operators and filters by reconstruction. IEEE Trans. Image Process. **3**(8), 1153–1160 (1995)
84. Sato, Y., Nakajima, S., Atsumi, H., Koller, T., Gerig, G., Yoshida, S., Kikinis, R.: 3D multi-scale line filter for segmentation and visualization of curvilinear structures in medical images. In: Computer Vision, Virtual Reality and Robotics in Medicine and Medical Robotics and Computer-Assisted Surgery – CVRMed-MRCAS 1997. Lecture Notes in Computer Science, vol. 1205, pp. 213–222. Springer, Berlin (1997)
85. Schaap, M., Manniesing, R., Smal, I., van Walsum, T., van der Lugt, A., Niessen, W.: Bayesian tracking of tubular structures and its application to carotid arteries in CTA. In: Medical Image Computing and Computer-Assisted Intervention – MICCAI 2007. Lecture Notes in Computer Science, vol. 4792, pp. 562–570. Springer, Berlin (2007)
86. Serra, J.: Image Analysis and Mathematical Morphology. Academic, London, UK (1982)
87. Serra, J. (ed.): Image Analysis and Mathematical Morphology. Volume 2: Theoretical Advances. Academic, London, UK (1988)
88. Shahzad, R., Schaap, M., van Walsum, T., Klien, S., Weustink, A.C., van Vliet, L.J., Niessen, W.J.: A patient-specific coronary density estimate. In: International Symposium on Biomedical Imaging – ISBI 2010, pp. 9–12. IEEE (2010)
89. Soille, P.: Morphological Image Analysis. Springer, Heidelberg (2003)
90. Soille, P., Talbot, H.: Directional morphological filtering. IEEE Trans. Pattern Anal. Mach. Intell. **23**(11), 1313–1329 (2001)
91. Suh, J.W., Scheinost, D., Qian, X., Sinusas, A.J., Breuer, C.K., Papademetris, X.: Serial non rigid vascular registration using weighted normalized mutual information. In: International Symposium on Biomedical Imaging – ISBI 2010, pp. 25–28. IEEE (2010)
92. Sun, K.Q., Sang, N.: Morphological enhancement of vascular angiogram with multiscale detected by Gabor filters. Electron. Lett. **44**(2), 86–87 (2008)
93. Suri, J.S., Liu, K., Reden, L., Laxminarayan, S.: A review on MR vascular image processing algorithms: Acquisition and prefiltering: Part I. IEEE Trans. Inform. Technol. Biomed. **6**(4), 324–337 (2002)
94. Suri, J.S., Liu, K., Reden, L., Laxminarayan, S.: A review on MR vascular image processing: Skeleton versus nonskeleton approaches: Part II. IEEE Trans. Inform. Technol. Biomed. **6**(4), 338–350 (2002)
95. Talbot, H., Appleton, B.: Efficient complete and incomplete paths openings and closings. Image Vis. Comput. **25**(4), 416–425 (2007)
96. Tankyevych, O.: Filtering of thin objects, applications to vascular image analysis. Ph.D. thesis, University Paris-Est (2010)
97. Tatu, L., Moulin, T., Bogousslavsky, J., Duvernoy, H.: Arterial territories of the human brain: Brainstem and cerebellum. Neurology **47**(5), 1125–1135 (1996)
98. Tatu, L., Moulin, T., Bogousslavsky, J., Duvernoy, H.: Arterial territories of the human brain: Cerebral hemispheres. Neurology **50**(6), 1699–1708 (1998)

99. Thompson, P.M., Toga, A.W.: A framework for computational anatomy. Comput. Vis. Sci. **5**(1), 13–34 (2002)
100. Tizon, X., Smedby, Ö.: Segmentation with gray-scale connectedness can separate arteries and veins in MRA. J. Magn. Reson. Imaging **15**(4), 438–445 (2002)
101. Tsitsiklis, J.: Efficient algorithms for globally optimal trajectories. IEEE Trans. Autom. Control **40**(9), 1528–1538 (1995)
102. Tyrrell, J.A., di Tomaso, E., Fuja, D., Tong, R., Kozak, K., Jain, R.K., Roysam, B.: Robust 3-D modeling of vasculature imagery using superellipsoids. IEEE Trans. Med. Imaging **26**(2), 223–237 (2007)
103. van Bemmel, C.M., Spreeuwers, L.J., Viergever, M.A., Niessen, W.J.: Level-set-based artery-vein separation in blood pool agent CE-MR angiograms. IEEE Trans. Med. Imaging **22**(10), 1224–1234 (2003)
104. Vasilevskiy, A., Siddiqi, K.: Flux maximizing geometric flows. IEEE Trans. Pattern Anal. Mach. Intell. **24**(12), 1565–1578 (2002)
105. Vincent, L.: Grayscale area openings and closings, their efficient implementation and applications. In: International Symposium on Mathematical Morphology – ISMM 1993, pp. 22–27. Barcelona, Spain (1993)
106. Wilkinson, M.H.F., Westenberg, M.A.: Shape preserving filament enhancement filtering. In: Medical Image Computing and Computer-Assisted Intervention – MICCAI 2001. Lecture Notes in Computer Science, vol. 2208, pp. 770–777. Springer, Berlin (2001)
107. Wilson, D.L., Noble, J.A.: An adaptive segmentation algorithm for time-of-flight MRA data. IEEE Trans. Med. Imaging **18**(10), 938–945 (1999)
108. Wink, O., Frangi, A.F., Verdonck, B., Viergever, M.A., Niessen, W.J.: 3D MRA coronary axis determination using a minimum cost path approach. Magn. Reson. Med. **47**(6), 1169–1175 (2002)
109. Wink, O., Niessen, W.J., Viergever, M.A.: Fast delineation and visualization of vessels in 3-D angiographic images. IEEE Trans. Med. Imaging **19**(4), 337–346 (2000)
110. Wong, W.C.K., Chung, A.C.S.: Probabilistic vessel axis tracing and its application to vessel segmentation with stream surfaces and minimum cost paths. Med. Image Anal. **11**(6), 567–587 (2007)
111. Wörz, S., Rohr, K.: A new 3D parametric intensity model for accurate segmentation and quantification of human vessels. In: Medical Image Computing and Computer-Assisted Intervention - MICCAI 2004. Lecture Notes in Computer Science, vol. 3216, pp. 491–499. Springer, Berlin (2004)
112. Wörz, S., Rohr, K.: Segmentation and quantification of human vessels using a 3-D cylindrical intensity model. IEEE Trans. Image Process.**16**(8), 1994–2004 (2007)
113. Wyatt, C., Bayram, E., Ge, Y.: Minimum reliable scale selection in 3D. IEEE Trans. Pattern Anal. Mach. Intell. **28**(3), 481–487 (2006)
114. Yim, P.J., Choyke, P.L., Summers, R.M.: Gray-scale skeletonization of small vessels in magnetic resonance angiography. IEEE Trans. Med. Imaging **19**(6), 568–576 (2000)
115. Zahlten, C., Jürgens, H., Evertsz, C.J.G., Leppek, R., Peitgen, H.O., Klose, K.J.: Portal vein reconstruction based on topology. Eur. J. Radiol. **19**(2), 96–100 (1995)
116. Zitová, B., Flusser, J.: Image registration methods: A survey. Image Vis. Comput. **21**(11), 977–1000 (2003)
117. Zucker, S.W.: Region growing: Childhood and adolescence. Comput. Graph. Image Process. **5**(3), 382–399 (1976)

Chapter 7
Detecting and Analyzing Linear Structures in Biomedical Images: A Case Study Using Corneal Nerve Fibers

Mohammad A. Dabbah, James Graham, Rayaz A. Malik, and Nathan Efron

7.1 Introduction

Diabetic peripheral neuropathy (DPN) is one of the most common long-term complications of diabetes. The accurate detection and quantification of DPN are important for defining at-risk patients, anticipating deterioration, and assessing new therapies. Current methods of detecting and quantifying DPN, such as neurophysiology, lack sensitivity, require expert assessment and focus primarily on large nerve fibers. However, the earliest damage to nerve fibers in diabetic neuropathy is to the small nerve fibers. At present, small nerve fiber damage is currently assessed using skin/nerve biopsy; both are invasive technique and are not suitable for repeated investigations.

Recent research [1–3] using Corneal Confocal Microscopy (CCM) suggests that this noninvasive, and hence reiterative, test might be an ideal surrogate endpoint for human diabetic neuropathy. These studies demonstrate that measurements made by CCM accurately quantify corneal nerve fiber morphology. The measurements reflect the severity of DPN and relate to the extent of intra-epidermal nerve fiber loss seen in skin biopsy. However, the major limitation preventing extension of this technique to wider clinical practice is that analysis of the images using interactive image analysis is highly labor-intensive and requires considerable expertise to quantify nerve fiber pathology. To be clinically useful as a diagnostic tool, it is essential that the measurements be extracted automatically.

The chapter is organized as follows. Section 7.2 provides an overview of some linear structure detection methods. Section 7.3 focuses on the quantification of CCM imaging. Section 7.3.1 provides more detail on its relationship with diabetic neuropathy, while Sect. 7.3.2 discusses the characteristics of CCM images and

M.A. Dabbah (✉)
The University of Manchester, Manchester, England
e-mail: m.a.dabbah@manchester.ac.uk

G. Dougherty (ed.), *Medical Image Processing: Techniques and Applications*, Biological and Medical Physics, Biomedical Engineering, DOI 10.1007/978-1-4419-9779-1_7,

Sect. 7.3.3 presents the metrics used to quantify the images. The dual-model detection algorithm is then presented in Sect. 7.4. The foreground and background models are described in Sect. 7.4.1 and the method for estimating local orientation is presented in Sect. 7.4.2. The final outcome of the dual-model and its postprocessing are discussed in Sects. 7.4.3 and 7.4.4, respectively. Section 7.5 describes the method and results of the comparative evaluations together with the assessment of clinical utility. The conclusion is provided in Sect. 7.6.

7.2 Linear Structure Detection Methods

The first critical stage in analysis of CCM images is the detection of nerve fibers. This is challenging as the nerve fibers often show poor contrast in the relatively noisy images. The literature on this topic is not extensive, although the problem has a superficial similarity to other, more widely investigated, applications, such as detection of blood-vessels in retinal images. Ruggeri et al. [4] describe a heuristic method that was adapted from retinal analysis. Linear structures occur in a number of imaging applications, in biomedicine and other fields. Below, we briefly review some that are particularly relevant to this application.

In [5], we conducted a preliminary comparison of methods for contrast enhancement of nerve fibers, comparing a Gabor wavelet with a well-established line detector. This method (Line Operator – LinOp), originally developed for detection of asbestos fibers [6] has also been used in other contexts, such as the detection of cracks in metal castings [7] and has been shown to be effective in detecting ducts and other linear structures in mammograms [8]. LinOp exploits the linear nature of the structures to enhance their contrast by computing the average intensity of pixels lying on a line passing through the reference pixel for multiple orientations and scales. The largest values are chosen to correspond to the line, the strength of which is determined by the difference with the average intensity of the similarly oriented square neighborhood. In the original LinOp implementation, processing was conducted at a single scale, but later versions used a multi-scale analysis.

In our comparative study [5], the 2D Gabor filter [9] was used to detect nerve fibers in CCM images. The filter is a band-pass filter that consists of a sinusoidal plane wave with a certain orientation and frequency, modulated by a Gaussian envelope. This spatial domain enhancement is based on the convolution of the image with the even-symmetric Gabor filter that is tuned to the local nerve fiber orientation. The comparison indicated that the oriented Gabor response, gave slightly improved enhancement of nerve fibers over that provided by LinOp.

Inspired by [10, 11], Frangi et al. [12] used a multi-scale decomposition of the Hessian matrix (matrix of second order image derivatives) to detect and measure blood vessels in Digital Subtraction Angiography images in 2D and 3D Magnetic Resonance Angiography images. Local second order structure of the image can be decomposed by extracting the principal directions using the eigenvalues of the Hessian. Unlike Lorenz et al. [10] and Sato et al. [11], Frangi et al. [12] simultaneously used the eigenvalues and eigenvectors of the Hessian to derive a discriminant function that has maximum response for tube-like structures. The method uses the norm of the Hessian to distinguish between background and foreground based on the observation that the magnitude of the derivatives (and thus the eigenvalues) is small at background pixels. Several other models based on second derivatives have been widely used for linear structure detection in medical image analysis [13].

The dual-tree complex wavelet transform (DTCWT) [14] is an extension of the discrete wavelet transform (DWT), which provides a sparse representation and characterization of structures and texture of the image at multiple resolutions. The DTCWT utilizes two DWT decompositions (trees) with specifically selected filters that give it the properties of approximate shift-invariance and good directionality. The key feature of the DTCWT operation lies in the differences between the filters in the two trees. DTCWT has been used in extracting and decomposing information from images to obtain rich feature descriptors of key-points [15]. It was also used to detect linear structures in retinal image [16] and mammograms [17].

The Monogenic Signal [18] (a variant of a 2D analytic signal) is an extension of the analytic signal using quaternionic algebra in an attempt to generalize the method to enable it to analyze intrinsically 2D signals, for example, structures within images. The Monogenic Signal is based on the Riesz transform, which is a 2D generalization of the Hilbert transform used in the conventional analytic signal. The Monogenic Signal is defined as the combination of the original signal and the Riesz-transformed one in the algebra of quaternions. It has been used in extracting structure information (such as edge, ridge, etc.) from images in several medical image analysis applications [19, 20].

7.3 Quantification of Nerve Fibers in Corneal Confocal Microscopy Imaging

CCM imaging is a new technology based on confocal laser scanning microscopes (CLSM), which captures high-resolution images with depth selectivity from the cornea of the human eye. However, confocal microscopy itself is not new, having been first introduced in 1961 [21] and later adopted as a standard technology in the late 1980s.

Confocal microscopy provides highly improved image quality over "conventional" transmitted-light microscopy due to its highly controlled and limited depth of focus. Images are reconstructed by detecting light from the focal plane in a point-by-point fashion, with the result that light from out-of-focus parts of the sample does not contribute to image background. Confocal imaging is widely used for constructing three dimensional images of prepared samples. A number of instruments are available that apply the same imaging procedure to the cornea in vivo (such as *Tomey Confoscan*,[1] *Nidek*,[2] *HRT-III*.[3])

7.3.1 CCM for Imaging Diabetic Peripheral Neuropathy

Diabetic neuropathy is one of the commonest long-term complications of diabetes and is the main initiating factor for foot ulceration, Charcot's neuroarthropathy, and lower extremity amputation. As 80% of amputations are preceded by foot ulceration, an effective means of detecting and treating neuropathy would have a major medical, social, and economic impact. The development of new treatments to slow, arrest, or reverse this condition is of paramount importance but is presently limited due to difficulties with end points employed in clinical trials [22].

Recent studies suggest that small unmyelinated c-fibers may be the earliest to be damaged in diabetic neuropathy [23–25]. The only technique which allows a direct examination of unmyelinated nerve fiber damage are those of sural nerve biopsy with electron microscopy [25, 26], and the skin-punch biopsy [27–29], but both are invasive procedures. However, our previous studies in patients with diabetic neuropathy have shown that CCM can be used to quantify early small nerve fiber damage and accurately quantify the severity of diabetic neuropathy [1, 2]. These observations led us to suggest that CCM may be an ideal noninvasive surrogate marker for detecting small fiber damage in diabetic and other peripheral neuropathies [3]. Moreover, we have shown that corneal nerve damage assessed using CCM relates to the severity of intra-epidermal nerve fiber loss in foot skin biopsies [30] and the loss of corneal sensation [31] in diabetic patients. CCM also detects early nerve fiber regeneration following pancreas transplantation in diabetic patients [32]. Recently, we have also shown that CCM detects nerve fiber damage in patients with Fabry disease [33] and idiopathic small fiber neuropathy [34] in the presence of normal electrophysiology and quantitative sensory testing (QST). CCM offers considerable potential as a surrogate marker, and hence as an end-point for clinical trials in diabetic neuropathy [35].

7.3.2 CCM Image Characteristics and Noise Artifacts

CCM images are captured at different depths in the cornea by manual focusing of the CLSM. Due to different capturing conditions such as saccadic eye movements

---

[1]Tomey Corporation, http://www.tomey.com/.

[2]Nidek Inc. http://usa.nidek.com/.

[3]Heidelberg Engineering, Inc. http://www.heidelbergengineering.com/.

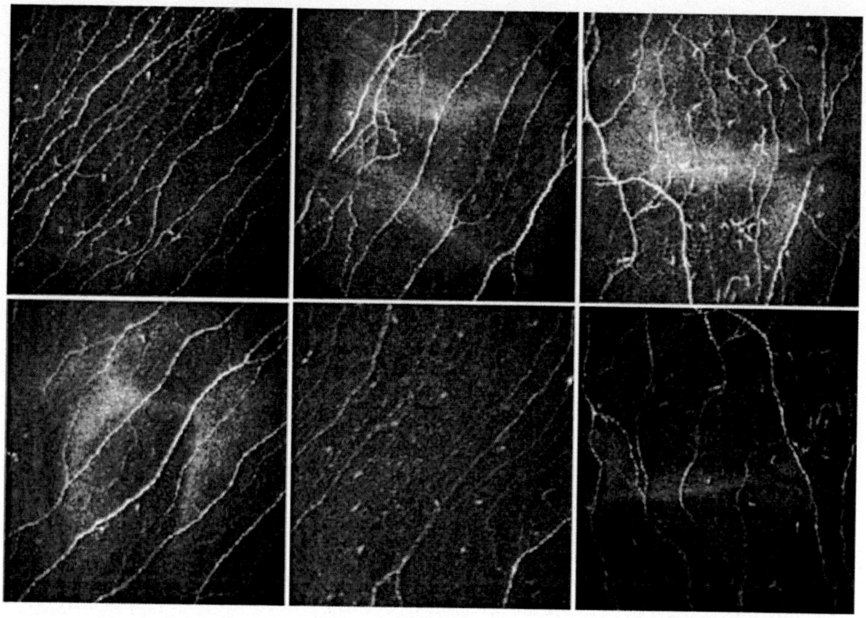

Fig. 7.1 CCM images of nerve fibers obtained from the Bowman's membrane. The images exhibit different varieties of artifacts due to acquisition conditions

in the eye, the degree of physical pressure on the eye, the spherical shape of the cornea, etc. images can suffer a variety of artifacts (Fig. 7.1).

One of the common artifacts seen in CCM images is uneven illumination: low frequency intensity variations across the entire CCM image with no correspondence to the feature of interest of the image (nerve fibers, cells, etc.). These illumination artifacts are caused by the physical pressure of the CLSM lens on the spherically shaped cornea. It is difficult to avoid such artifacts while capturing the images as the acquisition requires constant physical contact through a medium that matches the refractive index of the cornea. These artifacts could not be defined by a certain orientation or frequency. This makes it plausible for pass-band directional filters to eliminate them from the image. Two-dimensional wavelets are an example of such filters.

Motion artifacts arise from the fact that it is difficult for patients to keep their eyes still with the microscope lens in contact with the cornea, despite the application of an anesthetic drop. The result is blurring of the nerve fibers.

The human cornea has a diameter of about 11.5 mm and a thickness of 0.5–0.6 mm in the centre and 0.6–0.8 mm at the periphery. It is transparent, has no blood supply and gets oxygen directly from the atmosphere. It consists of five layers: Epithelium, Bowman's membrane, Stroma, Descemet's membrane, and Endothelium. The Bowman's membrane is 8–12 μm thick, helps the cornea maintain its shape and is composed of nerve fibers. It is in this layer that CCM images are acquired for analysis.

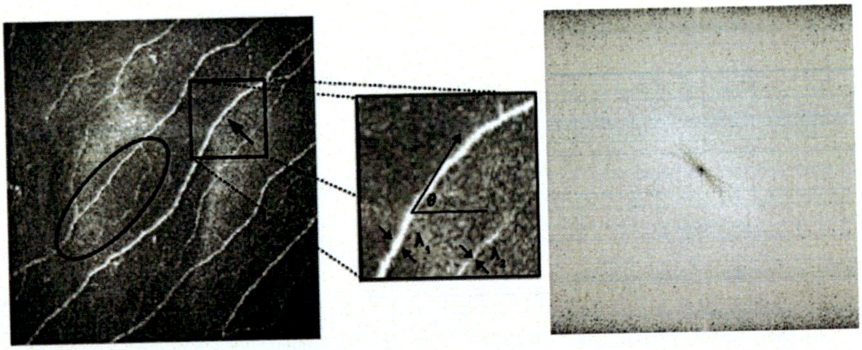

Fig. 7.2 CCM image characteristics. Nerve fibers flow in a predominant direction everywhere in the image with some minor variation in the direction and size. The size of these nerve fibers can be attributed to high-frequency components in the Fourier domain as shown in the right-hand side of the figure

These nerve fibers extend into different depths of the membrane causing them to disappear when they move out of the scanning focal plane. The spherical shape of the cornea also contributes to this depth-of-field artifact, especially around the periphery of the field of view.

Nerve fibers in CCM images display different lengths, widths, patterns, and orientations (Fig. 7.2). However, within an image they exhibit a predominant global orientation with minor variations across the image.

This predominant orientation depends on the part of the cornea the image is captured from. Across the cornea, the nerve fibers tend to converge toward the centre, in front of the pupil. The 2D Fourier Transform of the image in Fig. 7.2 shows the high frequency components (i.e., edges or nerve fibers) lying in an oriented pattern around the origin. These image characteristic can be used to detect nerve fibers and filter out noise.

7.3.3 *Quantified Metrics of Nerve Fibers in CCM Images*

For quantitative measurement of nerve fibers four main metrics are used: nerve fiber length (NFL), nerve fiber density (NFD), nerve-branch density (NBD), and nerve fiber tortuosity (NFT). The nerve fibers are considered as being the main trunks while braches are the secondary fibers which originate from the main nerve fibers (see Fig. 7.2).

NFL is defined as the total length of all nerve fibers visible in the CCM image per square millimeter. The total length is computed by tracing all the nerve fibers and nerve-branches in the image. This number is then divided by the area of the field-of-view provided by the microscope to produce the NFL [mm/mm^2]. According to

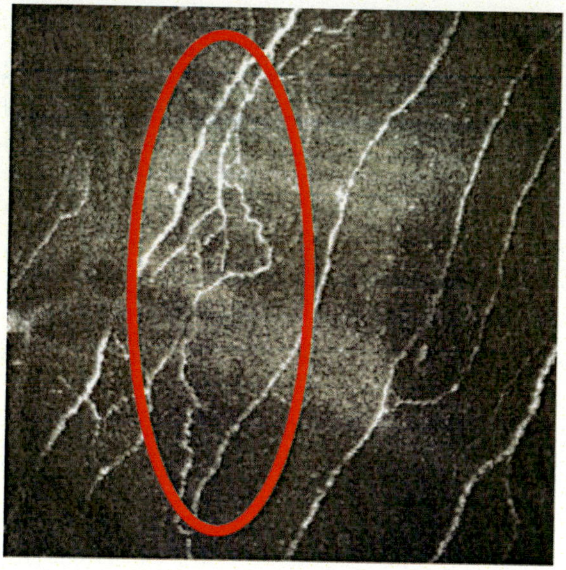

Fig. 7.3 An example of a CCM image with a tortuous nerve fiber

ongoing research and previously published data, this metric seems to be the most significant measure in categorizing and analyzing the diabetic neuropathy. The NFD and nerve-branch densities are the number of the major nerves per square millimeter and the number of branches emanating from those major nerve trunks per square millimeter of corneal tissue, respectively.

NFT is a metric indicating the degree of curvature or tortuousness of the main nerve fibers. There have been several attempts to quantify this property in medical image analysis such as in retinopathy [36], corneal neuropathy [2], etc. Clinical research [2] has shown that NFT (Fig. 7.3) can differentiate between differing severities of neuropathy.

7.4 A Dual-Model Detector for Linear Structures in CCM Images

To quantify the CCM images, the nerve fibers have to be reliably detected. The quality of captured images is often low due to the imaging effects outlined in Sect. 7.3.2 and nerve fibers may appear faint due to either their small size or being only partly in the focus plane. Therefore, a nerve fiber contrast enhancement algorithm is needed to exploit the linear structure of the nerve fibers and distinguish them from the background noise. All of the methods described in Sect. 7.2 are capable of providing this enhancement. In the next section, we describe our approach, the dual-model [37].

The dual-model consists of a 2D Gabor wavelet (foreground model) and a Gaussian envelope (background model), which are applied to the original CCM images. The detection relies on estimating the correct local and dominant orientation of the nerve fibers. We evaluate our dual-model in comparison with feature detectors described in Sect. 7.2 that are well established for linear and more general image features. In addition to the evaluation of the nerve fiber detection responses, we have also evaluated the clinical utility of the method by a comparison with manual analysis.

7.4.1 Foreground and Background Adaptive Models

For this purpose, the foreground model M_{F} is an even-symmetric and real-valued Gabor [9,38] wavelet and the background model M_{B} is a two-dimensional Gaussian envelope, Fig. 7.4.

$$M_{\mathrm{F}}(x_\theta, y_\theta) = \left(\cos\left(\frac{2\pi}{\lambda}x_\theta + \varphi\right)\right)\exp\left\{-\frac{1}{2}\left(\frac{x_\theta^2}{\sigma_x^2} + \frac{\gamma^2 y_\theta^2}{\sigma_y^2}\right)\right\} \tag{7.1}$$

$$M_{\mathrm{B}}(x_\theta, y_\theta) = \alpha\exp\left\{-\frac{1}{2}\left(\frac{x_\theta^2}{\sigma_x^2} + \frac{\gamma^2 y_\theta^2}{\sigma_y^2}\right)\right\} \tag{7.2}$$

$$x_\theta = x\cos\theta + y\sin\theta \tag{7.3}$$

$$y_\theta = -x\sin\theta + y\cos\theta \tag{7.4}$$

The x and y axes of the dual-model coordinate frame x_θ and y_θ are defined by a rotation of θ, which is the dominant orientation of the nerve fibers in a particular region within the image (see Sect. 7.4.2). This dual-model is used to generate the positive response $R_{\mathrm{P}} = M_{\mathrm{F}} + M_{\mathrm{B}}$ and the negative response $R_{\mathrm{N}} = M_{\mathrm{F}} - M_{\mathrm{B}}$ that are applied to the original CCM image and can be represented as in (7.5) and (7.6), respectively.

$$R_{\mathrm{P}}(x_\theta, y_\theta) = \left[\cos\left(\frac{2\pi}{\lambda}x_\theta + \varphi\right) + \alpha\right]\exp\left\{-\frac{1}{2}\left(\frac{x_\theta^2}{\sigma_x^2} + \frac{\gamma^2 y_\theta^2}{\sigma_y^2}\right)\right\} \tag{7.5}$$

$$R_{\mathrm{N}}(x_\theta, y_\theta) = \left[\cos\left(\frac{2\pi}{\lambda}x_\theta + \varphi\right) - \alpha\right]\exp\left\{-\frac{1}{2}\left(\frac{x_\theta^2}{\sigma_x^2} + \frac{\gamma^2 y_\theta^2}{\sigma_y^2}\right)\right\} \tag{7.6}$$

The equations of R_{P} and R_{N} assume that the Gaussian envelope of both responses are identical, that is, they have the same variances $\sigma^2(x,y)$ and the same aspect ratio γ. The magnitude of the Gaussian envelope α defines the threshold in which a

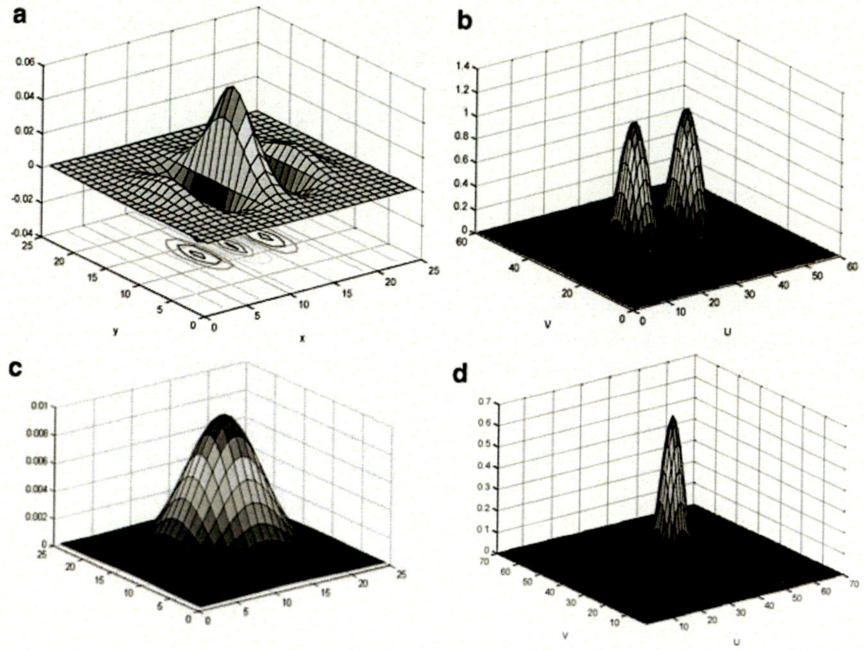

Fig. 7.4 Foreground and background models for the nerve fibers. (**a**) the two-dimensional Gabor wavelet at a particular orientation and frequency. It represents the foreground model of the nerve fibers. (**b**) the Fourier transforms of (**a**). (**c**) the two-dimensional Gaussian envelope that represents the background model and (**d**) its Fourier transform

nerve fiber can be distinguished from the background image. The value of α can be set empirically to control sensitivity and accuracy of detection. The wavelength λ defines the frequency band of the information to be detected in the CCM image, and is related to the width of the nerve fibers (see Fig. 7.2). Its value might be computed for a sub-region within the image that has significant variability of nerve fiber width. However, for simplicity, λ is chosen to be a global estimate of the entire image based on empirical results.

7.4.2 Local Orientation and Parameter Estimation

In CCM images, the nerve fibers flow in locally constant orientations. In addition, there is a global orientation that dominates the general flow. The orientation field describes the coarse structure of nerve fibers in the CCM images and has been proven to be of a fundamental importance in many image analysis applications [39, 40]. Using the least mean square algorithm [41], the local orientation $\theta(i,j)$ of the block centered at pixel (i,j) (7.9), is computed using the following equations [39].

$$V_x(i,j) = \sum_{u=i-\frac{\omega}{2}}^{i+\frac{\omega}{2}} \sum_{v=j-\frac{\omega}{2}}^{j+\frac{\omega}{2}} \left(\partial_x^2(u,v) - \partial_y^2(u,v)\right) \tag{7.7}$$

$$V_y(i,j) = \sum_{u=i-\frac{\omega}{2}}^{i+\frac{\omega}{2}} \sum_{v=j-\frac{\omega}{2}}^{j+\frac{\omega}{2}} 2\partial_x(u,v)\partial_y(u,v) \tag{7.8}$$

$$\theta(i,j) = \pi/2 + \frac{1}{2}\tan^{-1}\left(\frac{V_y(i,j)}{V_x(i,j)}\right) \tag{7.9}$$

The gradients $\partial_x(u,v)$ and $\partial_y(u,v)$ are computed at each pixel (u,v) and may vary from the simple *Sobel* operator to the more complex *Canny* operator depending on the computational requirements. ω is the width of the block centered at pixel (i,j). The orientation field is then smoothed by convolving the x and y vector field components in (7.7) and (7.8), respectively, with a low-pass Gaussian filter. This smoothed orientation field is calculated by (7.14), where $\widehat{\Phi}_x(i,j)$ and $\widehat{\Phi}_y(i,j)$ are the smoothed continuous x and y vector field components.

$$\Phi_x(i,j) = \cos(2\theta(i,j)) \tag{7.10}$$

$$\Phi_y(i,j) = \sin(2\theta(i,j)) \tag{7.11}$$

According to the original algorithm [41], the low-pass 2-dimensional Gaussian filter G is applied on the block level ω of the orientation field computed earlier in (7.9). The filter has a unit integral and a kernel size of $\omega_\Phi \times \omega_\Phi$. However, since the orientation in CCM images varies at a slow rate, the low-pass filter is applied globally to further reduce errors at near-nerve fiber and nonnerve fiber regions. The estimated orientation is not always correct, hence, the low-pass filter tries to rectify the error given that the orientation in the local neighborhood varies slowly;

$$\widehat{\Phi}_x(i,j) = \sum_{u=-\frac{\omega_\Phi}{2}}^{\frac{\omega_\Phi}{2}} \sum_{v=-\frac{\omega_\Phi}{2}}^{\frac{\omega_\Phi}{2}} G(u,v)\Phi_x(i-u,j-v) \tag{7.12}$$

$$\widehat{\Phi}_y(i,j) = \sum_{u=-\frac{\omega_\Phi}{2}}^{\frac{\omega_\Phi}{2}} \sum_{v=-\frac{\omega_\Phi}{2}}^{\frac{\omega_\Phi}{2}} G(u,v)\Phi_y(i-u,j-v) \tag{7.13}$$

$$O(i,j) = \frac{1}{2}\tan^{-1}\left(\frac{\widehat{\Phi}_y(i,j)}{\widehat{\Phi}_x(i,j)}\right) \tag{7.14}$$

The least square estimate produces a stable smooth orientation field in the region of the nerve fibers. However, when applied on the background of the image, that is, between fibers, the estimate is dominated by noise due to the lack of structure and uniform direction, which is expected and understandable. Figure 7.5 shows a CCM image and its orientation field estimate.

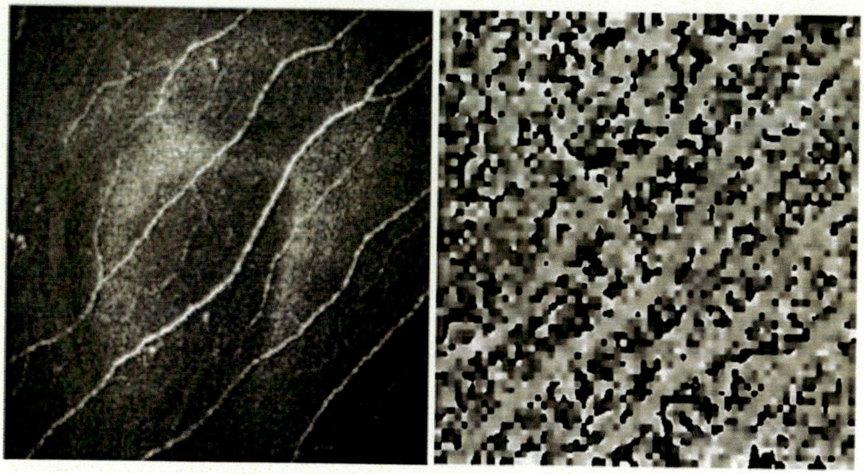

Fig. 7.5 An illustration of the orientation field (*right*) of the original CCM image (*left*). The orientations on the nerve fibers and their surrounding are similar and follow the predominant orientation in the image, while orientations everywhere else (background) are random and noisy

7.4.3 Separation of Nerve Fiber and Background Responses

The models are applied on the image pixel-wise. During this operation, they are adjusted to suit the local neighborhood characteristics of the reference pixel at $f(i,j)$ by modifying their parameters of the foreground and background separately in (7.5) and (7.6). The dot products of the models and the reference pixel's neighborhood ((7.15) and (7.16)) are then combined to generate the final enhanced value of this particular reference pixel $g(i,j)$ (7.17).

$$\Gamma_p(i,j) = \langle f_\omega(i,j), R_\mathrm{P} \rangle \tag{7.15}$$

$$\Gamma_n(i,j) = \langle f_\omega(i,j), R_\mathrm{N} \rangle \tag{7.16}$$

$$g(i,j) = \frac{\Gamma_p(i,j)}{1 + \exp\{-2k\Gamma_n(i,j)\}} \tag{7.17}$$

The neighborhood area of the reference pixel is defined by the width ω. The transition from foreground to background at a particular pixel $g(i,j)$ occurs at $\Gamma_n = 0$. The sharpness of this transition is controlled by k: larger k results in sharper transition. This in turn enhances the nerve fibers that are oriented in the dominant direction, and decreases noisy structures that are oriented differently by increasing the contrast between the foreground and the noisy background, whilst effectively reducing noise around the nerve fiber structure as shown in Fig. 7.6.

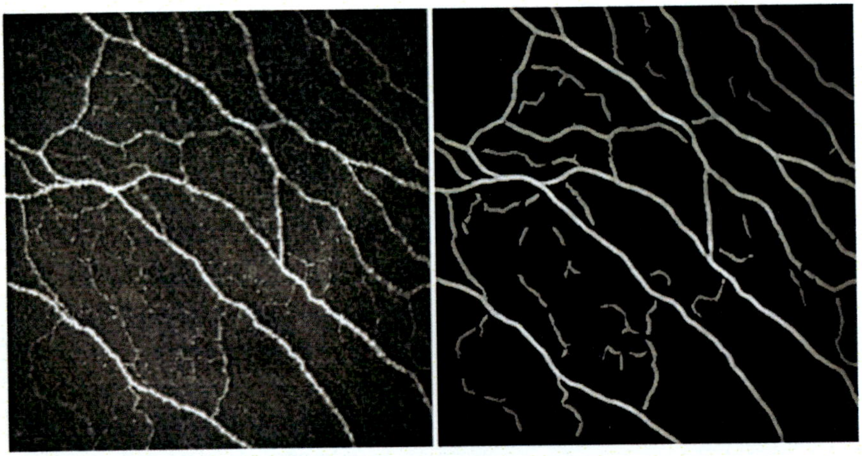

Fig. 7.6 An illustration of the dual-model enhancement results. The dual-model algorithm was applied on the original CCM image on the *left* resulting in the response image on the *right*. Even the structures of small and faint nerve fibers were enhanced

7.4.4 Postprocessing the Enhanced-Contrast Image

Once the CCM image is enhanced, the detection of the nerve fibers becomes a trivial task. The response image of the dual-model has a zero value for a background pixel and a value that is greater than zero, up to unity for everything else. This makes global thresholding of intensities an effective technique of separating background and foreground.

Changing the threshold value of the dual-model detector α in (7.5) and (7.6) will change the sensitivity of the detection as shown in Fig. 7.7. A lower threshold value will produce sensitive detection even for the very faint nerve fibers. However, this will also cause a more noisy response image due to the false positive detected nerve fibers that are represented as small fragments. These fragments can then be easily filtered out by simple postprocessing techniques as shown in Fig. 7.7.

The values that correspond to foreground pixels, that is, pixels on a nerve fiber, represent a confidence measure. The higher the value the more likely this is a nerve fiber pixel. Once the image is thresholded and turned into a binary form, zeros for background and ones for foreground, nerve fibers appear as thick ridges flowing across the image. This is followed by morphological operators to eliminate islands (separate pixels) between nerve fibers and to reduce the number of spurs in the thinned image.

To be able to estimate the center of these ridges, that is, the one-pixel line detection of nerve fibers, the binary linear structures are thinned using [42]. The skeletonized image (e.g., Fig. 7.8) provides a straightforward representation for

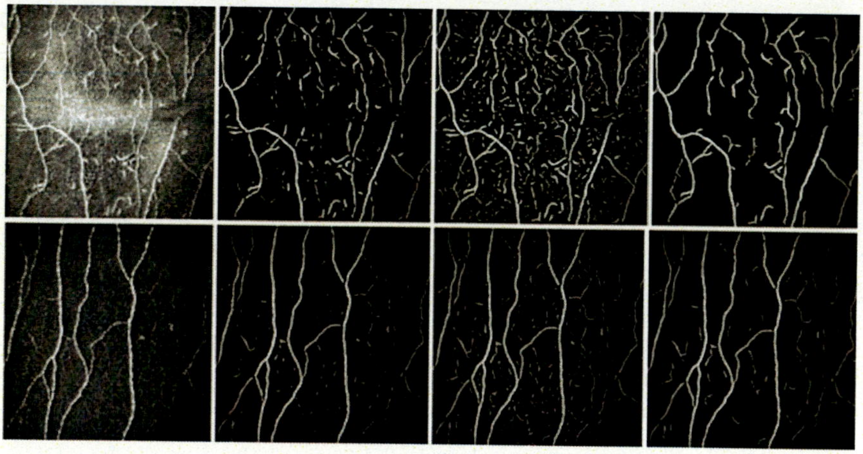

Fig. 7.7 An illustration of the dual-model enhancement threshold α on the response image. The first column contains the original CCM; the images in the second column are enhanced with a relatively high threshold. The images in the third and the last column are enhanced with the same low threshold but the images in the last column are also postprocessed to remove small fragmented nerve fibers

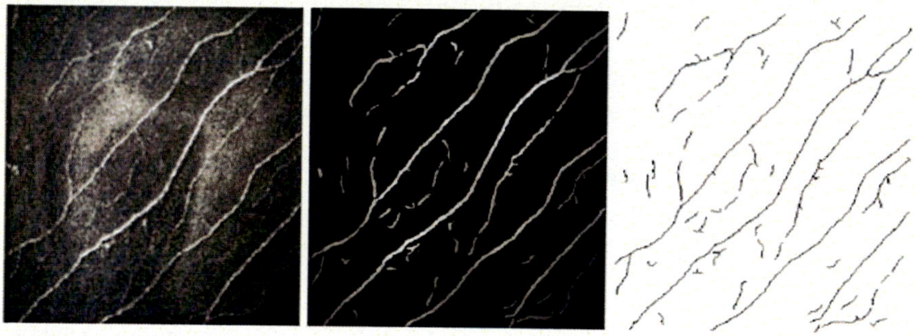

Fig. 7.8 After the original CCM image is enhanced to exploit the nerve fiber structures, the image is thresholded to produce a binary image which is then thinned to a skeleton image as shown in the last image on the *right*

defining the image features described in Sect. 7.3.3. NFL is simply the count of pixels with binary value of one. Fully connected lines of detected pixels that have a length greater than a certain threshold are counted to give the number of major nerve fibers in the CCM image and used to compute the NFD.

At each pixel on the skeleton, we can calculate the crossing number [43], which defines the number of neighbors the pixel has on the skeleton. One defines an end point, two is a ridge point, and three or more indicate a branch point. This allows us to recognize and count branch points.

7.5 Quantitative Analysis and Evaluation of Linear Structure Detection Methods

7.5.1 Methodology of Evaluation

The performance of the dual-model detector and the other methods described in Sect. 7.2 is obtained by validating the extracted nerve fibers in comparison with an expert manual delineation using *CCMetrics*.[4] Only the raw response of each method is taken into account without any further postprocessing operations or shade correction methods as shown in Fig. 7.9. Binary images are obtained by a simple uniform thresholding operation that is followed by a thinning operation to achieve a one-pixel-wide skeleton image.

To be consistent in this comparison of different methods, the detection algorithm did not include any pixel classifications. Responses from techniques with multi-scale analysis, such as LinOp, Hessian, DTCWT, and Monogenic Signal, were considered by taking the maximum magnitude of all levels.

Three measures have been used to quantify the evaluation: the false-positive (FPR), the true-positive (TPR), and the equal-error rate (EER), which is the average of optimal FPR and false-negative rate at minimal difference between both. A receiver operating characteristic (ROC) analysis was conducted by comparing the generated skeleton at different threshold intervals of the methods' responses with the manually delineated ground-truth. A tolerance of $\pm 3.141\,\mu\text{m}$ (3 pixels) was allowed in determining coincidence between the ground-truth and the detected nerve fibers.

The peak signal to noise ratio (PSNR) in (7.18) is also used to evaluate the performance of all methods.

$$\text{PSNR}_{\text{dB}} = 20\log\left(\frac{\text{MAX}_\text{I}}{\sqrt{e}}\right) \tag{7.18}$$

The PSNR is computed with respect to the mean squared error e, which is the mean square difference between the detected nerve fibers and the ground-truth manual delineation. MAX_I is the maximum possible intensity (fixed) and e is the mean square error. The practical implementations of the Hessian, the DTCWT, and the Monogenic Signal were obtained from public domain sources [44–46], while the rest were implemented by our research group.

[4] *CCMetrics* is a purpose built interactive graphical interface which helps in the analysis undertaken by experts to manually delineate nerve fibers in CCM images.

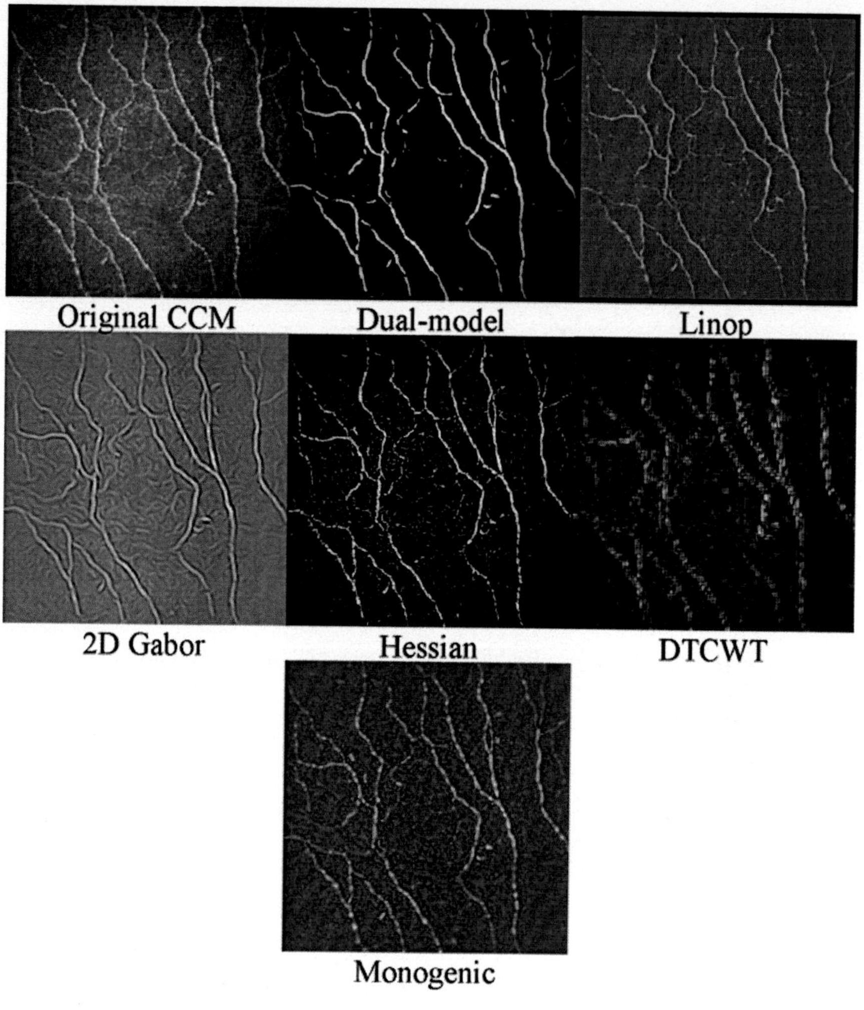

Fig. 7.9 Example response images for all different detection methods. The responses were taken as a raw output from the detector without any postprocessing and converted to binary images and then to skeleton images for fair comparison

7.5.2 Database and Experiment Setup

The evaluation has been conducted on a database of 525 CCM images captured using the HRT-III[5] microscope from 69 subjects (20 controls and 49 diabetic patients). The pixel size is 1.0417 μm and the field of view is $400 \times 400\,\mu m^2$ of

[5]Heidelberg Engineering Inc. modified to acquire corneal confocal images.

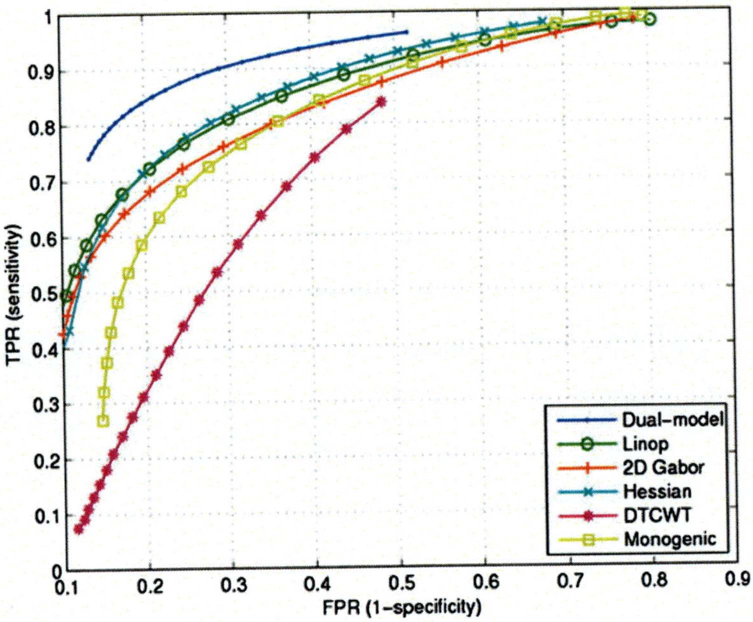

Fig. 7.10 The receiver operating characteristic (ROC) curves of all five detectors. The dual-model performance of detecting nerve fibres has clearly outperformed the other methods

the cornea. For each individual, several fields of view are selected manually in the centre of the cornea from the Bowman's layer showing recognizable nerve fibers.

Using the neuropathy disability score (NDS) [47], 48 patients were categorized into four groups according to severity of neuropathy (nonneuropathic: $0 \leq NDS \leq 2 (n = 26)$, mild: $3 \leq NDS \leq 5 (n = 9)$, moderate: $6 \leq NDS \leq 8 (n = 10)$ and severe: $9 \leq NDS \leq 10 (n = 3)$.

7.5.3 Nerve Fiber Detection Comparison Results

The superior performance of the dual-model is borne out by the ROC curves of Fig. 7.10 in which the dual-model shows improved detection at all operation points. The EER and PSNR values for all the methods are presented in the box-plots in Fig. 7.11 and Table 7.1. Each data point in Fig. 7.11 corresponds to the evaluation on one of the 525 CCM images in the database.

The dual-model shows lower EER and higher PSNR than all other methods (Table 7.1). These improvements are statistically significant ($p \approx 0$ using three different nonparametric tests). The table also shows that the standard deviations of both EER and PSNR are low for the dual-model, which indicates a more stable and robust behavior.

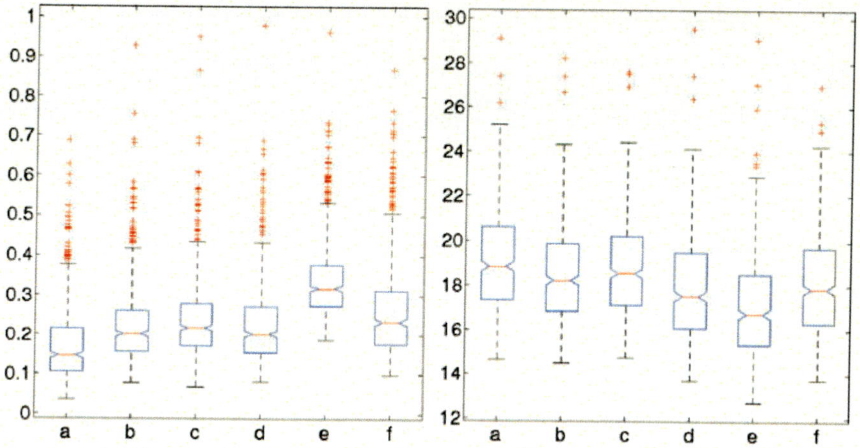

Fig. 7.11 The box-plots of the EER (*left*) and the PSNR (*right*) are shown for all methods. The box-plots indicate the upper and the lower quartiles as well as the median (*the bar*) of the EER and PSNR values respectively; whiskers show the extent of the rest of the data while crosses indicate outliers for **(a)** dual-model, **(b)** LinOp, **(c)** 2D Gabor, **(d)** Hessian, **(e)** DTCWT, and **(f)** Monogenic

Table 7.1 A comparison of mean EER and PSNR and their standard deviations for all five detection methods; the dual-model has achieved the lowest EER and the highest PSNR

	EER(%)		PSNR(dB)	
	μ	σ	μ	σ
Dual-model	17.79	10.58	19.0774	2.16
LinOp	22.65	10.76	18.5132	2.09
2D Gabor	24.15	10.74	18.8042	2.11
Hessian	23.14	11.53	17.9269	2.27
DTCWT	34.17	10.43	17.0045	2.23
Monogenic	26.50	12.58	18.1084	2.20

The closest performance to the dual-model has been achieved by LinOp, which has 4.86% greater EER on average. The performance of the Hessian methods is also similar with an average EER of 23.14% (Table 7.1). The poorest performance is obtained with the DTCWT and Monogenic Signal, as these are general-purpose methods. The dual-model has also shown a superior performance in terms of achieving higher PSNR values for the response images. As shown in the box-plot (Fig. 7.11), the average PSNR of the dual-model is 19.08 dB, while all PSNR groups have means smaller than the dual-model as indicated by Table 7.1, which shows a summary of the comparison. The closest PSNR is at 18.80 dB.

7.5.4 *Evaluation of Clinical Utility*

Of the several features listed in Sect. 7.3.3, which may be used to quantify the nerve fibers, NFL has been shown to be the most discriminating, and it is that feature that

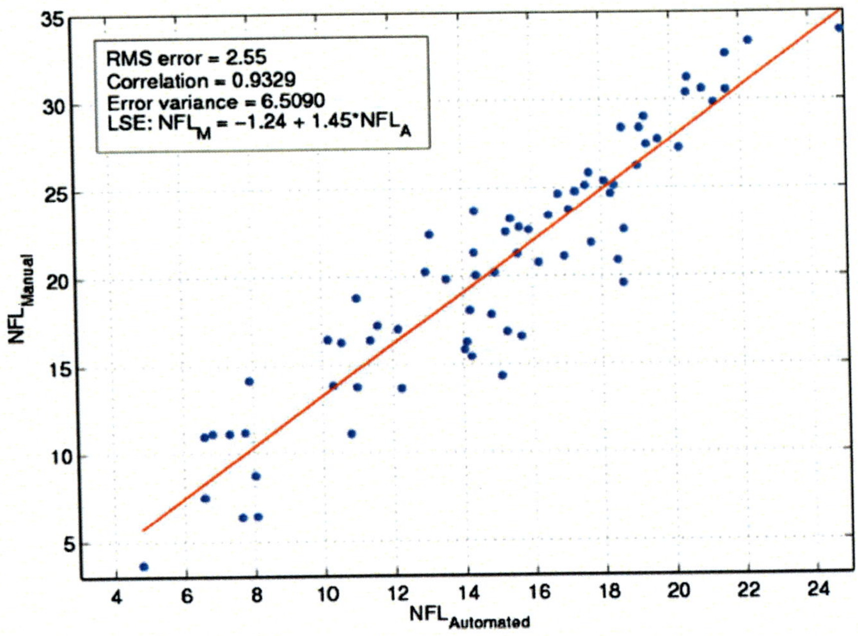

Fig. 7.12 The scatter plot of the manually and the automatically computed NFL metrics. There is clearly a very strong correlation ($r = 0.93$)

Table 7.2 A comparison of the manual and the automated analysis; unlike manual analysis, the automated analysis is insensitive to observer variability and can be much quicker

	Manual	Automated
p-value$(\times 10^{-8})$	0.03	2.03
Coefficient of variation	0.34	0.29
Observer variability	Yes	No
Processing time	5–10 min	≈ 5 s

we use to compare automatic detection with expert manual analysis (ground-truth). NFL is measured as the total number of pixels in the nerve fiber skeleton after the postprocessing of Sect. 7.4.4.

Figure 7.12 shows a scatter plot of manual vs. automatic measurements of NFL. There is clearly a strong correlation ($r = 0.93$) indicating that the automated system is successfully identifying the correct nerve fibers. The coefficient of variation $cv = \sigma/\mu$ of the manual analysis is 0.34, reducing for the automated analysis to 0.29, which indicates more reliability and robustness of the results (Table 7.2).

The box-plots in Fig. 7.13 shows NFL measured manually and automatically for the stratified group of subjects. There is a strong similarity between the manual and the automated analysis. However, the scale of the NFL has slightly changed from

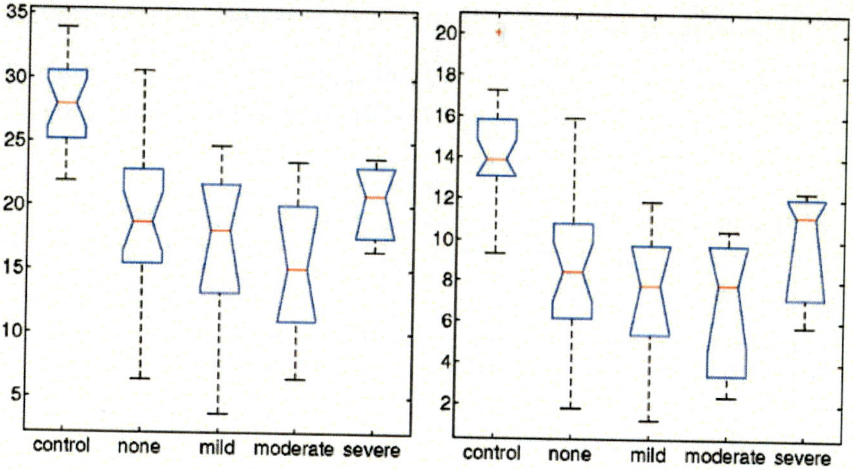

Fig. 7.13 Boxplots showing the NFL scores for each of the NDS groups calculated manually (*left*) and automatically (*right*)

(3.68–33.91) for the manual analysis to (1.22–20.03) for the automated analysis. ANOVA analysis results in a *p*-value for discrimination among these groups which is slightly higher for the automated than the manual analysis, though both are highly significant ($p \approx 0$) (Table 7.2).

7.6 Conclusion

The analysis of CCM images requires the identification of fiber-like structures with low contrast in noisy images. This is a requirement shared by a number of imaging applications in biology, medicine, and other fields. A number of methods have been applied in these applications, and we have compared some of these, and more generic methods, with a dual-model detection algorithm devised for this study. The comparison used a large set of images with manual ground-truth. In terms of both error-rates (pixel misclassification) and signal-to-noise ratio, the dual model achieved the highest performance. It seems reasonable to propose that this filter is likely to prove equally useful in applications of a similar nature.

The question of the clinical utility of the method was also addressed. The evaluation has shown that the automatic analysis is consistent with the manual ground-truth with a correlation of $r = 0.93$. Similarity in grouping control and patient subjects between manual and automated analysis was also achieved with ($p \approx 0$). Therefore, we conclude that automated analysis of corneal nerve fibers is a much quicker and potentially more reliable practical alternative to manual analysis due to its consistency and immunity to the inter/intra-observer variability. These prosperities will help deliver CCM from a research tool to a practical and potentially large scale clinical technique for the assessment of neuropathic severity.

Acknowledgments The authors thank *Dr. Mitra Tavakoli* and *Mr. Ioannis Petropoulos* for capturing the CCM images and providing the manually delineated ground-truth. The authors also thank the *Juvenile Diabetes Research Foundation* (JDRF) for the financial support under the scholar grant *17–2008–1031*.

References

1. Malik, R.A., Kallinikos, P., Abbott, C.A., van Schie, C.H.M., Morgan, P., Efron, N., Boulton, A.J.M.: Corneal confocal microscopy: A non-invasive surrogate of nerve fibre damage and repair in diabetic patients. Diabetologia **46**, 683–688 (2003)
2. Kallinikos, P., Berbanu, M., O'Donnell, C., Boulton, A., Efron, N., Malik, R.: Corneal nerve tortuosity in diabetic patients with neuropathy. Invest. Ophthalmol. Vis. Sci. **45**, 418–422 (2004)
3. Hossain, P., Sachdev, A., Malik, R.A.: Early detection of diabetic peripheral neuropathy with corneal confocal microscopy. Lancet **366**, 1340–1343 (2005)
4. Ruggeri, A., Scarpa, F., Grisan, E.: Analysis of corneal images for the recognition of nerve structures. In: IEEE Conference of the Engineering in Medicine and Biology Society (EMBS), pp. 4739–4742, September 2006
5. Dabbah, M.A., Graham, J., Tavakoli, M., Petropoulos, Y., Malik, R.A.: Nerve fibre extraction in confocal corneal microscopy images for human diabetic neuropathy detection using gabor filters. In: Medical Image Understanding and Analysis (MIUA), pp. 254–258, July 2009
6. Dixon, R.N., Taylor, C.J.: Automated asbestos fibre counting. Mach. Aided Image Anal. 178–185 (1979)
7. Bryson, N., Dixon, R.N., Hunter, J.J., Taylor, C.J.: Contextual classification of cracks. Image Vis. Comput. **12**, 149–154 (1994)
8. Zwiggelaar, R., Astley, S., Boggis, C., Taylor, C.: Linear structures in mammographic images: Detection and classification. IEEE Trans. Med. Imag. **23**, 1077–1086 (2004)
9. Jain, A.K., Farrokhnia, F.: Unsupervised texture segmentation using Gabor filters. Pattern Recogn. **24**, 1167–1186 (1991)
10. Lorenz, C., Carlsen, I., Buzug, T.: Fassnacht, C., Weese, J.: Multi-scale line segmentation with automatic estimation of width, contrast and tangential direction in 2D and 3D medical images. In: CVRMed-MRCAS'97, pp. 233–242 (1997)
11. Sato, Y., Nakajima, S., Atsumi, H., Koller, T., Gerig, G., Yoshida, S., Kikinis, R.: 3D multi-scale line filter for segmentation and visualization of curvilinear structures in medical images. In: CVRMed-MRCAS'97, pp. 213–222 (1997)
12. Frangi, A.F., Niessen, W.J., Vincken, K.L., Viergever, M.A.: Multiscale vessel enhancement filtering. In: Medical Image Computing and Computer-Assisted Interventation (MICCAI), pp. 130–137, July 1998
13. Krissian, K., Malandain, G., Ayache, N., Vaillant, R., Trousset, Y.: Model-based detection of tubular structures in 3D images. Comput. Vis. Image Underst. **80**, 130–171 (2000)
14. Kingsbury, N.: Complex wavelets for shift invariant analysis and filtering of signals. Appl. Comput. Harmonic Anal. **10**, 234–253 (2001)
15. Kingsbury, N.: Rotation-invariant local feature matching with complex wavelets. In: European Conference on Signal Processing (EUSIPCO), Florence, pp. 4–8, 2006
16. Sadeghzadeh, R., Berks, M., Astley, S., Taylor, C.: Detection of retinal blood vessels using complex wavelet transforms and random forest classification. In: Proceedings of Medical Image Understanding and Analysis (MIUA), pp. 127–131 (2010)
17. Chen, Z., Berks, M.: Astley, S., Taylor, C., Classification of linear structures in mammograms using random forests. In: Digital Mammography, pp. 153–160 (2010)
18. Felsberg, M., Sommer, G.: The Monogenic Signal. IEEE Trans. Signal Process. **49**, 3136–3144 (2001)

19. Pan, X.B., Brady, M., Highnam, R., Declerck, J.: The use of multi-scale monogenic signal on structure orientation identification and segmentation. In: Digital Mammography, pp. 601–608 (2006)
20. Ali, R., Gooding, M., Christlieb, M., Brady, M.: Advanced phase-based segmentation of multiple cells from brightfield microscopy images. In: IEEE International Symposium on Biomedical Imaging: From Nano to Macro (ISBI), pp. 181–184 (2008)
21. Minsky, M.: Microscopy apparatus US Patent 3013467. U. S. P. Office, Ed. US, 1961
22. Dyck, P., Norell, J., Tritschler, H., Schuette, K., Samigullin, R., Ziegler, D., Bastyr, E., Litchy, W., O'Brien, P.: Challenges in design of multicenter trials: end points assessed longitudinally for change and monotonicity. Diabetes Care **30**, 2619–2625 (2007)
23. Umapathi, T., Tan, W., Loke, S., Soon, P., Tavintharan, S., Chan, Y.: Intraepidermal nerve fiber density as a marker of early diabetic neuropathy. Muscle Nerve **35**, 591–598 (2007)
24. Loseth, S., Stalberg, E., Jorde, R., Mellgren, S.: Early diabetic neuropathy: thermal thresholds and intraepidermal nerve fibre density in patients with normal nerve conduction studies. J. Neurol. **255**, 1197–1202 (2008)
25. Malik, R., Tesfaye, S., Newrick, P., Walker, D., Rajbhandari, S., Siddique, I., Sharma, A., Boulton, A., King, R., Thomas, P., Ward, J.: Sural nerve pathology in diabetic patients with minimal but progressive neuropathy. Diabetologia **48**, 578–585 (2005)
26. Malik, R., Veves, A., Walker, D., Siddique, I., Lye, R., Schady, W., Boulton, A.: Sural nerve fibre pathology in diabetic patients with mild neuropathy: relationship to pain, quantitative sensory testing and peripheral nerve electrophysiology. Acta Neuropathol. (Berl.) **101**, 367–374 (2001)
27. Novella, S., Inzucchi, S., Goldstein, J.: The frequency of undiagnosed diabetes and impaired glucose tolerance in patients with idiopathic sensory neuropathy. Muscle Nerve **24**, 1229–1231 (2001)
28. Singleton, J., Smith, A., Bromberg, M.: Increased prevalence of impaired glucose tolerance in patients with painful sensory neuropathy. Diabetes Care **24**, 1448–1453 (2001)
29. Sumner, C., Sheth, S., Griffin, J., Cornblath, D., Polydefkis, M.: The spectrum of neuropathy in diabetes and impaired glucose tolerance. Neurology **60**, 108–111 (2003)
30. Quattrini, C., Tavakoli, M., Jeziorska, M., Kallinikos, P., Tesfaye, S., Finnigan, J., Marshall, A., Boulton, A.J.M., Efron, N., Malik, R.A.: Surrogate markers of small fiber damage in human diabetic neuropathy. Diabetes **56**, 2148–2154 (2007)
31. Tavakoli, M., Kallinikos, P.A., Efron, N., Boulton, A.J.M., Malik, R.A.: Corneal sensitivity is reduced and relates to the severity of neuropathy in patients with diabetes. Diabetes Care **30**, 1895–1897 (2007)
32. Mehra, S., Tavakoli, M., Kallinikos, P.A., Efron, N., Boulton, A.J.M., Augustine, T., Malik, R.A.: Corneal confocal microscopy detects early nerve regeneration after pancreas transplantation in patients with type 1 diabetes. Diabetes Care **30**, 2608–2612 (2007)
33. Tavakoli, M., Marshall, A., Thompson, L., Kenny, M., Waldek, S., Efron, N., Malik, R.A.: Corneal confocal microscopy: A novel noninvasive means to diagnose neuropathy in patients with Fabry disease. Muscle and Nerve **40**, 976–984 (2009)
34. Tavakoli, M., Marshall, A., Pitceathly, R., Fadavi, H., Gow, D., Roberts, M.E., Efron, N., Boulton, A.J.M., Malik, R.A.: Corneal confocal microscopy: A novel means to detect nerve fibre damage in idiopathic small fibre neuropathy. Exp. Neurol. **223**, 245–250 (2010)
35. Tavakoli, M., Quattrini, C., Abbott, C., Kallinikos, P., Marshall, A., Finnigan, J., Morgan, P., Efron, N., Boulton, A.J.M., Malik, R.A.: Corneal confocal microscopy: A novel non-invasive test to diagnose and stratify the severity of human diabetic neuropathy. Diabetes Care **33**, 1792–1797 (2010)
36. Dougherty, G., Johnson, M.J., Wiers, M.D.: Measurement of retinal vascular tortuosity and its application to retinal pathologies. Med. Biol. Eng. Comput. **48**, 87–95 (2010)
37. Dabbah, M.A., Graham, J., Tavakoli, M., Petropoulos, Y., Malik, R.A.: Dual-model automatic detection of nerve-fibres in corneal confocal microscopy images. In: The International Conference on Medical Image Computing and Computer Assisted Intervention (MICCAI), vol. 6361, pp. 300–307 (2010)

38. Daugman, J.G.: Two-dimensional spectral analysis of cortical receptive field profiles. Vis. Res. **20**, 847–856 (1980)
39. Rao, A.R.: A taxonomy for texture description and identification. Springer, New York (1990)
40. Kass, M., Witkin, A.: Analyzing oriented patterns. Comp. Vis. Graph. Image Process. **37**, 362–385 (1987)
41. Hong, L., Wan, Y., Jain, A.: Fingerprint image enhancement: algorithm and performance evaluation. IEEE Trans. Pattern Anal. Mach. Intell. **20**, 777–789 (1998)
42. Zhang, T.Y., Suen, C.Y.: A fast parallel algorithm for thinning digital patterns. Commun. ACM **27**, 236–239 (1984)
43. Rutovitz, D.: Pattern recognition. J. R. Stat. Soc. A (General) **129**, 504–530 (1966)
44. Kingsbury, N.: Dual-tree complex wavelet transform pack. http://www-sigproc.eng.cam.ac.uk/\~ngk/. Accessed June 2002
45. Kovesi, P.: An implementation of Felsberg's monogenic filters. http://www.csse.uwa.edu.au1pk/research/matlabfns/. Accessed August 2005
46. Kroon, D.J., Schrijver, M.: Hessian based Frangi Vesselness filter. http://www.mathworks.co.uk/. Accessed October 2009
47. Abbott, C.A., Carrington, A.L., Ashe, H., Bath, S., Every, L.C., Griffiths, J., Hann, A.W., Hussein, A., Jackson, N., Johnson, K.E., Ryder, C.H., Torkington, R., Ross, E.R.E.V., Whalley, A.M., Widdows, P., Williamson, S., Boulton, A.J.M.: The North-West diabetes foot care study: Incidence of, and risk factors for, new diabetic foot ulceration in a community-based patient cohort. Diabetic Med. **19**, 377–384 (2002)

Chapter 8
High-Throughput Detection of Linear Features: Selected Applications in Biological Imaging

Luke Domanski, Changming Sun, Ryan Lagerstrom, Dadong Wang, Leanne Bischof, Matthew Payne, and Pascal Vallotton

8.1 Introduction

Psychovisual experiments support the notion that a considerable amount of information is contained in region boundaries such as edges and linear features [1]. Thus, as long as these elements are preserved, it is possible to simplify images drastically with no apparent loss of content. Linear features also underlie the organization of many structures of interest in biology, remote sensing, medicine, and engineering. Examples include rivers and their deltas, road networks, the circulatory system, and textile microstructure (see [2] for a more extensive list and Chapters 6, 7, and 11 in this book).

Given that linear features play a central role in image analysis, a major aim is to develop fast and sensitive implementations for identifying them in digital images. Section 8.2 describes an efficient approach based on nonmaximum intensity suppression. The method systematically probes the image intensity along short segments in the image. By sampling a few transect directions (typically 8) at each image pixel in turn, linear features can be detected very rapidly. Although the algorithm in its original form is unable to detect linear features in close proximity, small modifications can deal with this issue while sacrificing very little in terms of speed. This is described in Sect. 8.2.3.

The output of our linear feature detector sometimes contains artifacts that need treatment. Thus, one typically removes isolated pixels and attempts to restore skeleton continuity when the latter is broken. Our approach to these issues is presented in Sect. 8.2.4.

Linear features can be considered intermediate representations on which to apply further computations to obtain morphological information, such as the number of

P. Vallotton (✉)
CSIRO (Commonwealth Scientific and Industrial Research Organisation), North Ryde, Australia
e-mail: pascal.vallotton@csiro.au

G. Dougherty (ed.), *Medical Image Processing: Techniques and Applications*, Biological and Medical Physics, Biomedical Engineering, DOI 10.1007/978-1-4419-9779-1_8,

branches, their length, or the number of branching points. Branching structures typically possess a hierarchical organization, and it is important to capture and deliver information at each level. This higher order processing is described in Sect. 8.2.5.

There is renewed excitement nowadays around computational platforms such as Graphics Programming Units (GPUs) to speed up algorithms. This is particularly the case for high-throughput applications and for 3D and 4D datasets, where execution time still represents a major bottleneck. Section 8.3 describes the implementation of our linear feature detection algorithm on the GPU.

The technology presented in this contribution has been implemented in a Windows™ package called HCA-Vision. For the user's benefit, Sect. 8.4 outlines the software architecture and the main capabilities of this user-friendly tool (www.hca-vision.com).

Section 8.5 describes several representative applications in biological imaging. The first example demonstrates that even subtle phenotypic changes in the arbors of neurons in culture can be characterized if a sufficient number of neurons are analyzed. In this type of application, automation is central to reaching statistical significance. The importance of maintaining linear feature continuity is also highlighted by this example.

We have also used HCA-Vision with success to characterize the phenotype of astrocytes in response to drug-induced stress. In this application, fluorescently stained bundles of intermediate filaments were traced rather than neurites. Intermediate filaments are typical of the many polymers present within the cells, and we expect this application to grow in importance in the future.

In our final example, we describe how we used our sensitive linear feature detector to separate closely adjoining bacteria under minimal contrast, thus complementing the capability of more mainstream edge detectors.

Section 8.6 concludes this chapter by giving insight into recent developments, as well as by speculating about future directions.

8.2 Methods

There are many different algorithms suitable for linear feature detection. Good reviews are provided in [3,4]. Sun and Vallotton {Sun, 2009 #2}developed a fast linear feature detection method using multiple directional nonmaximum suppression (MDNMS). This approach is particularly intuitive and suitable for a nonspecialist audience. We will summarize the main steps of this algorithm in this section and describe its implementations on the GPU in Sect. 8.3.

8.2.1 Linear Feature Detection by MDNMS

Linear features are thin objects across which the image presents an intensity maximum in the direction of the largest variance, gradient, or surface curvature

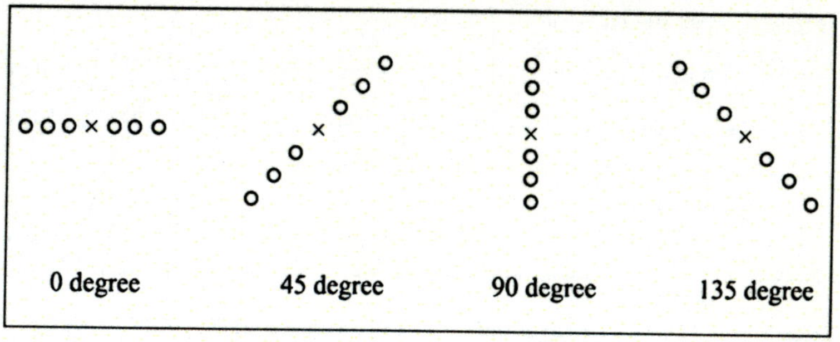

Fig. 8.1 Illustration of linear windows at four different directions. "x" indicates the centre of the linear windows. The other pixels in the window are shown as *small circles*. The length of the linear window is 7 pixels in this illustration

(i.e., perpendicular to the linear feature). This direction may be obtained by the use of computationally expensive Hessian-based detectors or by using matched or steerable filters. Rather than searching for the local direction of linear features, we use 1D nonmaximum suppression (NMS) [5] in multiple directions to identify candidate pixels on a linear feature.

NMS is the process of removing all pixels whose intensity is not the largest within a certain local neighborhood, that is searching for local maxima. The shape of this local neighborhood is usually a square or rectangular window. For our purpose, we choose this local neighborhood as an orientated linear window. A number of approaches are available to obtain the directional local maximum within a one-dimensional window. For example, a local maximum can be obtained by using basic morphological operators to check whether a pixel value in the input image is equal to that in the dilated image [6].

Figure 8.1 shows four examples of linear windows at angles equal to 0, 45, 90, and 135°. Additional directions, such as 22.5, 67.5, 112.5, and 157.5° can also be used. NMS is performed successively for each direction at every pixel. The result is independent of the order of the scanning process. The longer the linear window, the lower the number of candidate pixels detected in an image. The ideal scenario occurs when the linear window crosses the linear feature at right angles as this will produce maximum contrast.

The outputs of the directional NMS are binary images, as opposed to grayscale images produced by most directional filter methods [3]. This eliminates the need for an arbitrary thresholding step. Linear features are detected by the union of MDNMS responses, that is we combine the NMS responses at each orientation.

The algorithms for linear feature detection using NMS can be easily extended for 3D images by using 3D linear windows. We used either 3 or 9 directions in 3D, although additional directions may be necessary, depending on the data.

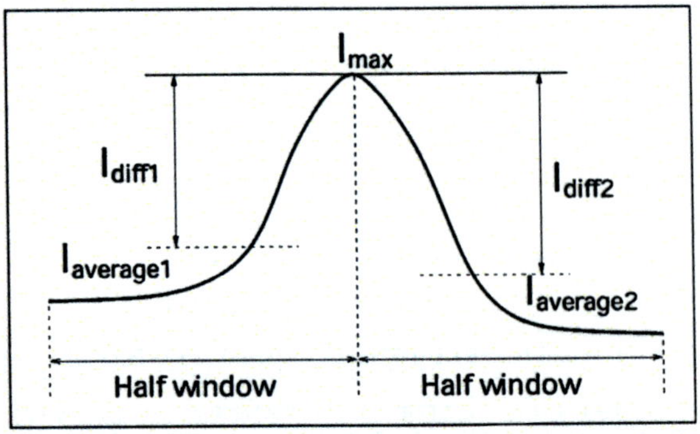

Fig. 8.2 Cross section through a linear feature

8.2.2 *Check Intensities Within 1D Window*

Bona fide linear features are characterized by an approximately symmetric intensity profile across the feature, as opposed to edges which show a step-like profile. This translates into approximately equal intensity values around the central pixel on both sides of the linear window. Figure 8.2 illustrates the parameters that characterise the shape of a profile. I_{max} is the value of a local directional maximum at the centre pixel of the linear window. I_{average1} and I_{average2} are the average intensity values over the two sides of the local maximum within the local window; and I_{diff1} and I_{diff2} are the differences between the maximum value and these two average values. For a genuine linear feature, both I_{diff1} and I_{diff2} should be large. The parameter I_{diff1} and I_{diff2} can be used to control the sensitivity of the algorithm. Lowering the values of I_{diff1} and I_{diff2} increases the number of linear features detected.

8.2.3 *Finding Features Next to Each Other*

Some images present linear features in close proximity to one another. The NMS process will detect only one feature in each linear window. Figure 8.3 shows an example where four local maxima lie within the linear window. We can extend the NMS process so that multiple local maxima are detected for any particular linear window. Once a local maximum is found at the center of a linear window, a second local maximum search is performed within the original window using a smaller window size. The use of a smaller window size allows additional local maxima to be found within the original window. The newly found local maxima are ordered based on their maximum intensity as shown in Fig. 8.3. To prevent detecting false peaks due to noise, we also check that the intensity does not deviate significantly from I_{max} and that it conforms to the symmetry condition defined in Sect. 8.2.2. For example, in Fig. 8.3, $I_{\text{max}3}$ and $I_{\text{max}4}$ are two local maxima, but they are not strong enough to be retained.

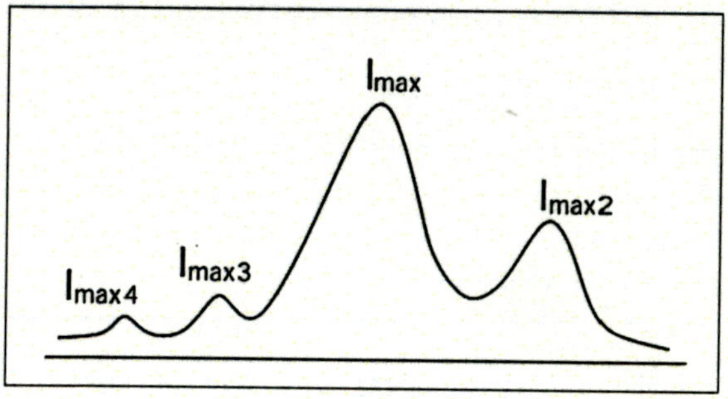

Fig. 8.3 Multiple local maxima within a linear window

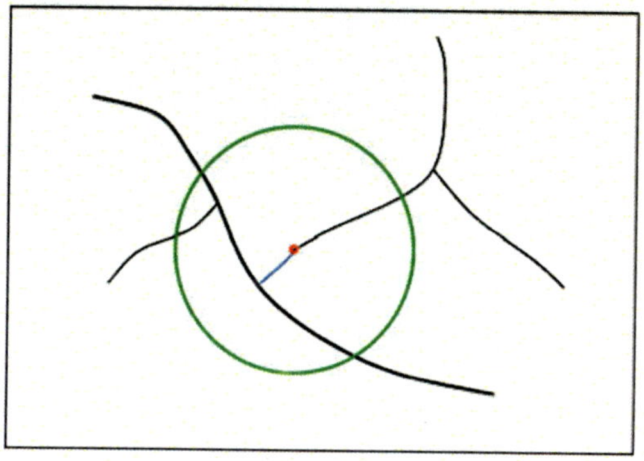

Fig. 8.4 Endpoint (*red*) and its neighborhood (*green*) for gap linking. The shortest path is shown in blue

8.2.4 Gap Linking for Linear Features

The union of the multiple responses of NMS at different directions generates many small objects, which may not belong to genuine linear features. We fit an ellipse to each object (set of connected pixels) and remove them if the major axis is too small. Alternatively, a simple pixel count can be used.

The continuity of the linear features may be broken due to noise or weakness of linear features. This results in small gaps in the combined responses of NMS. We restore continuity by joining linear feature endpoints to neighboring linear features through a shortest path. Here, a shortest path should be understood as a connection, which displays a high average intensity along its length. A standard technique called dynamic programming limits the combinatorial explosion that would otherwise be associated with the exploration of all candidate paths [7]. The computational overhead also remains small because the operation is only performed on small gaps – typically below 20 pixels in length. Figure 8.4 illustrates the process of gap linking from an endpoint to the skeleton in its neighborhood. Note that the domain shown in green in Fig. 8.4 is first transformed into polar coordinates prior to calculating the shortest path.

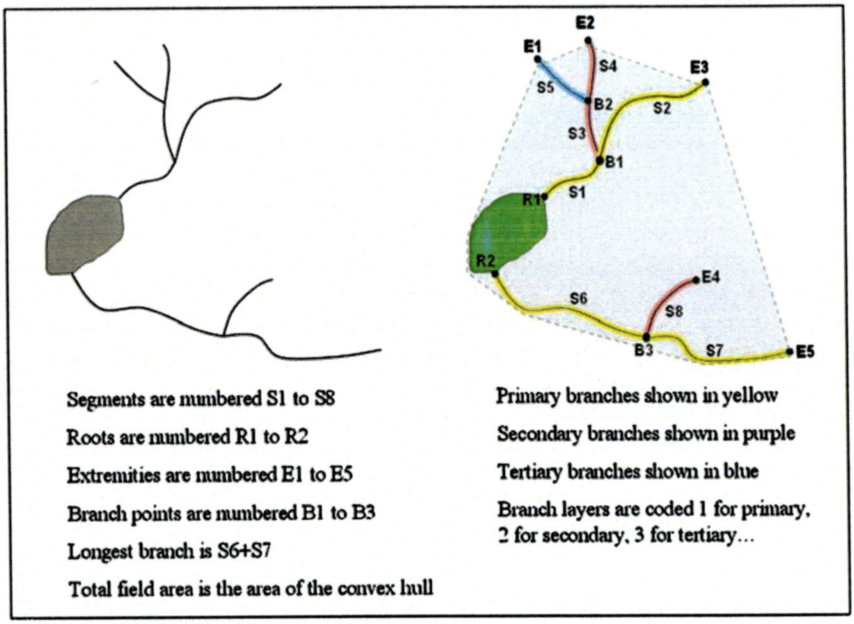

Fig. 8.5 Illustration of the measurements derived by quantifying linear feature detection results on neuronal cells

8.2.5 Quantifying Branching Structures

Linear features are produced as intermediate representations towards image quantification and image understanding. The conversion of a full image into a set of linear feature entails a considerable compression of the original information. In this section, we describe how to reduce the feature size further while preserving as much information as possible. To this end, the linear features are processed to generate measures of length, branching, and complexity (see Fig. 8.5). This framework is quite general and several applications are described in Sect. 8.5.

A) Feature representation
After preprocessing, the linear features are paths of width equal to only one pixel, often connected in complex ways. The sensitivity of the feature detection process typically leads to false positive detection events. These inaccuracies manifest themselves as small barbs in the skeleton, which can be pruned by a process, where small lateral branches below a chosen length are removed. The skeleton is then divided into unique segments, defined as sections of linear feature between two intersections, or branching points. This division process first requires identifying the branching points as having more than two 4-connected neighbors. Branching points are then removed from the skeleton. In doing so, the skeleton is divided into segments which remain 4-connected and each segment is given a unique label.

A graph of neighborhood relationships for segments then has to be built. We first morphologically dilate [6] uniquely labeled intersection points with a 3×3 structuring element, so that they overlap with the extremities of segments. Segments which overlap a common intersection point are considered neighbors. This information is initially contained in a bivariate histogram of segments versus intersection points. A linked list is then created by scanning across each row of the histogram and locating nonzero entries in the histogram indicating neighborhood relationships among segments.

When the image corresponds to a structure possessing an organizing centre from which branches are protruding (such as the mitotic organizing centre, the trunk of a tree, or the cell body of a neuron), we identify segments that are in contact with that organizing centre as "root" segments. To identify them, we first thicken the labelled centre, so that it overlaps root segments (a thickening is a dilation that preserves an object's label [6]). Again, we use a bivariate histogram to store the overlap information. Nonzero entries correspond to root segments for a particular cell body.

B) Tree growing using the watershed algorithm

At this stage, we must associate all segments with a particular tree. A tree is a connected network extending from a single root segment. We use the watershed algorithm to derive the association. Typically, the watershed is performed on an image called the segmentation function, which highlights object boundaries. A set of unique seeds are grown on the segmentation function using a priority queue. Seeds are placed in the queue and neighboring pixels are added with priority given to those with the lowest value in the segmentation function. Pixels are repeatedly taken from the top of the queue and added to the object defined by the pixel's neighboring seed.

We use the watershed methodology to grow all trees from their root segment. The framework for the watershed in our case is different to that which is used for 2D images: we are dealing with graph nodes instead of pixels. The nodes are the individual segments and our seeds are the root segments as found in the previous section. Root segments are initially put in the priority queue and neighboring segments are added with priority given to segments with the highest average brightness. The average brightness is calculated over the pixels that form the segment. Brightness was chosen as priority feature because it is generally found to be preserved along branches. Other criteria for the prioritization could be used such as the relative orientation of the segments. Segments are repeatedly taken from the top of the queue and associated with their neighboring neurite tree until all segments have been removed from the queue.

C) Measures on segments and trees

Various measurements can be accumulated for each branch during the tree growing process. These measurements can in turn be aggregated on a per-tree or per-organizing centre basis. It is also common to report measurements on a per-image basis. There are two groups of measurements collected during the watershed process: those relating to length, width or brightness and those relating to complexity. Fig. 8.5 illustrates these measurements.

Length and width measurements seem particularly important in applications (see Sect. 8.5 for some examples). Before the tree growing process is initiated, the length of each segment is estimated [8]. The average width of each segment is also computed using the method proposed by Lagerstrom et al. [9]. The average brightness of the segment is computed not only to guide the watershed process, but also as a reportable measure in itself. As each segment is removed from the queue, we accumulate the length back to the centre for the segment, the longest path back to the centre for the tree and the total length of the tree. In a similar fashion, the average width of the tree, the total area of the tree, the average brightness and integrated intensity of the tree are also accumulated. Once the trees have been grown, the total field area is calculated, defined by the area of the convex hull of all trees associated with a single organizing centre.

A variety of complexity measures for capturing additional morphological measures are also collected via the tree growing process. Often trees display behaviour where a dominant or primary branch extends from the centre, with secondary branches projecting from the primary branches, and recursively. On a per-line basis, we refer to this as branching layer. Root segments are given a primary branching layer coded as "1." As segments are removed from the queue, they inherit their parent's branching layer if they represent a child segment with the highest average brightness. The remaining child segments inherit an incremented branching layer. The average branching layer per tree, the number of branching points per tree and the number of extreme segments (i.e., those with no children) are accumulated as the tree is grown.

8.3 Linear Feature Detection on GPUs

While the algorithm presented in Sect. 8.2 is generally considered fast, execution time can become an issue in the case of large images or those containing complex linear structures. In this context, complexity refers to both density of linear structures and their branching rate, as well as the variation of intensity along linear features. In high-throughput biological experiments, where thousands of images may need to be processed in batch during a single experiment, the overall increase in processing time can be significant, thus motivating attempts to improve algorithm performance further.

In this section, we will look at using many-core processing units and parallel programming techniques to help accelerate parts of the linear feature detection algorithm described in Sect. 8.2. This problem will serve as an interesting example of how these methods can be used to accelerate image processing problems in general. We will utilize Graphics Processing Units (GPUs) as the many-core processors in our discussions and tests. These commodity processing chips are now a widely available and popular parallel processing platform, with most personal and office computers containing one or more of them.

8.3.1 Overview of GPUs and Execution Models

Originally designed to accelerate the rendering of 3D computer graphics, GPUs are now used widely as architecture for executing general purpose parallel programs [10]. Modern GPUs consist of hundreds of light-weight processor cores, capable of executing thousands of parallel threads concurrently. GPUs are coupled to dedicated off-chip RAM through a high-bandwidth memory interface. Data is transferred between this dedicated GPU RAM and a host processor's memory via an expansion card bus (usually PCI-Express). GPUs also provide a number of on-chip storage spaces including register files, unmanaged shared memory, and various memory caches. Accessing these on-chip storage spaces can be orders of magnitude faster than accessing GPU RAM which, while being high-bandwidth, can incur significant access latencies.

On the NVIDIA GPUs [4] used in our tests, the processors are grouped into a number of *streaming multi-processors* (SMs), which can concurrently execute a large number of assigned threads by switching execution between different groups of these threads. On-chip storage is arranged such that each SM has its own private register file and shared memory space, which cannot be accessed by threads executing on other SMs.

Threads are logically grouped into n-dimensional *blocks* whose sizes are customisable. A regular *grid* of common sized blocks is used to parameterize threads over a problem domain. Each thread is assigned unique n-dimensional grid and block level IDs to distinguish it from other threads. A block is assigned to a single SM for its lifetime, and its threads can synchronize their execution and share data via SM shared memory. Each thread in the grid of blocks executes the same program, which is defined by a parallel *kernel* function. A grid of threads only executes a single kernel at a time, and on-chip storage does not remain persistent between kernel launches.

For example, the process of executing a simple image operation on the GPU that subtracts image A from B and places the result in image C can be performed as follows:

1. Transfer image A and B from host RAM to GPU RAM.
2. Assign a single thread to each output pixel by constructing a grid of 2D thread blocks to cover the image domain.
3. Launch a subtraction kernel where each thread reads corresponding pixel values from image A and B in GPU RAM and writes the subtraction result to image C in GPU RAM.
4. Transfer image C from GPU RAM to host RAM.

Data transfer between host and GPU can be a performance bottleneck, so it should be avoided where possible. For example, when performing a number of dependent parallel image operations in succession on the GPU, it may not be necessary to transfer the result of each operation back to the host.

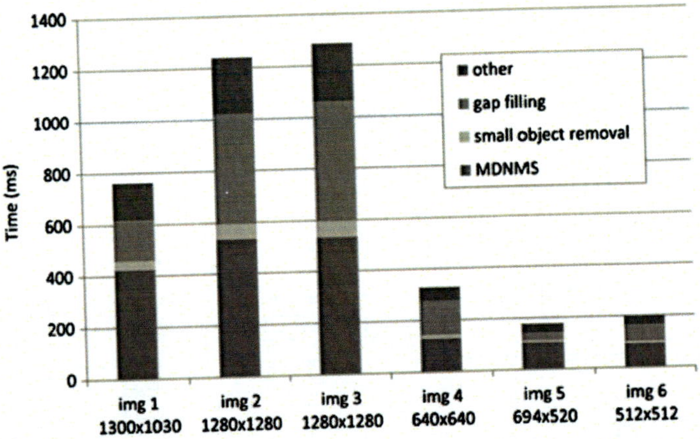

Fig. 8.6 Execution times for different stages of linear feature detection for images shown in Fig. 8.7. MDNMS, small object removal, and gap filling steps are described in Sects. 8.2.1–8.2.4. Aggregated time taken by short utility operations performed between steps, such as labelling and skeletonization, is represented as "other"

8.3.2 *Linear Feature Detection Performance Analysis*

Before attempting to parallelize the algorithm on a GPU, one should analyze performance issues and determine how different changes might take effect. For example, the MDNMS algorithm from Sect. 8.2 is a classic neighborhood filter that computes the value for each pixel in the output image by analyzing only pixels within a small neighborhood around the pixel. Although the number of symmetry checks performed (Sect. 8.2.2) may vary with image content, its performance is primarily determined by the size of the image, as well as by the size and number of linear windows used for filtering. Accelerating this portion of the algorithm should provide performance improvement irrespective of the image complexity. However, the number of false objects and feature mask endpoints in the MDNMS output can increase significantly with input image complexity. This places a higher workload on the steps that remove small objects and bridge gaps in the resulting feature masks. The performance of these steps is, therefore, affected more strongly by image complexity than by size.

Figure 8.6 shows the breakdown of linear feature detection processing time for a number of images with varying size and linear structure complexity (Fig. 8.7). In general, we see that MDNMS takes the largest portion of overall execution time for each image. Gap filling also consumes a large amount time for complex images (Fig. 8.7, img 2 and img 3) due to an increase in the number of feature mask endpoints produced by the MDNMS step, and the need to perform costly shortest path calculations for each endpoint (Sect. 8.2.4). Although small object removal performance also appears to be affected by image complexity, as hypothesized above, its execution time is relatively low compared to the other two steps. The same remark applies to the utility functions.

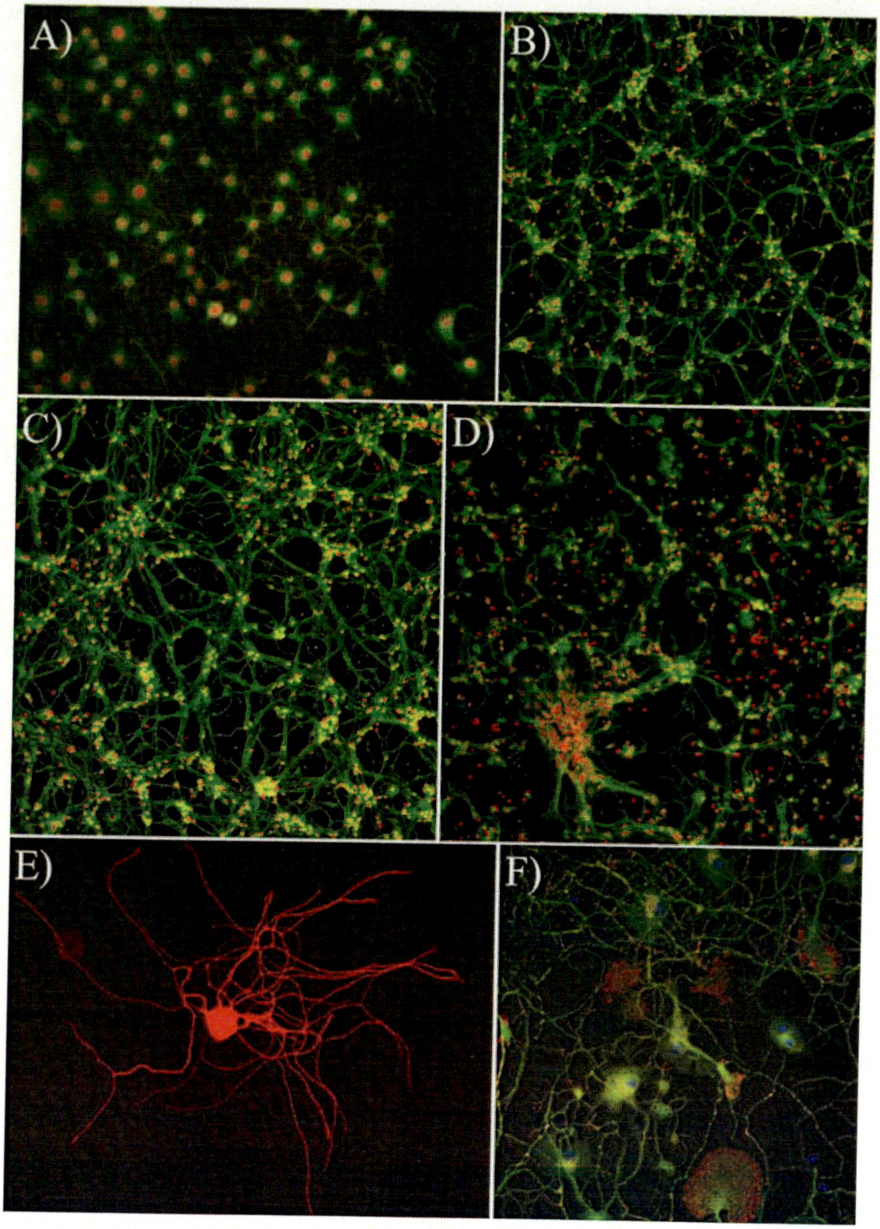

Fig. 8.7 Neurite images used in performance tests. (**a**) img 1: 1,300 × 1,030 pixels. (**b**) img 2: 1,280 × 1,280 pixels. (**c**) img 3: 1,280 × 1,280 pixels. (**d**) img 4: 640 × 640 pixels. (**e**) img 5: 694 × 520 pixels. (**f**) img 6: 512 × 512 pixels

Intuitively, one would expect that efforts would be best directed towards improving the performance of those steps, which consume the largest percentage of time, assuming they can be accelerated. Therefore, we will focus our attention on the implementation of the initial linear feature detection on the GPU.

8.3.3 Parallel MDNMS on GPUs

The MDNMS algorithm with extensions to support symmetry checks and dual local maxima detection (Sects. 8.2.2 and 8.2.3) consists of four steps for each window orientation:

1. Detecting primary maxima by NMS with given window size.
2. Detecting candidate secondary maxima by NMS with a smaller window size.
3. Symmetry check on primary maxima.
4. Secondary maxima search and symmetry check in presence of positive primary detection.

NMS is achieved most simply using a brute force neighborhood filter. In this case, each pixel is compared directly to every pixel within its local linear window to determine whether it is the maximum in the window. Although it is possible to reuse the max operator observations across nearby pixels to improve performance in a serial context [2], this would not work effectively in a parallel implementation where nearby pixels can be processed concurrently and where the order of operations is not predictable. The symmetry checks are performed using similar brute force filters, but carry out different operations on the values within a pixel's linear window.

In each of these brute force neighborhood filters, the result for a single pixel is not dependent on the output of other pixels, and can be performed in parallel on the GPU using one thread to calculate each output pixel. Examples of the parallel filter kernels are shown in Listing 8.1 through Listing 8.3. Note that the NMS kernel is executed with two different windows sizes to produce the primary and secondary

```
nms (){

i = get pixel index for this thread
val = in[i]
is_max = true

for each pixel j (≠ i) in linear window do
        if val ≤ in[j] then
            is_max = false
end

if is_max then maxima[i] = val
}
```

Listing 8.1 Parallel NMS kernel

```
symmetry_check (){

i = get pixel index for this thread

if prim[i] is a maxima then
        for each pixel k (≠ i) in first half of linear window do
                sum1 = sum1+ in[k]
        end

        for each pixel j (≠ i) in second half of linear window do
                sum2 = sum2+ in[j]
        end

        avg1 = sum1/((window_size-1)/2)
        avg2 = sum2/((window_size-1)/2)

        if (in[i],avg1,avg2) doesn't pass symmetry check then
                prim[i] = 0
}
```

Listing 8.2 Parallel symmetry check kernel

```
2nd_maxima_search (){

i = get pixel index for this thread
max = first pixel in linear window around i
out[i] = prim[i]

if prim[i] is a maxima then
        for each pixel k (≠ i) in linear window around i do
                if sec[k] > sec[max] then
                        max = k
        end

        for each pixel j (≠ max) in first half of linear window around max do
                sum1 = sum1+ in[j]
        end

        for each pixel j (≠ max) in second half of linear window around max do
                sum2 = sum2+ in[j]
        end

        avg1 = sum1/((window_size-1)/2)
        avg2 = sum2/((window_size-1)/2)

        if (sec[max],avg1,avg2) passes symmetry check then
                out[max] = sec[max]
}
```

Listing 8.3 Parallel secondary maxima search and symmetry check maxima images "prim" and "sec." Each of these parallel kernels can be executed using a simple 2D grid and block configuration that assigns one thread to each output pixel.

8.3.4 Combining Steps for Efficiency

In Sect. 8.3.1, we discussed the high latency of GPU RAM accesses (where the input image resides) compared to on-chip data access to registers, shared memory or cache memories. Because of these latencies, it is important to minimize transfers to GPU RAM where practical. The steps outlined in Sect. 8.3.3 can easily be performed using separate parallel image filters by executing separate GPU kernel functions for each step. However, a solution with fewer GPU RAM accesses can be developed if we consider the following properties of the filters:

1. *NMS*: A pixel can be compared with the pixels in its linear window in *any order* to determine its suppression status.
2. *NMS*: Testing a pixel's suppression status in a given linear window is a *subtask* of doing the same for a larger linear window at the same orientation.
3. *Symmetry check*: Performing a symmetry check on a maxima pixel requires the *same set of values* as its primary suppression test

With these properties identified, we can combine the two NMS steps and the primary maxima symmetry check into a single kernel shown in Listing 8.4. The code in this kernel visits the first and second halves of a pixel's linear window separately, allowing it to calculate the average value in each half-window while simultaneously checking whether the pixel is a maximum. It also assesses the secondary NMS result for a pixel at the same time as the primary NMS. This allows everything except the secondary maxima symmetry check to be calculated after reading a pixel's primary linear window values only once from GPU RAM. In contrast, using a separate kernel for each filter requires these values to be read multiple times from RAM, since on-chip storage is not persistent between kernel launches. It would also require the values of the primary NMS kernel to be communicated to the primary symmetry check kernel via GPU RAM. It should be noted that this combination of steps might also help speed up the CPU algorithm by reducing overall workload and memory accesses.

The secondary maxima search and symmetry check around a primary maximum cannot be performed efficiently in the same kernel as the other filters. This is because the thread responsible for a pixel requires both the secondary NMS and half-window average values calculated by other threads to avoid recalculating them. It is non-trivial to share these values using shared memory, as many threads with data interdependency relationships will belong to different processing blocks. These threads will not be able to access each other's shared memory space or synchronize their execution. All NMS and half-window averages are, therefore, written to global memory by the kernel in Listing 8.4, and a different kernel is utilized to facilitate the exchange of values and perform the necessary processing.

```
nms_dual_peak (){

i = get pixel index for this thread
sum1 = sum2 = val = in[i]
w_max = sw_max = true

for each pixel k (≠ i) in first half of small linear window do
        if val ≤ in[k] then
                w_max = sw_max = false
        sum1 = sum1+ in[k]
end

for each pixel k (≠ i) in first half of large linear window but not small window do
        if val ≤ in[k] then
                w_max = false
        sum1 = sum1+ in[k]
end

for each pixel j (≠ i) in second half of small linear window do
        if val ≤ in[j] then
                w_max = sw_max = false
        sum2 = sum2+ in[j]
end

for each pixel j (≠ i) in second half of large linear window but not small window do
        if val ≤ in[j] then
                w_max = false
        sum2 = sum2+ in[j]
end

if w_max and (val,sum1,sum2) passes symmetry check then
        prim[i] = val

if sw_max then
        sec[i] = val
        avg1[i] = sum1/((large_window_size-1)/2)
        avg2[i] = sum2/((large_window_size-1)/2)
}
```

Listing 8.4 Parallel kernel calculating primary and secondary NMS, and primary symmetry check

Table 8.1 Speedup of linear feature detection algorithm after GPU acceleration. Results indicate how many times faster the NMS and gap filling ran on the GPU in isolation, as well as how many times faster the process ran overall

	Speedup on GPU (times ×)		
Img	NMS	Gap filling	Overall
1	13.3	4.8	3.0
2	13.7	8.3	3.3
3	13.6	8.6	3.3
4	9.0	8.6	3.3
5	7.9	3.5	2.7
6	8.0	6.9	3.0

8.3.5 Results for GPU Linear Feature Detection

We have also sped up the gap closing steps on the GPU, which involved parallelizing polar image generation and shortest path calculations. Various lower level code and memory optimisations were applied throughout. For example, loop unrolling has been applied to the NMS kernels [11], while texture fetches, a special kind of cached [12, 13]. GPU RAM access operation used in computer graphics, have been used to accelerate polar image generation. Discussion of these topics is beyond the scope of an introductory text. The reader may want to check [14] if his/her curiosity has been aroused.

Tests were performed on a Geforce GTX 260 GPU (NVIDA Corp., Santa Clara, CA, USA). The host computer consisted of a Xeon E5520 2.3GHz CPU (Intel Corp., Santa Clara, CA, USA) with 3GB of RAM running 32-bit Windows XP (Microsoft Corp., Redmond, WA, USA). Table 8.1 shows the performance of the linear feature detection algorithm after GPU accelerations for the images displayed in Fig. 8.7. The speedup for NMS is quite high, and is more significant for larger images (Fig. 8.7, img 1–3). The speedup for gap filling is also significant in the case of complex images (img 2–4, 6), but reduces for less complex images. These types of performance inconsistencies are common in parallel programs, where higher workloads help offset the costs of setup and communication that tend to take a constant amount of time or do not parallelize well.

Note that the overall speedup of the algorithm is much lower than the speedup of the accelerated steps in isolation due to the portion of the process consumed by the steps that were not accelerated. Since these steps consumed around 21–24% of the original CPU algorithm's execution time for the test images, the maximum overall speedup we could hope for by accelerating NMS and gap filling is less than 4.7 times. The described GPU algorithms have achieved up to 70% of this theoretical speedup. Greater overall speedup will come from accelerating other parts of the algorithm.

8.4 HCA-Vision, A Software Platform for Linear Feature Detection and Analysis

The algorithms detailed in Sects. 8.2 and 8.3 have been integrated in standalone software that rapidly and reproducibly quantifies linear features and their organization. It can be used both for linear feature detection and for more dedicated morphological analyses, such as neurite analysis. HCA-Vision was originally aimed for cell phenotyping and thus also provides functions for cell and nucleus detection and counting, cell scoring and sub-cellular analysis. It supports batch processing with a built-in database to store results, a batch result viewer and an ad hoc query builder for users to retrieve events of interest (e.g., identify all cells with neurite field above 50 μm^2 and projecting in excess of 5 root neurites). Results from HCA-Vision have been shown to be more objective, reliable and reproducible than those from manual or semi-automated approaches [15].

8.4.1 Architecture and Implementation

HCA-Vision has been designed with three layers: the image processing layer written in C, a C++ wrapper layer, and the data and image presentation layer, written in C#. It also includes an optional web service for users to remotely view their batch processing results.

The GUI provides an assistant to guide the user in the process of choosing optimal parameters for the automated segmentation of cells and sub-cellular structures. It delivers quantitative measures and statistics that are biologically relevant, together with result images.

8.4.2 HCA-Vision Features

HCA-Vision enables a range of various activities including the detection of neurons, neurites, astrocytes, and other glial cells, even in the presence of varying background brightness, variable neurite staining, high cell densities, and high levels of neurite interconnections. It delivers quantitative measures and statistics that are biologically relevant, including measurements of neurite features at various levels of branching, neuron "webbiness," and astrocyte "starriness" and roundness (see Table 8.3 for definitions).

To accommodate a wide range of applications, a step-by-step wizard guides the user through a process that fine-tunes individual parameter required for obtaining best results. At each step of the process, the user is given a semantic description of each parameter and controls their effect through user-friendly interfaces. As the processing has been optimized for speed, intermediate results are immediately presented on screen as each parameter is adjusted (Fig. 8.8).

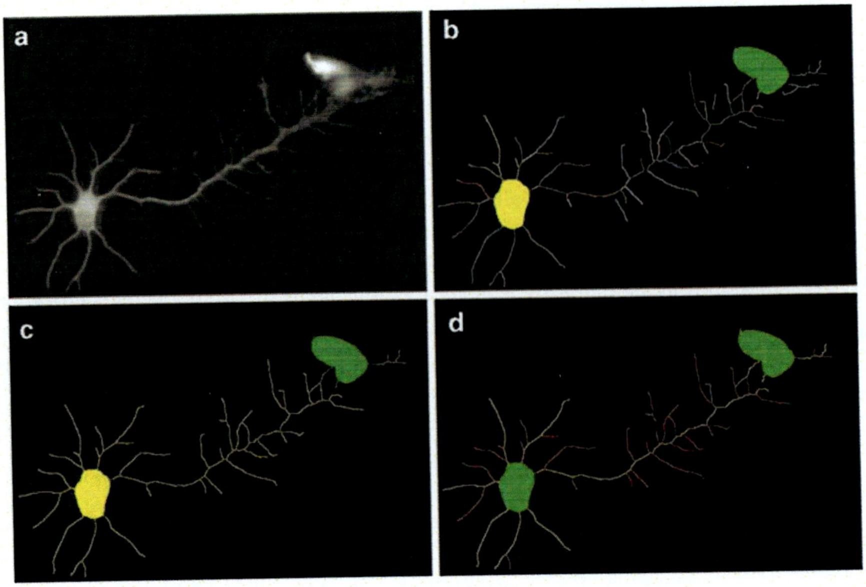

Fig. 8.8 Results produced by HCA-Vision. (**a**) Original image. (**b**) Detected neurite segments. (**c**) Assignment of neurites to neuron bodies. (**d**) Labelled neurite branching structure

Once parameters are tuned, they can be saved into a profile. The user can load the saved profile to process individual images or batch process all images generated in an experiment. Summaries of batch processing results can be produced using the query facilities provided.

8.4.3 Linear Feature Detection and Analysis Results

HCA-vision detects linear features and reports their properties such as length, width, area and complexity. For example, with the neurite analysis module, neurites can be detected and their branching behaviour can be analysed, including the primary neurites in contact with the cell body and the number of layers of branching from these primary neurites into secondary, tertiary and quaternary branches (see Fig. 8.9).

Users can query and view the batch processing results using the batch result viewer (Fig. 8.10), including both the result images and measured features.

When HCA-Vision is used with microplates, the batch processing also generates a plate summary with normalized features for each well. The features extracted from the images captured from the same well are averaged to produce the well-based normalized statistics. A sample plate summary is shown in Table 8.2.

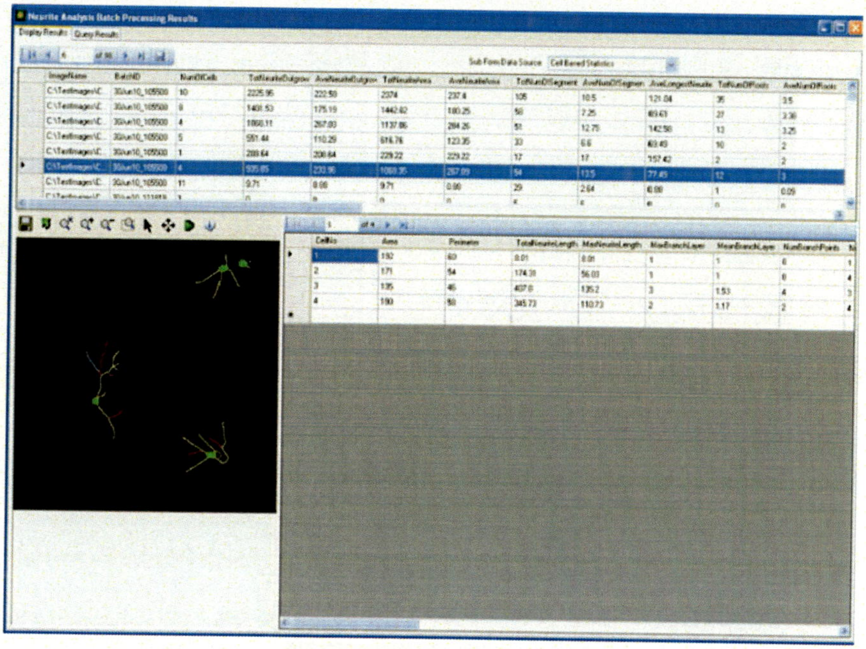

Fig. 8.9 Batch processing result viewer

8.5 Selected Applications

8.5.1 Neurite Tracing for Drug Discovery and Functional Genomics

High Content Screening or Analysis (HCS, respectively HCA) has virtually become an obligatory step of the Drug Development process. Cells in small transparent wells in a 96-, 384-, or 1,536-microplate format are exposed in a fully automated manner to thousands of different candidate compounds (see Table 8.2). They are then imaged and analysed using computer vision algorithms for evidence of drug action. In the case of neuronal cells, such evidence includes the growth of neuronal projections (neurites), but it can also include receptor trafficking, apoptosis, motility, as well as many other assays [16]. Measuring neurite dynamics is a particularly direct and informative approach but it is also challenging because neurites tend to be very thin, long, and may present extensive branching behaviour.

Some drugs trigger spectacular effects on neurites (e.g., nocodazole destabilizes microtubules thus inducing neurite retraction). More often however, dendritic arbours are altered in subtle ways only. Pharma is generally interested in changes to the length, shape and complexity of neurites. In fact, most of the general image features described in Sect. 8.2.5 are directly relevant for this particular application.

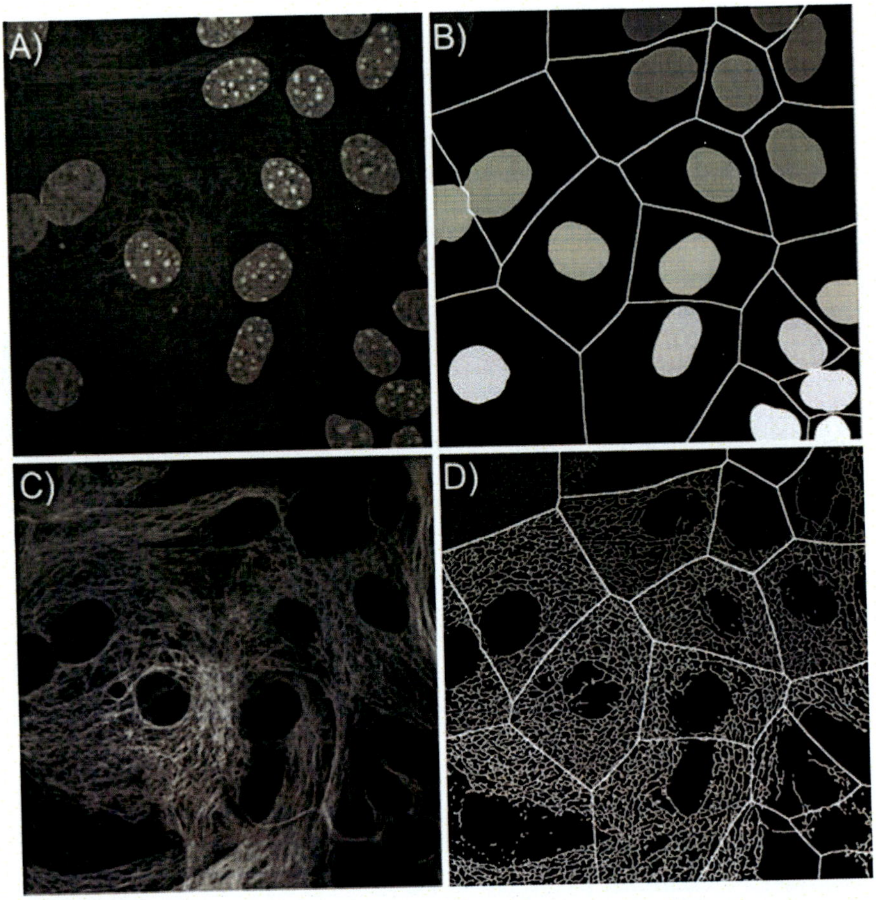

Fig. 8.10 (**a**) Original image showing astrocyte nuclei. (**b**) Nuclei identified by the software are *gray* coded, with surrogate cellular region boundaries overlaid in *white*. (**c**) Original image showing staining of GFAP fibres of the cytoskeleton. (**d**) Linear features identified by the software and gray coded as per nuclei, with surrogate cellular region boundaries overlaid in *white*

Generally, one does not know in advance how the phenotype will be altered. Therefore, it is desirable to apply as wide a spectrum of quantitative features as possible. Neuronal phenotype may also be altered by mutations, or by changes in the protein expression level elicited, for example, by small inhibitory RNAs. In collaboration with the Group of S.S. Tan and J.M. Gunnersen at the Howard Florey institute, we have been particularly interested in uncovering the role of the Seizure-related protein type 6 (Sez-6) [15]. From mouse behavioural studies, Sez-6 had already been implicated in cognitive processes but it was not clear yet whether the cell morphology was affected. In general, the biological variability across cells precludes drawing definitive conclusions from observing by eye a limited number of cells. Indeed, an individual knockout cell (lacking Sez-6) may appear more similar

Table 8.2 Plate summary showing well-based normalized features

Well number	A1	A2	–	–	–	H11	H12
Number of cells	625	425	–	–	–	899	648
Total neurite outgrowth	32,124	19,801	–	–	–	30,883	11,887
Average neurite outgrowth	51.4	46.59	–	–	–	34.35	18.34
Total neurite area	36,466	23.348	–	–	–	36,432	14,025
Average neurite area	58.35	54.94	–	–	–	40.52	21.64
Total number of segments	3,826	2,110	–	–	–	4,820	2,119
Average number of segments	6.12	4.96	–	–	–	5.36	3.27
Average longest neurite length	25.68	27.58	–	–	–	18.97	12.97
Total number of roots	1,213	583	–	–	–	1,479	571
Average number of roots	1.94	1.37	–	–	–	1.65	0.88
Total number of extreme neurites	1,353	747	–	–	–	1,492	615
Average number of extreme neurites	2.16	1.76	–	–	–	1.66	0.95
Total number of branching points	455	251	–	–	–	438	101
Average number of branching points	0.73	0.59	–	–	–	0.49	0.16
Average branching layers	1.28	1.13	–	–	–	1.17	0.71

to a wild type cell (possessing Sez-6) than to another knockout cell (Fig. 8.9). It is only when large numbers of cells are systematically analysed that statistically significant differences can be uncovered. In conducting these comparisons, it is extremely important to ensure that the analysis is performed identically on both the knockout and the wild-type images.

Our results demonstrated clearly that while the neurite field area was not affected, the mutation both increased branching and diminished the mean branch length. The full biological significance of these findings is not yet appreciated but these experiments clearly indicate that the geometry of neurite arbours is in large part under genetic control.

8.5.2 *Using Linear Features to Quantify Astrocyte Morphology*

In this example, we show how linear feature detection can be used to characterise morphological changes in the cytoskeleton of astrocytes, as induced by kinase inhibitors. This was part of a larger study into the role played by astrocytic glutamate transporters in maintaining brain homeostasis [17].

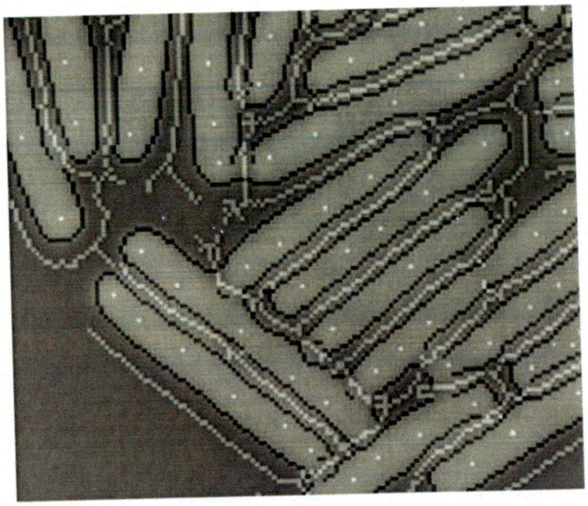

Fig. 8.11 The sensitivity of our linear feature proved helpful in this bacterial segmentation problem. By themselves, the detected edges (in *gray*) would not be sufficient to segment cells successfully. Together with the detected edges, our linear features (in *white*) form a double barrier system that enables accurate segmentation

It is often important to make measurements on a per-cell basis rather than on a per-image basis. To achieve this, one needs some way to identify the extent of each cell. This is done either directly by acquiring an additional image of a labelled cytoplasmic protein, or indirectly by generating a surrogate for the cell extent. The surrogate commonly used in cellular screening requires the capture of an additional image of labelled nuclei, the segmentation of those nuclei and the placing of a doughnut or ring around each nucleus. If cells are isolated, the surrogate cell region will appear roughly elliptical and the approximation to the actual cell shape tends to be crude. However, if cells are closely packed (as they often are in screening assays), the surrogate cell regions from neighboring nuclei deform to the midpoint between the two nuclei. This gives rise to regions which are close to the actual cell shapes (Fig. 8.11).

Within these surrogate cell regions, we quantify the features of the linear structures forming the astrocyte cytoskeleton (see Table 8.3). These measures have been used by our collaborators, O'Shea et al. of the Howard Florey Institute, to quantify the changes induced by the Rho-kinase inhibitor HA1077 in primary cultures of mouse astrocytes.

The astrocyte cytoskeleton was labelled using immunocytochemical staining for the astrocytic intermediate filament protein GFAP. Nuclei were labelled using Hoechst 33342. Figure 8.11 shows a sample nucleus and cytoskeleton image, along with the detected nuclei, the calculated surrogate cell extent and the detected lines in the cytoskeleton. Treatment with HA1077 (100 μM) produced rapid (<1h) and persistent changes in astrocytic morphology. The lineAngleVar feature (variance of the orientation angles of the "lines" within the cells) was significantly reduced by HA1077 (to $81 \pm 4\%$ of control, $p < 0.05$). A low lineAngleVar means that the

Table 8.3 Features developed specifically to characterize linear structures in the astrocyte cytoskeleton

lineNo	Number of lines detected within cell
lineMean	Mean brightness of lines within cell
lineLength	Length of lines within cell
lineDensity	Density of lines within cell
lineAngle	Mean orientation angle of lines Within cell
lineAngleVar	Variance of orientation angle of lines within cell
lineWidthMean	Mean width of lines within cell
lineWidthMedian	Median width of lines within cell
lineStar	"Starriness" measure based on how much linear structure is removed by an opening of a specific radius
lineWeb	"Webbiness" measure based on how much the gaps between lines are filled in by a closing of specific radius

linear structures within a cell are generally aligned in a particular direction, whereas a high value indicates that the linear structures are randomly oriented. In addition, the lineDensity measure (the density of these "lines" within cells) also decreased (to $44 \pm 5\%$ of control, $p < 0.05$), demonstrating a decrease in the cellular area labelled by GFAP.

8.5.3 Separating Adjacent Bacteria Under Phase Contrast Microscopy

In bacterial cultures, cells tend to come in very close proximity to each other, such that the contrast between individual cells is sometimes minute (see Fig. 8.11). This makes bacterial counting and measurement of size and shape difficult. We have found that our linear feature detector was capable of detecting the very weak linear features that separate adjacent bacteria (see Fig. 8.11). By combining this tool with a standard Canny edge detector, a system was put together that permitted counting and segmenting bacteria with over 97% reliability [18]. We expect this new capability to be generally useful for microbial studies and we are planning to extend this work to allow tracking of bacterial cells with a comparable reliability. This work is motivated by the need to understand how bacterial films are able to spread quickly over tissue surfaces and thus cause infections.

8.6 Perspectives and Conclusions

Our approach to linear feature is both conceptually simple and very fast. It is fairly general – independent of any model assumption about linear features. This also means that the approach is not recommended for very noisy images. In the rare instances where noise was an issue, we found that it was often possible to use a preliminary processing step to increase the contrast of linear features, for example, using the technique of anisotropic filtering [19]. Another promising tool to preprocess linear feature images in this manner is afforded by so-called flexible path openings, which aim to identify paths along which the intensity remains high on average [20].

There are also instances where the contrast mechanism of the optical instrument makes the analysis difficult. This is the case with Differential Interference Contrast (DIC) microscopy. For such images, the Hilbert transform can be used to create images that are suitable for analysis.

In the case of very noisy images, the potential of more global methods, such as "shortest paths" [7], or even "linear paths" [9], has been explored. While useful, these more complex approaches may also represent warnings that the experimenter should go back to the bench to produce better data. With the availability of EMMCD camera and the availability of bright and photostable dyes, such as the Alexa^{TM} dyes, this is often the best course of action.

This is an exciting time for image analysis, with a growing number of applications that can be automated. The samples presented in this chapter only touch the surface, as illustrated by the content of other exciting chapters in this book.

Acknowledgements The authors would like to thank the following people for allowing us to use their images as sample images: Marjo Götte, Novartis Institutes for BioMedical Research; Dr. Myles Fennell, Wyeth Research, Princeton, NJ, USA; Dr. Xiaokui Zhang, Helicon Therapeutics, Inc., USA; Prof. Pat Doherty, Kings College, London, UK; Dr. Jenny Gunnersen, Prof. Seong-Seng Tan, and Dr. Ross O'Shea Howard Florey Institute, Melbourne; Ass. Prof. Cynthia Whitchurch, UTS, Sydney.

References

1. Palmer, S.E.: Vision science. MIT, MA (1999)
2. Sun, C., Vallotton, P.: Fast linear feature detection using multiple directional non-maximum suppression. J. Microsc. **234**, 147–57 (2009)
3. Kirbas, C., Quek, F.: A review of vessel extraction techniques and algorithms. ACM Comput. Surv. **36**, 81–121 (2004)
4. Meijering, E.: Neuron tracing in perspective. Cytometry **77**, 693–704 (2010)
5. Neubeck, A., Van Gool, L.: Efficient non-maximum suppression. In: Tang, Y.Y., Wang, S.P., Lorette, G., Yeung, D.S., Yan, H. (eds.) Proceedings of 18th International Conference on Pattern Recognition, vol. 3, pp. 850–855. IEEE Computer Soc, Los Alamitos (2006)
6. Soille, P.: Morphological image analysis, 2nd edn. Springer, Heidelberg (2004)
7. Sun, C., Pallottino, S.: Circular shortest path in images. Pattern Recogn. **36**, 709–719 (2003)
8. Dorst, L., Smeulders, A.W.M.: Length estimators for digitized contours. Comput. Vis. Graph. Image Process. **40**, 311–333 (1987)
9. Lagerstrom, R., Sun, C., Vallotton, P.: Boundary extraction of linear features using dual paths through gradient profiles. Pattern Recogn. Lett. **29**, 1753–1757 (2008)
10. Owens, J.D., Luebke, D., Govindaraju, N., Harris, M., Kruger, J., Lefohn, A.E., Purcell, T.J.: A survey of general-purpose computation on graphics hardware. Comput. Graph. Forum **26**, 80–113 (2007)
11. Harris, M.: Optimizing parallel reduction in CUDA. NVIDIA SDK white paper (2007)

12. Hakura, Z.S., Gupta, A.: The design and analysis of a cache architecture for texture mapping. In: 24th Annual International Symposium on Computer Architecture, Conference Proceedings, pp. 108–120. Assoc Computing Machinery, New York (1997)
13. Cox, M., Bhandari, N., Shantz, M.. Multi-level texture caching for 3D graphics hardware. In: Proceedings of the 25th Annual International Symposium on Computer Architecture. IEEE Computer Soc, Los Alamitos, pp. 86–97 (1998)
14. Domanski, L.: Linear feature detection on GPUs. In: International Conference on Digital Image Computing: Techniques and Applications (DICTA), Sydney (2010)
15. Vallotton, P., Lagerstrom, R., Sun, C., Buckley, M., Wang, D., De Silva, M., Tan, S.S., Gunnersen, J.M.: Automated analysis of neurite branching in cultured cortical neurons using HCA-Vision. Cytometry A **71**, 889–895 (2007)
16. Conrad, C., Gerlich, D.W.: Automated microscopy for high-content RNAi screening. J. Cell Biol. **188**, 453–461 (2010)
17. Lau, C., O'Shea, R., Broberg, B., Bischof, L., Beart, P.: The Rho kinase inhibitor Fasudil up-regulates astrocytic glutamate transport subsequent to actin remodelling in murine cultured astrocytes. https://nsw.owa.csiro.au/pubmed/21309758. Br. J. Pharmacol. **163**, 533–545 (2011)
18. Vallotton, P., Sun, C., Wang, D., Ranganathan, P., Turnbull, L. Whitchurch, C.: Segmentation and tracking of individual Pseudomonas aeruginosa bacteria in dense populations of motile cells. In: Image and Vision Computing New Zealand Wellington, New Zealand (2009)
19. Orkisz, M.M., Bresson, C., Magnin, I.E., Champin, O., Douek, P.C.: Improved vessel visualization in MR angiography by nonlinear anisotropic filtering. Magn. Reson. Med. **37**, 914–919 (1997)
20. Heijmans, H., Buckley, M., Talbot, H.: Path-based morphological openings. In: ICIP: 2004 International Conference on Image Processing, vol. 1–5, pp. 3085–3088. IEEE, New York (2004)

Chapter 9
Medical Imaging in the Diagnosis of Osteoporosis and Estimation of the Individual Bone Fracture Risk

Mark A. Haidekker and Geoff Dougherty

Abstract Osteoporosis is a degenerative disease of the bone. In an advanced state, bone weakened by osteoporosis may fracture spontaneously with debilitating consequences. Beginning osteoporosis can be treated with exercise and calcium/vitamin D supplement, whereas osteoclast-inhibiting drugs are used in advanced stages. Choosing the proper treatment requires accurate diagnosis of the degree of osteoporosis. The most commonly used measurement of bone mineral content or bone mineral density provides a general orientation, but is insufficient as a predictor for load fractures or spontaneous fractures. There is wide agreement that the averaging nature of the density measurement does not take into account the microarchitectural deterioration, and imaging methods that provide a prediction of the load-bearing quality of the trabecular network are actively investigated. Studies have shown that X-ray projection images, computed tomography (CT) images, and magnetic resonance images (MRI) contain texture information that relates to the trabecular density and connectivity. In this chapter, image analysis methods are presented which allow to quantify the degree of microarchitectural deterioration of trabecular bone and have the potential to predict the load-bearing capability of bone.

9.1 Introduction

Osteoporosis *is defined as a skeletal disorder characterized by compromised bone strength predisposing a person to an increased risk of fracture. Bone strength primarily reflects the integration of bone density and bone quality* [1].

The official definition of osteoporosis further specify *bone density* as referring to specify mineral content and *bone quality* as referring to architecture, turnover, damage accumulation, and mineralization [1]. Bone density peaks at an age between

M.A. Haidekker (✉)
University of Georgia, Faculty of Engineering, Athens, GA 30602, Georgia
e-mail: mhaidekk@uga.edu

G. Dougherty (ed.), *Medical Image Processing: Techniques and Applications*, Biological and Medical Physics, Biomedical Engineering, DOI 10.1007/978-1-4419-9779-1_9,

20 and 30 and declines as people age. Hormonal changes, most notably menopause, accelerate this decline. For the purpose of diagnosis, individual bone density is commonly compared to an age-matched reference collective. The World Health Organization defines osteopenia as a loss of bone density to one standard deviation below the age-matched mean (T-score of -1) and osteoporosis as a loss of bone density to below 2.5 standard deviations (T-score of -2.5). The major health concern is the risk of fracture. The relationship between reduced bone density and the incidence of fractures is well known [2–5].

Bone loss can be slowed or prevented. A diet rich in calcium and vitamin D, or dietary supplements thereof, reduce the risk of osteopenia and osteoporosis [6]. Strength-building exercise stimulates bone formation (see [7] for a critical review). Whereas calcium intake and exercise primarily improve the baseline, patients with a low T-score need to be treated with drugs that reduce bone deterioration, such as calcitonin or bisphosphonates.

The primary goal of the diagnostic procedures is to assess the degree of bone loss for a decision on possible treatment. Whereas calcium and vitamin D supplementation are widely recommended, the type and vigorousness of a possible exercise regimen strongly depends on the degree of bone deterioration. The use of drugs also depends on the diagnosis. In advanced stages of bone deterioration it is, therefore, crucial to establish the *individual fracture risk*.

Presently, the diagnostic process most commonly involves the measurement of bone density (see Sect. 9.2). However, bone deterioration that leads to osteopenia and osteoporosis is a complex process [2, 8] that affects bone microarchitecture. In fact, early studies show that osteoporosis is associated with a deterioration of the complex three-dimensional network of trabeculae, which form the weight-bearing component of spongy bone [9]. There is a discrepancy between the relatively low bone density gain of around 1% by exercise [10] and the strong reduction of fracture incidence [11]. The benefits of exercise clearly include improved muscular strength, dexterity, and range of motion, thus directly contributing to a lower incidence of falls, accidents, or fracture-causing motions. Conversely, treatment with fluorides has been shown to strongly increase bone density while not decreasing [12] or even increasing fracture incidence [13]. Similarly, observations have been made for drugs that enhance bone formation. Moreover, bone density has been shown to strongly overlap between patients with and without osteoporosis-related fractures. Clearly, bone density alone is not a sufficiently specific predictor of the individual fracture risk [14].

Bone is heterogeneous and biomechanically complex. Fracture-prone sites, such as vertebrae, wrist, femoral head, and calcaneus are composed of spongy bone, which is a three-dimensional strut-like network of trabeculae, and the surrounding cortical shell, which is composed of compact bone. Both parts contribute to the weight-bearing capacity of bone. The loss of bone density reflects both the deterioration of the cortical shell and thinning of the trabeculae in spongy bone. An early study by Rockoff et al. found that the compact bone of the cortical shell carried between 45% and 75% of the total mechanical load, and that the weight-bearing contribution of the cortical shell increased with decreasing ash content [15].

A more recent study indicated that both vertebral stiffness and vertebral strength is almost completely attributable to spongy bone, and the cortical shell plays only a very small role in the weight-bearing capacity [16]. Further biomechanical studies suggested a power–law relationship between apparent bone density ρ and elastic modulus E and maximum compressive stress $\sigma_{\max}$ in the form of (9.1),

$$E = \rho^A = \left(\frac{V_b}{V_t}\right)^A; \qquad \sigma_{\max} = \rho^B = \left(\frac{V_b}{V_t}\right)^B \tag{9.1}$$

where V_b is the apparent bone volume, V_t is the total volume, and A and B are experimentally-determined constants [17]. Although a case can be made that microstructural deterioration is reflected in a loss of bone density [18], the deterioration of trabeculae is not isotropic. Rather, deterioration of vertical trabeculae occurs more rapidly than that of horizontal trabeculae [9]. Non-isotropic deterioration may in part explain differences in failure load at the same bone density [19]. Consequently, analysis of bone microstructure remains under active investigation [20].

The focus of recent research has been three-pronged. On the treatment side, scientists are striving to understand the cellular mechanisms that determine the balance between bone-resorbing cells (osteoclasts) and bone-forming cells (osteoblasts), with the long-term goal to influence this balance in favor of bone formation. On the diagnostic side, researchers are striving to obtain information about the bone microarchitecture, because the combined measurement of bone density and microstructural parameters promise to improve the prediction of the fracture load and therefore the individual fracture risk. Finally, basic research efforts are aimed at understanding the complex biomechanical behavior of bone. In all three cases, imaging methods play a central role.

9.2 Bone Imaging Modalities

The clinical modalities for imaging bone include X-ray imaging and the related dual energy X-ray absorptiometry (DEXA), computed tomography, magnetic resonance imaging, and ultrasound imaging.

9.2.1 X-Ray Projection Imaging

X-ray imaging provides excellent bone-tissue contrast, with a spatial resolution of about 30–40 μm. Dual-Energy X-ray Absorptiometry (DEXA) reduces the influence of soft tissue, such as muscle or marrow, which surrounds the bone. Since X-ray attenuation coefficients are energy-dependent, the X-ray intensity is measured at two different energies along the same path to eliminate the contribution of the soft tissue. Like conventional X-ray imaging, DEXA is a projection imaging method. It is typically applied to the thoracic or lumbar spine, the femoral neck, or the calcaneus.

The accuracy of the DEXA method is limited because the X-ray beam is polychromatic and because the soft tissue may be composed of muscle and adipose tissue, with varying absorption coefficients between individuals. It is possible to obtain the soft tissue composition from the DEXA image [21] to correct for the error. DEXA is best known for the measurement of bone density. Typical DEXA scanners feature a spatial resolution in the millimeter range, which is not sufficient to image structural details.

9.2.2 Computed Tomography

Computed tomography (CT) is an X-ray based technique that provides cross-sectional images of the X-ray absorption coefficient. Unlike projection imaging methods, computed tomography provides bone density as a true volumetric value that can be calibrated in $\mathrm{mg/cm^3}$. Accuracy and reproducibility of bone density measurements can be further increased by introducing a calibration phantom into the image. With a phantom that provides representative image values of bone I_B and of soft tissue I_S, bone density D, calibrated in milligrams of hydroxyapatite per milliliter of bone volume, can be computed from the average image value $<I>$ in the bone region and the known specific density of bone, ρ_B, through (9.2):

$$D = \frac{<I> - I_\mathrm{S}}{I_\mathrm{B} - I_\mathrm{S}} \rho_\mathrm{B} \tag{9.2}$$

The computed tomography method that provides calibrated bone density values is often referred to as *quantitative CT*. It is further possible to use a dual-energy principle similar to DEXA to eliminate artifacts caused by soft tissue and bone marrow [22]. Dual-energy quantitative CT is often regarded as the gold-standard for the noninvasive measurement of bone density.

Whereas a low spatial resolution, dominantly in the form of a wide slice thickness, is used for bone density measurement, CT can be used to image the bone microarchitecture when a high resolution is selected. Slice thicknesses of 1 mm with in-plane pixel sizes of 0.2×0.2 mm are possible with many clinical CT scanners. Micro-CT scanners are available that feature isotropic voxel sizes in the micron range, but these devices can hold only small samples, and are therefore reserved for biopsies or for in vivo imaging of, for example, the wrist (Fig. 9.1). The interior of the radius and ulna show a clear texture that can be related to the trabecular microarchitecture.

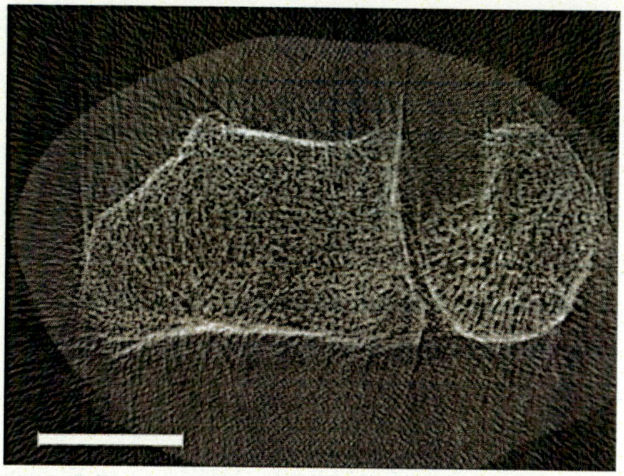

Fig. 9.1 Cross-sectional Micro-CT image of a human forearm. The voxel size is 70 μm, small enough to make the trabecular structure visible. Note the presence of reconstruction artifacts (the pseudo-texture that is particularly prominent in the air region surrounding the arm) and beam-hardening artifacts (straight lines extending from bone edges). Note also the low tissue-tissue contrast that makes tissue structures (blood vessels, muscle, tendons) indiscernible. The scale bar represents 10 mm

9.2.3 Magnetic Resonance Imaging

Magnetic resonance imaging (MRI) is presently not clinically used for imaging bone because of the low water content of bone, which leads to a very weak resonance signal. Figure 9.2 shows a T_1-weighted high-resolution spin-echo image of the wrist, acquired with a conventional clinical 1.5T scanner. Inside the ulna and radius areas, texture becomes apparent that is related to the trabecular architecture, analogous to Fig. 9.1. In-plane resolution is approximately 120 μm, with pixels almost twice as large as in Fig. 9.1. The long T_1 relaxation of bone marrow makes spongy bone appear bright in this image, although compact bone itself appears dark. Although MRI is not currently a method of choice for bone densitometry, there is a rising popularity of MRI methods in research to examine bone microstructure and its changes in osteoporosis.

9.2.4 Ultrasound Imaging

Ultrasound is widely used for bone densitometry. Sound waves travel much faster in bone than in soft tissue, with the speed of sound being approximately 4,080 m/s in compact bone and 1,600 m/s in muscle tissue [23]. A broadband ultrasound signal is attenuated in tissue in a frequency-dependent manner. Broadband ultrasound attenuation is commonly measured in transverse transmission mode by computing

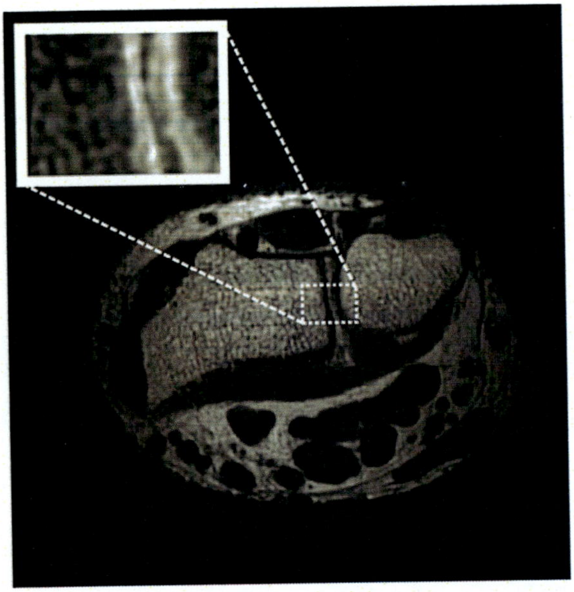

Fig. 9.2 Cross-sectional magnetic resonance image of a human forearm. Similar to Fig. 9.1, radius and ulna can easily be recognized. In contrast to the CT image, however, compact bone appears dark, and the MR image clearly reveals soft tissue features, such as muscle, tendons, and blood vessels. Spongy bone appears particularly bright because of the long relaxation times of bone marrow. The inset shows the section inside the white rectangle magnified and inverted to match the higher image values for bone in CT images. In the spongy area, texture that is related to trabecular structure is discernible, although it appears more blurred than in the corresponding CT image. Image courtesy of Dr. Kai Haidekker

the ratio of the transmitted spectra with and without the sample. A single value, often referred to as BUA (the broadband ultrasound attenuation value), is obtained from the slope of the attenuation over the frequency. No theoretical relationship between ultrasound attenuation and the mechanical properties of cancellous bone has been established [24], but both the speed of sound and the BUA value are higher in healthy bone than in osteoporotic bone.

9.3 Quantifying the Microarchitecture of Trabecular Bone

In Sect. 9.1, we discussed the need to estimate the individual fracture risk and the role that bone microstructure plays in that estimation. A very rigorous approach is computerized modeling of the biomechanical behavior of bone under a defined load. For this purpose, the exact three-dimensional microscopic geometry of the trabeculae and their interface with cortical bone need to be available. It is feasible to extract the geometry with sufficient precision by micro-CT or microscopy or histology in a slice-by-slice manner [25], yet these methods are normally restricted to *ex vivo* samples. Other three-dimensional imaging methods with high resolution, for example, micro-MRI, change the apparent geometry due to the system's point-spread function. At lower resolution, individual trabeculae cannot be imaged accurately. In such cases, the image shows a distinct texture that is more or less related to trabecular microarchitecture. Even two-dimensional cross-sectional slices and projection images can provide such texture information, but other influences (noise, reconstruction artifacts, pseudo-texture) become more dominant as the resolution becomes lower. Below a certain resolution, artifacts dominate, and any texture in the image is unrelated to trabecular microarchitecture, at which the image can only be used to measure average bone density. The most important methods to quantify image texture and relate it to trabecular microarchitecture are discussed in this section.

9.3.1 Bone Morphometric Quantities

Morphometric quantities are directly related to the microstructure of the trabecular network. From the segmented three-dimensional bone image, total volume and bone surface area can be obtained immediately. In the literature [26], these are abbreviated *TV* and *BS*, respectively. Bone volume (*BV*) is the space occupied by actual bone mineral. *BV* can be determined by measuring the total volume of the trabeculae after segmentation of those voxels that lie above a certain density. These primary indices can be used in normalized form, that is, relative bone volume *BV/TV*, relative bone surface *BS/TV*, and bone surface–volume-ratio *BS/BV*, to allow comparison between individuals.

Microstructural analysis of the trabeculae leads to the derived indices of trabecular thickness (*Tb.Th*) and trabecular spacing (*Tb.Sp*). In a simplified model, trabeculae can be interpreted as thin plates of constant thickness and width. In this case, the following relationships can be found [27]:

$$\begin{aligned} Tb.Th &= 2BV/BS \\ Tb.Sp &= 2(BV - TV)/BS \\ Tb.N &= BS/2BV \end{aligned} \tag{9.3}$$

Tb.N is the number of trabecular plates. Bone mineral content (BMC) was found to directly correlate with the normalized bone volume [28],

$$BMC = \left(\frac{BV}{TV}\right)\rho_{\mathrm{B}}\alpha \tag{9.4}$$

where ρ_{B} is the specific weight of bone mineral and α is the ash fraction. Values for α range from $\alpha = 0$ for osteoid to $\alpha = 0.7$ for fully mineralized bone [28].

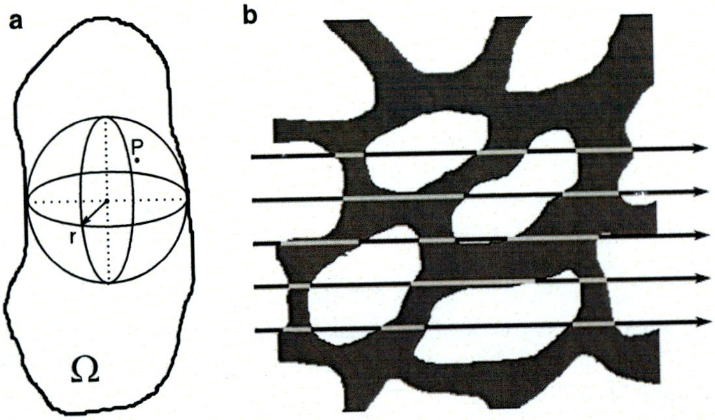

Fig. 9.3 Analysis of the trabecular microstructure. (**a**): Structures can be quantitatively analyzed by fitting a maximum-radius sphere into the structure Ω so that a point $P \in \Omega$ is an element of the maximum-radius sphere (adapted from [29]). (**b**): Scan-line method to quantify the microstructure. Parfitt [26] suggested counting the intersections of the bone structure with a rectangular grid of lines to obtain *Tb.N*. The method can be extended to obtain the distribution of runs along bone (*light gray*) and runs along marrow spaces (*black*)

These values correspond to a dry tissue density of $1.41\,\mathrm{g/cm^3}$ and $2.31\,\mathrm{g/cm^3}$, respectively. Furthermore, an experimental relationship of these values to elasticity E and ultimate stress σ_{ult} was found (9.5) with the constants a, b, c, and d determined by nonlinear regression as $a = 2.58 \pm 0.02$, $b = 2.74 \pm 0.13$, $c = 1.92 \pm 0.02$, and $d = 2.79 \pm 0.09$ [28].

$$E \propto (BV/TV)^a \alpha^b; \qquad \sigma_{\mathrm{ult}} \propto (BV/TV)^c \alpha^d \tag{9.5}$$

The assumption of homogeneous plates is a very rough approximation, and actual measurements from images provide more accurate estimations. Hilebrand and Rüegsegger proposed an image analysis method where a maximum-size sphere is fitted into the space between the segmented trabeculae [29]. By examining each point P inside the structure (Fig. 9.3a), statistical analysis of the void spaces can be performed. For each structure, the mean thickness (the arithmetic mean value of the local thicknesses taken over all points in the structure), the maximum thickness, average volume, and similar metrics can be found and examined over the entire bone segment, which can then be characterized by statistical methods. This method is capable of analyzing 3D volumetric data, but by using a circle rather than a sphere, it can be adapted to 2D slices.

An alternative method, known as the *run-length method*, can produce similar quantitative parameters with relatively low computational effort. Originally, Parfitt [26] suggested to use a rectangular grid of scanlines and count the intersections with trabecular bone in microscopic images to determine *Tb.N*. This idea can be extended to the run-length method (Fig. 9.3b): In an image that contains the segmented bone, linear runs of bone and marrow are measured. The average length of the marrow runs is related to *Tb.Sp*, and the average length of the bone runs is related to *Tb.Th*. The number of bone runs relative to the total number of runs provides *Tb.N*. With progressing bone deterioration, where trabeculae become thinner and disconnected, we can expect fewer and longer marrow runs, and shorter bone runs. Typically, runs are calculated at 0°, 45°, 90°, and 135°, and the resulting run lengths are averaged to obtain a measurement that is widely orientation-independent. Alternatively, runs in different directions can be used to determine orientational preferences. The run length method is a 2D method.

Two representative examples for the application of the run-length method for the quantification of trabecular microstructure on CT images [30, 31] demonstrate that this technique does not require an accurate microscopic representation of trabecular bone. A relationship between morphometric quantities and bone strength has been established [32]. On the other hand, noise and partial-volume effects have a strong influence on which voxels are classified as bone. Furthermore, image degradation by the point-spread function of the device makes the selection of a suitable threshold for the separation of bone and soft tissue difficult. Even with micro-CT and micro-MRI techniques, the thickness of individual trabeculae is on the order of a single voxel.

Topological quantities, such as the number of struts or the number of holes, are popular in the analysis of the trabecular network. The interconnected trabeculae can be represented as a graph, and the number of links (i.e., trabecular struts), the number of nodes, the number of holes (i.e., the marrow spaces), and related metrics, such as the Euler characteristic can be determined. The Euler characteristic can be seen as the number of marrow cavities completely surrounded by bone. To obtain the topological quantities, the image needs to be segmented into bone and non-bone areas followed by thinning of the structures (skeletonization, Fig. 9.4). One of the most interesting features of the topological quantities is the invariance under affine transformations. Therefore, these quantities should be robust – within limitations of the pixel discretization – against changes in scale, rotations, and even shear and perspective distortions.

Analysis of the skeleton is traditionally a 2D method, but extensions to 3D have been reported [33]. In two and three dimensions additional parameters have been defined. These include the connectivity index by Le et al. [34]; the marrow star volume by Vesterby et al. [35], which also reflects connectivity; the trabecular bone pattern factor (often abbreviated *TBPf*) by Hahn et al. [36], which decreases with bone deterioration; the ridge number density by Laib et al. [37]; and the structure model index (SMI) by Hildebrand and Rüegsegger [38], which is designed to characterize a 3D shape as being plate-like or rod-like and requires a 3D image of microscopic resolution.

9.3.2 Texture Analysis

Whereas morphometric analysis is based on the assumption that individual trabeculae are resolved in the image (thus, requiring high-resolution images with voxel

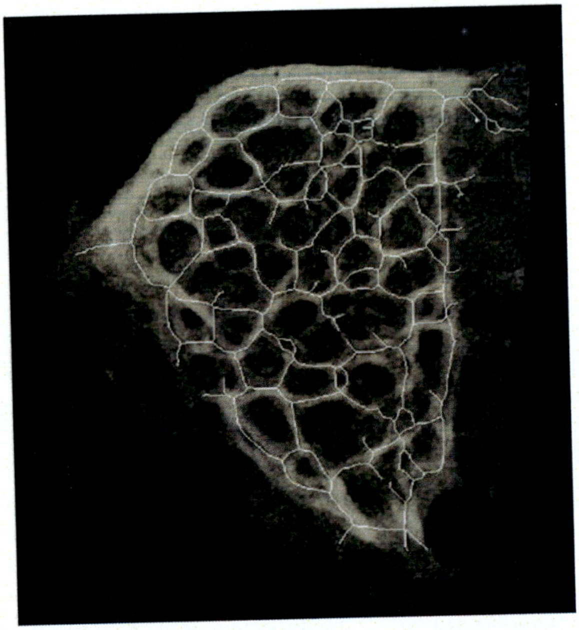

Fig. 9.4 X-ray image of an excised section of trabecular bone with its skeleton superimposed. Skeletonization is a popular method to quantify the connectivity. Microstructural deterioration of bone increases the number of branches (links that are connected to the network on only one end), and decreases the number of loops

sizes generally smaller than trabecular width), such an assumption is not needed for the analysis of the texture in image regions representing trabecular bone. Texture can be defined as a *systematic local variation* of the image values [39]. This definition normally implies the existence of multiple image values (gray values) as opposed to the purely binary values used in the morphometric analysis. Moreover, an assumption must be made that the image texture is related to the microstructure of trabecular bone. This is a reasonable assumption, as can be demonstrated in a simple experiment (Fig. 9.5). Let us consider an imaging device with a non-ideal point-spread function, for example, X-ray projection imaging where the size of the focal spot of the X-ray tube and the bone–film distance introduce blur. This process can be simulated by additive superposition of blurred images of some defined binary structure. A structure that somewhat resembles the distribution of trabecular bone is shown in Fig. 9.5a. The purely binary image represents trabecular bone (white) and marrow space (black). This image has been generated by a suitable random generator. Six such images, blurred with a second-order Butterworth filter adjusted to a random cutoff frequency of 10 ± 3 pixel^{-1} were created and added on a pixel-by-pixel basis (Fig. 9.5b). The similarity of the resulting texture to an actual CT image (Fig. 9.5c) is striking.

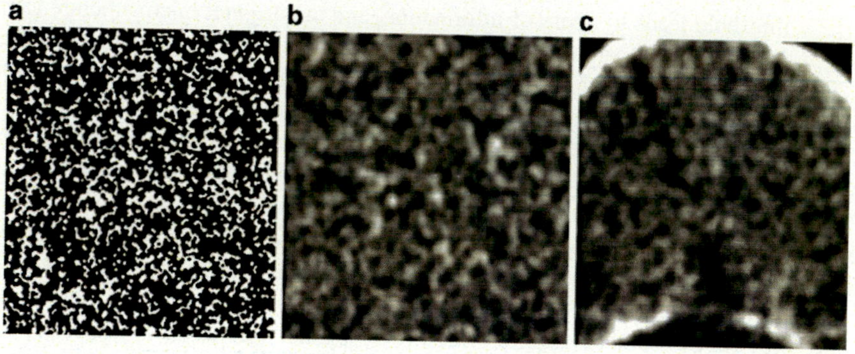

Fig. 9.5 Demonstration of the relationship between microarchitecture and texture in an imaging system with non-ideal point-spread function. (**a**): A synthetically generated pattern that somewhat resembles segmented and binarized trabeculae in a microscopic image. (**b**): Six patterns similar to the one shown in (**a**), blurred and superimposed. (**c**): Magnified computed tomography slice of the trabecular area in a lumbar vertebra. The trabecular area has been slightly contrast-enhanced, leading to saturation of the image values that correspond to cortical bone

Texture can be analyzed in two or three dimensions. Because most imaging modalities have anisotropic voxel sizes with much lower resolution in the axial direction than in-plane, texture analysis normally takes place as a two-dimensional operation. In 3D imaging modalities (CT and MRI), texture analysis can take place slice-by-slice. When an analysis method is extended into three dimensions, a possible voxel anisotropy needs to be taken into account.

The simplest method for texture analysis is the computation of the statistical moments of the histogram inside a region of interest. Most notably, the standard deviation and its normalized equivalent, the coefficient of variation, contain information about the irregularity of the structure. Healthy bone can be expected to have a more irregular structure with a higher coefficient of variation than osteoporotic bone with larger and more homogeneous marrow spaces. Both variance and coefficient of variation can be computed locally inside a sliding window. A compact metric is the average local variance (ALV), which shows a similar trend as the global variance, i.e., declines with the lower roughness of the texture of osteoporotic bone. Basic statistical moments can also be computed on gradient images [40], and the first moment is sometimes referred to as *edgeness*. The use of edge enhancement to emphasize the roughness was applied by Caldwell et al. [41], who processed digitized radiographies of thoracic vertebrae with the Sobel operator and a thresholding algorithm to remove small gradients that are likely caused by noise. A histogram of the gradient values showed two maxima that were related to a preferredly horizontal and vertical orientation of the trabeculae, respectively. The two maxima, relative to the mean gradient, related to the fracture load.

The run length method, introduced in the previous section, can be adapted to analyze texture in a straightforward manner. The grayscale image (such as Fig. 9.5c can be binarized with the application of a threshold. Clearly, the application of a threshold leads to a loss of information, and the original microstructure cannot be restored. The binary run length method can be extended by allowing more than two gray level bins. In this case, a two-dimensional histogram $N(g,l)$ of the number of runs N as a function of the gray value g and the run length l is obtained. The run-length histogram is still relatively complex, and statistical quantities can be extracted that characterize the histogram. Examples are the short- and long-run emphasis (SRE and LRE), low and high gray value emphasis (LGRE and HGRE), combined metrics, such as the long-run low gray-value emphasis (LRLGE), and uniformity metrics, such as the gray-level and run-length nonuniformities (GNLU and RLNU). Three of these quantitative descriptors are listed in (9.6) as examples, and a complete list can be found in [39].

$$\begin{aligned} SRE &= \sum_{g=0}^{G-1} \sum_{l=1}^{L} \frac{P(g,l)}{l^2} \\ LGRE &= \sum_{g=0}^{G-1} \sum_{l=1}^{L} \frac{P(g,l)}{(g+1)^2} \\ GLNU &= \sum_{l=1}^{L} \left[\sum_{g=0}^{G-1} P(g,l) \right]^2 \end{aligned} \tag{9.6}$$

Here, L is the longest run, G is the number of gray bins, and $P(g,l)$ is the run probability where $P(g,l) = N(g,l)/n$ with n being the total number of runs. It can be expected that trabecular rarefaction due to osteoporosis would increase LRE and LRLGE because of the longer intratrabecular runs, and would decrease GLNU as an indicator of the loss of complexity. In fact, Chappard et al. found that X-ray based SRE and GLN showed a high negative correlation with 2D histomorphometric methods ex vivo. Lespessailles et al. found significant differences in the SRE value between patients with and without osteoporosis-related fractures in a multicenter study [42]. In radiography images, Ito et al. [31] reported a significant increase of a parameter termed I-texture with age and with the presence of fractures. The I-texture is the average length of dark runs in a binary run-length histogram and corresponds to intratrabecular spaces. Conversely, the *T-texture*, the white runs that correspond to the trabecular area, did not change significantly. This observation can be interpreted as the increase of marrow spaces with otherwise unchanged trabecular width.

Selection of the threshold, or placement of the gray-level bins, have a strong influence on the quantitative texture descriptors. Neither projection images nor CT images allow the application of a defined threshold to separate bone from soft tissue and marrow. Depending on the threshold, the runs can be almost arbitrarily shifted from black to white runs. Bin selection is less critical when a gray-level run-length histogram is computed. However, a large number of gray levels leads to predominantly short runs, whereas a small number of gray levels allows longer runs, but introduces a sensitivity against shifts in image value. Frequently, the bin size b is determined from the maximum and minimum image intensity as

$b = (I_{max} - I_{min})/G$. A few outlier pixels can strongly influence bin placement. In the example of Fig. 9.5b, clamping the brightest 0.1% of the pixels reduces SRE by 15%, LGRE by 25%, GLNU by 28%, SRLGE by 27%, and LRLGE by 16%. All other descriptors are also affected, albeit to a lesser degree. A more robust method to select a threshold or to place bins would be based on the gray-value histogram [43]. Histogram-based threshold and bin selection also ensures that the quantitative descriptors are independent of the image intensity. Bone deterioration leads to lower image values in X-ray and CT images. A change in exposure or a different CT calibration may have a similar effect, although the microstructure has not been altered. The quantitative descriptors directly reflect the image intensity rather than the microarchitecture. This potential fallacy applies to most other texture analysis methods as well. Any method can be subjected to a simple test. If all image values are linearly transformed such that $I'(x,y) = a \cdot I(x,y) + b$, where $a > 0$ and b are scalars, any algorithm that produces different descriptors for I' when a and b are modified does not reflect pure texture.

Other artifacts that may influence texture descriptors are image noise and pseudotexture, introduced in the image formation process. Noise can be seen as an artifactual texture on the pixel level. By using a low number of gray-value bins and by discarding the shortest runs, the run-length method can be made more robust against noise. Pseudo-texture (see, for example, Fig. 9.1) cannot be fully eliminated. In special cases, suitable filters can suppress the pseudo-texture to some extent. The presence of this artifact makes it difficult to compare quantitative parameters obtained with different instruments or reconstruction settings. Lastly, the application of a highpass filter is advisable before texture parameters are determined. A highpass filter removes broad trends in the image values, for example, inhomogeneous X-ray illumination or MR bias field artifacts.

Two other texture analysis methods, based on the co-occurrence matrix [44] and on Law's texture energy metrics [45], have been successfully applied in texture analysis of trabecular bone. The co-occurrence matrix is the histogram of probabilities that an image value i is accompanied by the image value j at the distance $\vec{r}$. To create the co-occurrence matrix, a copy of the image is shifted by $-\vec{r} = (\Delta x, \Delta y)$, and the two-dimensional joint histogram is computed. Analogous to the two-dimensional run-length histogram, single-value quantitative descriptors can be extracted. The most widely used set of quantitative descriptors are known as *Haralick's texture classification metrics* [44]. Haralick proposes 14 different classification metrics, such as the energy, entropy, contrast, and correlation. Each of these metrics can be determined for different $\vec{r}$. Texture analysis is often driven by a high-dimensional feature vector that is used as the input of some artificial-intelligence decision mechanism. In the analysis of bone, however, single values are normally used. The choice of $\vec{r}$ is critical. Since the texture features of interest are normally larger than a single pixel, single-pixel offsets reflect predominantly noise. A thorough analysis of the relevant size of the features is necessary to obtain meaningful values from the co-occurrence matrix. The influence of the offset $\vec{r}$ is shown in Fig. 9.6, where the co-occurrence matrices for $\Delta x = 1$ and $\Delta x = 15$

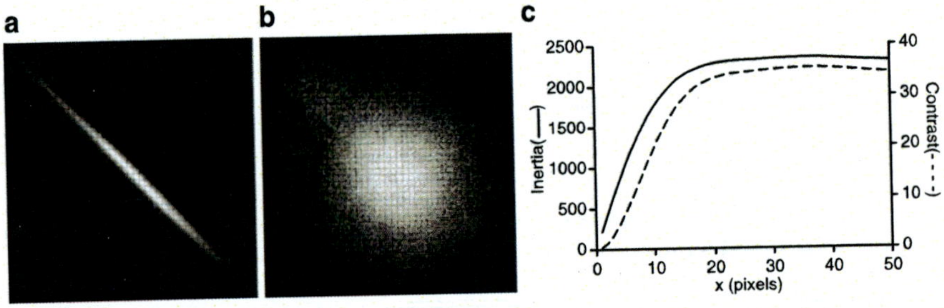

Fig. 9.6 Examples of the gray-level co-occurrence matrix of the image in Fig. 9.5b. A single-pixel offset (**a**) shows highly correlated pixel values with high-probability co-occurrences arranged along the diagonal $i = j$. With a larger offset ($\Delta x = 15$ pixels), a broader distribution of the co-occurrence matrix is seen (**b**). Analysis of the influence of the offset ((**c**), inertia and contrast shown as representative metrics) reveals that the values strongly depend on the choice of Δx until $\Delta x \gtrapprox 20$

are juxtaposed, and two representative metrics, inertia and contrast, are shown for different Δx. Neighboring pixels are highly correlated for small $\vec{r}$ (the diagonal $i = j$ dominates the histogram). The metrics become robust against changes of Δx for $\Delta x \gtrapprox 20$, which indicates that the features of interest coincide with this size. In fact, Euclidean distances between local maxima of the image fluctuate between 15 and 35. One example study be Lee et al. [46] examines the inverse difference moment *IDM* (9.7) of the co-occurrence matrix,

$$IDM(\theta, d) = \sum_{a,b,a \neq b} \frac{P_{\theta,d}(a,b)}{|a-b|^2} \tag{9.7}$$

where θ and d describe the displacement in polar coordinates, and a and b are the indices of the co-occurrence matrix. Lee et al. found a negative correlation of *IDM* in digitized radiographies of the femoral neck with bone strength, and found that *IDM* was significantly correlated with bone density, although highpass filtering and histogram normalization made *IDM* independent from radiographic density. On the other hand, Lee et al. [46] did not find a statistically significant difference of *IDM* between cases with and without fractures. One possible reason is the low resolution of approximately 8 pixels/mm, where the displacement $d = 7$ corresponds to 1 mm and is much larger than the trabecular size. Lespessailles et al. [47] found only a weak and nonsignificant correlation between the energy parameter and bone density. This observation can be interpreted in two ways. First, it should be expected that microarchitectural organization and density are independent quantities, and a high correlation cannot be expected. On the other hand, it is known that microarchitectural deterioration is seen as reduced bone density in large-volume averages, and a moderate correlation between the two quantities should be observed. Since Lespessailles et al. used single-pixel distances for r, the energy metric may have been dominated by pixel noise rather than actual microstructural information.

Unlike run-length classifiers and co-occurrence classifiers, Laws' energy metrics [45] are local neighborhood values obtained by convolving the image with two out of five one-dimensional convolution kernels to yield a feature vector with up to 25 elements. Laws proposes a three-step process where the image is first subjected to background removal (for example, a highpass filter step), subsequently convolved with the kernel pair, and finally lowpass filtered. The resulting image is referred to as the energy map. Typically, Laws' energy maps result in a high-dimensional feature vector per pixel and advertises itself for classification or segmentation with high-dimensional clustering techniques or artificial intelligence methods. In one example [48], classical statistical parameters were obtained from the texture energy maps, and the discrete first-order finite difference convolution kernel that approximates $\partial^2/\partial x \partial y$ showed some ability to discriminate between cases with and without fractures. However, Laws' texture maps are not in common use for this special application of texture analysis. One possible reason is the fixed scale on which the texture maps operate. Laws' convolution kernels are fixed to 5 by 5 pixels, and pixel noise dominates this scale. Thus, the method suffers from the same problems as the co-occurrence matrix with short displacements. However, it is conceivable that a combination of multiscale decomposition (such as a wavelet decomposition) with Laws' texture energy maps provides a more meaningful basis to obtain a quantitative description of the texture.

In summary, texture analysis methods in the spatial domain provide, to some extent, information that is related to actual trabecular structure. Therefore, texture analysis methods have the potential to provide information on bone architecture that is independent from bone density. However, texture analysis methods are sensitive towards the image formation function (e.g., projection image *versus* tomography), towards the point-spread function, the pixel size relative to trabecular size, and towards image artifacts, most notably noise.

9.3.3 Frequency-Domain Methods

The Fourier transform decomposes an image into its periodic components. A regular, repeating pattern of texture elements (sometimes referred to as *texels*), causes distinct and narrow peaks in the Fourier transform. A comprehensive discussion of the Fourier transform and its application in image analysis can be found in the pertinent literature [39, 40]. Since the Fourier transform reveals periodic components, i.e., the distances at which a pattern repeats itself, operations acting on the Fourier-transform of an image are referred to as *frequency-domain* operations in contrast to the *spatial-domain* operations that were covered in the previous section. The magnitude of the Fourier transform is often referred to as the *frequency spectrum*. In two-dimensional images, the frequency spectrum is two-dimensional with two orthogonal frequency components u and v. It is possible to analyze those frequency components separately and include properties such as texture anisotropy.

Alternatively, the spectrum can be reduced to one dimension by averaging all frequency coefficients at the same spatial frequency $\omega = \sqrt{u^2 + v^2}$. Intuitively, the frequency spectrum can be interpreted as a different representation of the spatial-domain image data that emphasizes a specific property, namely, the periodicity.

Trabecular bone does not exhibit any strict periodicity, because trabeculae are to some extent randomly oriented and have random size. The Fourier transform of random structures does not show distinct peaks. Rather, the frequency components decay more or less monotonically with increasing spatial frequency. This property is demonstrated in Fig. 9.7. The Fourier transform images (more precisely, the log-transformed magnitude of the Fourier transform) of a relatively regular texture and an irregular, bone-like structure are shown. Peaks that indicate the periodicity of the knit pattern do not exist in the Fourier transform of the bone-like structure. However, the decay behavior of the central maximum contains information about the texture. A fine, highly irregular texture, typically associated with healthy bone architecture, would show a broad peak with slow drop-off towards higher frequencies. Conversely, osteoporotic bone with large intratrabecular spacing would show a narrow peak with a faster drop-off towards higher frequencies.

In the analysis of trabecular bone texture, frequency-domain methods are often used to detect self-similar properties. These methods are described in the next section. Moreover, a number of single-value metrics can be derived from the Fourier transform. These include the root mean square variation and the first moment of the power spectrum (*FMP*, (9.8)):

$$FMP = \frac{\sum_u \sum_v \sqrt{u^2+v^2}|F(u,v)|^2}{\sum_u \sum_v |F(u,v)|^2} \tag{9.8}$$

Here, $F(u, v)$ indicates the Fourier coefficients at the spatial frequencies u and v, and the summation takes place over all u, v. The computation of the *FMP*-value can be restricted to angular "wedges" of width $\Delta\theta$, where the angle-dependent *FMP* (θ_i) is computed over all u, v with $\tan(\theta_i) < v/u \leq \tan(\theta_i + \Delta\theta)$. In this case, the minimum and maximum value of *FMP* (θ_i) provide additional information on the anisotropy of the texture. For example, if the texture has a preferredly horizontal and vertical orientation, $FMP(\theta)$ shows very distinct maxima for θ around $0°, \pm 90°$, and $180°$. Conversely, the *FMP*-index for a randomly oriented texture has less distinct maxima, and the coefficient of variation of $FMP(\theta)$ is lower. Two recent representative studies describe the use of frequency-domain metrics in radiographic images of the femur in patients with osteoprosis [49] and osteolysis [50]. Special use of the anisotropy was made in a study by Chappard et al. [51] and Brunet-Imbault et al. [52].

Frequency-domain methods have two major advantages over spatial-domain methods for the analysis of bone structure and its texture representation in images. First, frequency-domain methods are less sensitive against background irregularities and inhomogeneous intensity distributions. Trend-like background inhomogeneities map to very low spatial frequencies, and the corresponding Fourier coefficients

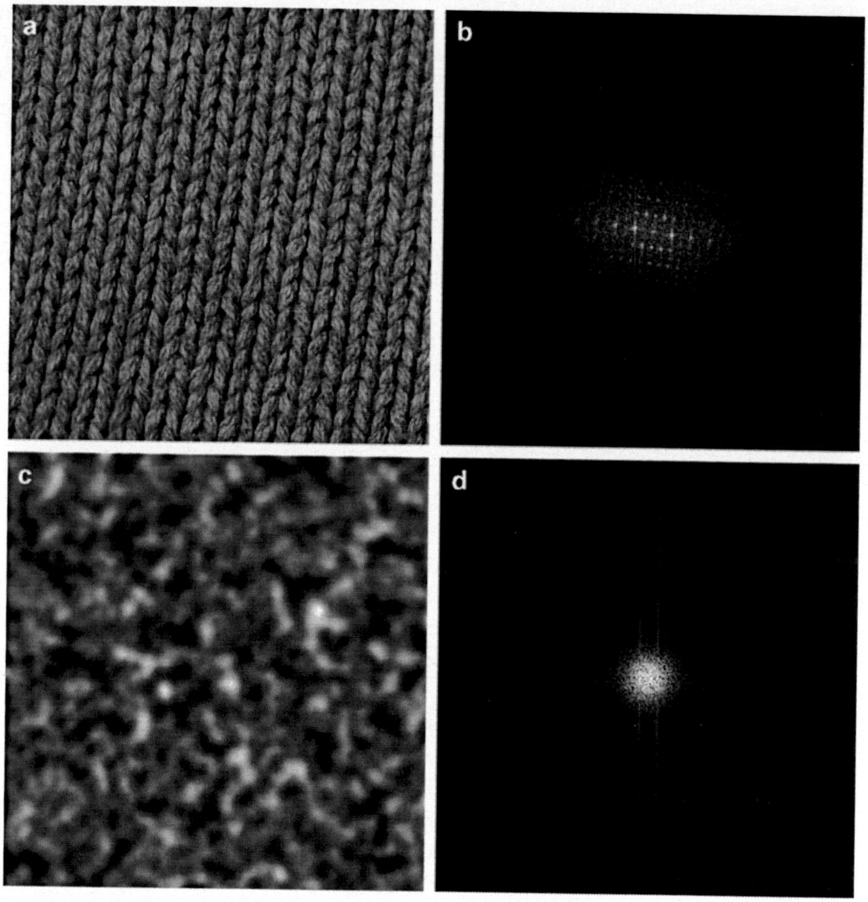

Fig. 9.7 Frequency-domain representation of texture. (**a**): Photography of a knit texture that contains a somewhat regular repeating pattern. The regularity of the pattern is represented by distinct peaks in the Fourier transform (**b**). The slight off-vertical orientation of the knit texture finds a correspondence of the off-horizontal orientation of the Fourier-transform pattern. (**c**): Synthetic trabecular pattern from Fig. 9.5. Such a pattern does not have repeated texture elements, and the frequency coefficients decay mostly monotonically with increasing frequencies (**d**). The decay behavior can be used to characterize any irregular texture

are close to the center of the Fourier spectrum image. By omitting the $F(u,v)$ from any single-value metric computation for small u,v, trends are automatically excluded from the metric. Second, frequency-domain methods are usually less noise-sensitive than spatial-domain methods. Many spatial-domain methods act on the pixel level (examples are Laws' texture energies or the co-occurrence matrix with low displacements), and pixel noise directly influences the measured values. In the frequency domain, the energy of the noise is broadly distributed over the entire frequency spectrum. High spatial frequencies, i.e., those frequencies that are above the inverse spatial distance of the image features of interest, can be omitted from the computation of any frequency-domain metric, leading to an immediate reduction of the influence of noise. Application of a moderate spatial-domain noise filter before performing the Fourier transform is none the less advisable, because it reduces aliasing artifacts.

9.3.4 Use of Fractal Dimension Estimators for Texture Analysis

Fractal models enjoy great popularity for modeling the texture in medical images, with the fractal dimension commonly used as a compact descriptor. The fractal dimension D describes how an object occupies space and is related to the complexity of its structure: it gives a numerical measure of the degree of boundary irregularity or surface roughness. In its original mathematical definition, a fractal object is created by a set of mapping rules that are applied iteratively on an object. After an infinite application of the mapping operation, the resulting object is invariant under the mapping operation and therefore referred to as the *attractor* of the mapping rules. The attractor is self-similar in the sense that any part, magnified, looks like the whole. Furthermore, the mapping rules determine the *self-similarity dimension* of the attractor, which is strictly less than, or equal to, the Euclidean dimension E of the embedding space ($E = 1$ for a line, $E = 2$ for a surface, and $E = 3$ for a volume). In nature, self-similarity occurs, but contains a certain degree of randomness. For this reason, no strict self-similarity dimension exists, and the apparent fractal dimension needs to be estimated by suitable methods. To cover the details of this comprehensive topic is beyond the scope of this chapter. A highly detailed overview of the subject with a mathematical focus can be found in [53], the groundbreaking work by Mandelbrot [54] deals with the aspect of randomness in natural fractal objects, and an in-depth overview of the methodology of estimating fractal dimensions in images can be found in [39].

Fractal analysis always involves the detection and quantification of self-similar behavior, i.e., to find a scaling rule under which a magnified part of the image feature is similar to the whole. This property can be illustrated with the example of the coastline of England. The length of the coastline can be estimated with calipers. However, as the caliper setting is reduced, the length of the coastline appears increased, because the caliper now covers more of the smaller detail, such as bays and headlands. In fact, nonlinear regression of the coastline length l as a function of the caliper setting s reveals a power law,

$$l = \frac{1}{s^D} \tag{9.9}$$

where D is the apparent fractal dimension. Equation (9.9) implies that the measured length exceeds all bounds as $s \to 0$, which is indeed a property of some mathematical fractals. In actual images, the scaling law fails when the limits of the resolution

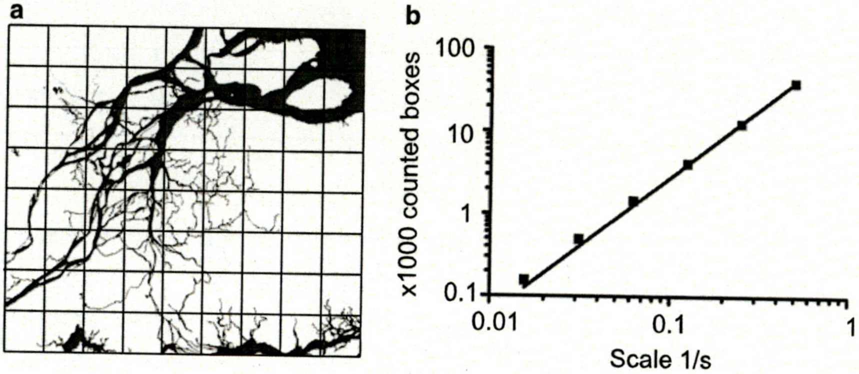

Fig. 9.8 Demonstration of the box-counting method to estimate the fractal dimension of binary images. The image (**a**) shows the water body (*black*) of the Amazon river. Superimposed is a square mesh of boxes. Every box that contains or touches the river is counted. Next, the mesh is refined (usually by powers of two), and the number of smaller boxes containing part of the river is counted again. If self-similar properties exist, the log-transformed number of boxes, plotted over the log-transformed inverse box size (**b**), will follow a straight line with slope D

are reached. In practice, the range of scales under which a scaling law similar to (9.9) can be found is even more limited. In the example of the coastline, values around $D = 1.2$ can be found [39, 53]. This is consistent with the notion of a very complex and convoluted line embedded in two-dimensional Euclidean space, where $1 \leq D \leq 2$. A surface, such as a grayscale elevation landscape, is embedded in three-dimensional space, where $2 \leq D \leq 3$ holds.

A very widespread method to estimate an apparent fractal dimension in binary images is the *box-counting method*, explained in Fig. 9.8. The feature needs to be segmented (the river in the example), and the number of boxes that contain part of the feature are counted as a function of the box size. If a power-law similar to (9.9) exists, the counted boxes N_B over the inverse scale $1/s$ lie on a straight line in a log–log plot, and nonlinear regression yields the box-counting dimension D_B:

$$\log N_B = -D_B \cdot \log s \tag{9.10}$$

In a grayscale extension of the box-counting method, the image intensities are interpreted as heights in an elevation landscape. The surface area is determined, whereby the scale is controlled by averaging neighboring pixel values inside boxes of a grid. Numerous other algorithms for estimating fractal dimension have been described [39, 55, 56]. They are all based on measuring an image characteristic, chosen heuristically, as a function of a scale parameter. Generally, these two quantities are linearly regressed on a log–log scale, and the fractal dimension obtained from the resulting slope, although nonparametric estimation techniques have also been used [57]. In all fractal estimation methods, the scale range needs to be chosen carefully. For large box sizes, to remain with the example of the

Table 9.1 A visual classification scheme for the assessment of the trabecular structure used for the determination of the degree of osteoporosis and its relationship to fractal properties

WHO classification	Bone strength	Spongiosa pattern	Marrow size	Fractal dimension of feature
1 (Healthy)	High	Homogeneously dense with granular structures	Small, homogeneous	Low, Unifractal
2 (Beginning demineral-ization)	Normal	Discrete disseminated intertrabecular areas	Medium, inho-mogeneous	High, Multifractal
3 (Osteopenia)	Low	Confluent intratrabecular areas <50% of the cross-sectional surface	Large, inhomo-geneous	High, Multifractal
4 (Osteoporosis)	Very low	Confluent intratrabecular areas ≥50% of the cross-sectional surface	Very large, homogeneous	Low, Multifractal

box-counting method, there is insufficient resolution to measure the feature area properly. In the presence of noise, the scaling law at the smallest scales may be dominated by noise, and partial-volume artifacts may result in misclassification of pixels in the segmented image [58]. Further critique of the method specifically for trabecular bone stems from the observation that trabeculae, on the microscopic scale, are not fractal [58]. This observation reinforces the notion that the scale range needs to be carefully considered [59, 60]. In spite of some criticism [58] and the limitations discussed above, estimation methods for the fractal dimension have been applied successfully in hundreds of studies (for representative reviews, see [61, 62]).

There has been considerable debate in the literature regarding the change in fractal dimension with decalcification. An early study of human calcaneous bone [63] during immobilization for fracture (causing a process similar to osteoporosis) found an increased fractal dimension during immobilization. Likewise, a study of mandibular alveolar bone reported an increased fractal dimension after decalcification [64]. On the other hand, a later study showed a reduction in fractal dimension of the ankle with immobilization and age [65], and a CT study of vertebral trabecular bone reported that osteoporotic patients had a smaller fractal dimension [66]. Furthermore, a study of dental radiographs [67] concluded that fractal dimension decreased with decalcification. In fact, the majority of studies agrees that the fractal dimension declines with the progression of the disease. Clearly, changes in fractal dimension need to be interpreted with care. We conclude that global fractal dimension does not change monotonically with decalcification (Table 9.1), but rather that it reflects the homogeneity of the spongiosa pattern of the trabecular bone. Studies with subjects who do not reflect the full range of this pattern can report either an increase (WHO classes 1,2, perhaps 3) or a decrease in fractal dimension (from class 2 or 3 to class 4) with osteoporosis.

9.3.4.1 Frequency-Domain Estimation of the Fractal Dimension

Self-similar properties of texture can be quantified in the frequency domain. The main advantage of frequency-domain methods is the better representation of the stochastic nature of images: Certain characteristics are less robust when applied to digitized data, especially when these are sparse, and algorithms that implicitly assume an exactly self-similar fractal model are inappropriate for medical images, because they are fractal only in a statistical sense and because pixel intensity and position are different physical properties and cannot be expected to scale with the same ratio. Thus, methods that do not meet the intensity–scale independency requirement [68], such as the surface area algorithm, may not be applicable. In contrast, the Fourier power spectrum method conveniently represents the statistical nature of real images by describing them in terms of a fractional Brownian motion model.

Roughness (or is opposite, smoothness) is an important feature of texture, and a commonly used method to estimate the smoothness of a one-dimensional function is from the decay of the Fourier power spectrum with increasing frequency f. For a two-dimensional image, the radial Fourier power spectrum should be used. For a rough image that adheres to the model of uncorrelated noise, the power spectrum falls off as $1/f^2$, whereas a smooth image (correlated or Brownian noise) has a power spectrum that decays with $1/f^4$. In a log-log plot of the spectral power over the frequency, a decay with a straight line of slope β indicates self-similar properties. Furthermore, β will be 2 and 4 for a rough- and a smooth-textured image, respectively. Similar to the scaling range in spatial-domain methods, the range in which a power-law decay of the power spectrum with frequency is found, may be limited. At very low spatial frequencies, corresponding to the bulk features of an object, the power spectrum may be fairly constant. At very high spatial frequencies, system noise will dominate and the power spectrum will become constant again. The power spectrum has been shown to estimate the fractal dimension of self-affine fractals reliably and accurately [69] and has been used to discriminate textures in conventional radiographs of osteoarthritic knees [70].

The link between power-law spectral decay β and fractal dimension is established by interpreting the image data as fractional Brownian motion (FBM), because FBM shows a statistical scaling behavior. FBM is an extension of the more familiar Brownian motion. It has been shown [57] that the decay exponent of the power spectrum, β, is related to the fractal dimension of a function modeled by FBM according to

$$D = 1 + \frac{1}{2}(3E - \beta) \tag{9.11}$$

where E is the Euclidean dimension of the embedding space (for a two-dimensional image, $E = 2$ and therefore $D = 4 - \beta/2$). In a two-dimensional image, the value of D will be constrained to be between 2 (smooth) and 3 (rough), and for a projected image generated by Brownian motion (a special case of FBM), the value of D will be 2.5.

In practice, images are degraded by noise and blurring within a particular imaging device. Image noise adds to the roughness and results in an overestimate of the fractal dimension, whereas blurring results in an underestimate of the fractal dimension. A very important advantage of the power spectrum method is that it allows for correction of these two effects. The noise power can be obtained by scanning a water phantom under the same conditions, and can then be subtracted from the power spectrum of the noisy image. Image blurring can be described by the modulation transfer function (MTF) of the system that typically attenuates higher frequencies in an image. The effect of system blurring can be eliminated by dividing the measured power spectrum by the square of the MTF, obtained by scanning a very small object approximating a point. With these corrections, accurate estimates of the fractal dimension of CT images of trabecular bone have been obtained, enabling very small difference in texture to be distinguished [71].

9.3.4.2 Lacunarity

Lacunarity (from *lacuna*, meaning gap or cavity) is a less frequently used metric that describes the complex intermingling of the shape and distribution of gaps within an image; specifically, it quantifies the deviation of a geometric shape from translational invariance. Lacunarity was originally developed to describe a property of fractals [54, 72] to distinguish between textures of the same fractal dimension. However, lacunarity is not predicated on self-similarity and can be used to describe the spatial distribution of data sets with and without self-similarity [73]. Lacunarity is relatively insensitive to image boundaries, and is robust to the presence of noise and blurring within the image.

Lacunarity is most frequently computed as a function of a local neighborhood (i.e., moving window) of size r. To compute the lacunarity, we first define a "score" $S(r,x,y)$ for each pixel, which is the sum of the pixel values inside the moving window centered on (x,y). The detailed derivation of the lacunarity $L(r)$ can be found in [73]. Simplified, we obtain $L(r)$ as

$$L(r) = \frac{\sigma_S^2(r)}{\bar{S}^2(r)} + 1 = \frac{\sum_{r=1}^{N}(\bar{S}(r) - S(r))^2}{\bar{S}^2(r)} + 1 \qquad (9.12)$$

where $\bar{S}(r)$ is the mean value of all $S(r)$ and σ_S^2 is the variance of $S(r)$.

Equation (9.12) reveals explicitly the relationship between lacunarity and the variance of the scores: Lacunarity relies on the variance of the scores, standardized by the square of the mean of the scores. The lacunarity, $L(r)$ of an image at a particular window size r uses all the scores obtained by exhaustively sampling the image. Thus, in general, as the window size r increases, the lacunarity will decrease, approaching unity whenever the window size approaches the image size (when there is only one measurement and the variance is consequently zero) – or for a spatially random (i.e., noisy) pattern, since the variance of the scores will be close to zero even for small window sizes. The lacunarity defined in (9.12) and its variants (including normalized lacunarity and grayscale lacunarity) are scale-invariant but are not invariant to contrast and brightness transformations, so that histogram equalization of images is a necessary pre-processing step.

A plot of lacunarity against window size contains significant information about the spatial structure of an image at different scales. In particular, it can distinguish varying degrees of heterogeneity within an image, and in the case of a homogeneous image it can identify the size of a characteristic substructure. Hierarchically structured random images can be generated using curdling [54]. Higher lacunarity values are obtained when the window sizes are smaller than the scale of randomness, and images with the same degree of randomness at all levels (*viz.* self-similar fractals) are close to linear, where the slope is related to the fractal dimension. Specifically, the magnitude of the slope of the lacunarity plot for self-similar fractals is equal to $D-E$, where D and E are the fractal and Euclidean dimensions, respectively.

One problem with the lacunarity metric defined in (9.12) is that the vertical scaling is related to the image density, with sparse maps having higher lacunarities than dense maps for the same window size. This complicates the comparison of plots between images of different density. It is possible to formulate a *normalized lacunarity* whose decay is a function of clustering only and is independent from image density. A normalized lacunarity, $NL(r)$, can be achieved by combining the lacunarity of an image, $L(r)$ with the lacunarity of its complement, $cL(r)$, which can assume values between 0 and 1 [71,74]:

$$NL(r) = 2 - \frac{1}{L(r)} + \frac{1}{cL(r)} \tag{9.13}$$

9.3.4.3 Lacunarity Parameters

Lacunarity plots, i.e., plots of $L(r)$ over r, show how the lacunarity varies with scale. The plots monotonically decay to a value of unity at large scales, unless there is considerable periodicity in the image; in which case it can pass through some minima (corresponding to the repeat distance) and maxima as it falls to unity. Most real images, as opposed to synthetic images, will show only the monotonic decay. In image features with strict self-similarity, $L(r)$ results in a straight-line plot from (0,1) to (1,0). If this line is seen as the neutral model, the deviation of the (normalized) lacunarity plots from the straight line, calculated as a percentage of the (normalized) lacunarity value, will emphasize subtle differences that are not conspicuous in the decay curves themselves and is useful in identifying size ranges for different tonal features [74]. Positive (negative) deviations indicate greater

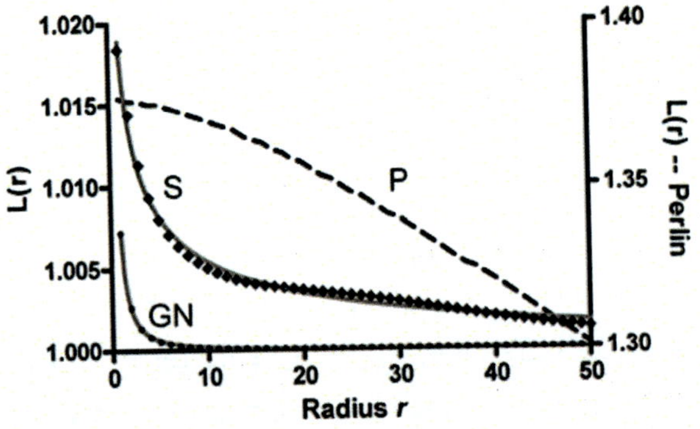

Fig. 9.9 Lacunarity plots for three sample textures, highly correlated noise (Perlin noise, *dashed line* labeled **P**), uncorrelated Gaussian noise (*black circles*, labeled **GN**), and the texture from a CT cross-section of spongy bone in a healthy lumbar vertebra (*black diamonds*, labeled **S**). Fitted curves (9.14) are shown in gray. The lacunarity for Perlin noise cannot be described by (9.14). For Gaussian noise, $\alpha = 1.5$ and $\beta = 0.007$ was found, and for the spongiosa texture, $\alpha = 0.45$ and $\beta = 0.021$

(lesser) spatial homogeneity than the underlying scale-invariant neutral (fractal) model. The presence of a prominent maximum would indicate the typical size of a structuring element in the image. Moreover, Lacunarity plots often resemble the plot of a power-law function, and Zaia et al. [75] have fitted non-normalized lacunarity plots from binary images to a function of the form

$$L(r) = \frac{\beta}{r^{\alpha}} + \gamma \tag{9.14}$$

where the parameters α, β, and γ are regression parameters that represent the order of the convergence of $L(r)$, the magnitude (vertical scaling) of $L(r)$, and the offset (vertical shift) of $L(r)$, respectively. We have explored the fitting to monotonic normalized lacunarity plots, where the parameter γ can be conveniently set to unity, which is the value that $NL(r)$ approaches at large scales. This simplifies the power-law fit to

$$NL(r) - 1 = \frac{\beta}{r^{\alpha}} \tag{9.15}$$

Examples for lacunarity plots $L(r)$ are shown in Fig. 9.9. The plots of $L(r)$ and the curve fits with (9.14) are shown. Highly correlated noise (Perlin noise) cannot be described by (9.14). Conversely, both the uncorrelated noise and the texture of spongy bone in a cross-sectional CT slice show a good fit with (9.14) with $R^2 > 0.99$ in both cases. The score $S(r)$ rapidly reaches statistical stability for increasing r, and the variance of S over the location of a sliding window becomes very low. The texture of trabecular bone has a higher variance with the location of S, and its decay with increasing window size is slower. It becomes obvious that window sizes with $r > 20$ do not carry additional information in this example, where a window size was used that coincides with the largest textural variations in the spongy bone image. Since bone density, the fractal dimension, and lacunarity are orthogonal metrics, they can be used for multidimensional clustering to better discriminate between degrees of osteoporosis (Table 9.1) [74].

9.3.5 *Computer Modeling of Biomechanical Properties*

Up to this point, Sect. 9.3 was primarily concerned with the empirical relationship between image properties and bone strength. Most notably X-ray attenuation, which is directly related to bone density, can be linked to bone strength. The texture analysis methods introduced in the previous sections aim at extracting information from biomedical images of trabecular bone that are independent from average density and therefore provide additional information. A less empirical approach is the modeling of bone biomechanical properties with finite-element models (FEM). In finite-element analysis, an inhomogeneous object is subdivided into a large number of small geometrical primitives, such as tetrahedrons or cuboids. Each element is considered homogeneous with defined mechanical properties (stress–strain relationship). External forces and displacements are applied by neighboring elements. External boundary conditions can be defined. Those include spatial fixation and external forces. The entire system of interconnected finite elements is solved numerically, and the forces, the shear tensor and the displacement for each element are known. Finite-element models also allow time-resolved analysis, providing information on motion and the response to time-varying forces.

The use of FEM for skeletal bone became popular in the late 1970s and has been extensively used to relate skeletal variation to function (for a general overview, see [76]). Since then, literally dozens of studies have been published each year where FEM were used for the functional understanding of bone, predominantly in the spine. Because of the large volume of available literature, we will focus on spinal vertebrae as one pertinent example. One of the most fundamental questions that can be approached with finite-element analysis is the contribution of compact bone to the overall weight-bearing capacity of bone. Studies by Rockoff et al. and Vesterby et al. indicate a major load-bearing contribution of the cortical shell [15, 77].

Two main approaches exist. A general vertebral model with representative geometry can be designed and used to study general spine biomechanics. Conversely, the geometry of individual vertebrae can be extracted from volumetric images (CT or MRI), approximated by finite elements, and subjected to load and deformation analysis. Although the second approach holds the promise to improve the assessment of the individual fracture risk, it has not found its way into medical practice, mainly because of the computational effort and because of uncertainties about the influence of the finite-element subdivision, material property assignment, and the exact introduction of external forces [78]. The two representative approaches are shown in Fig. 9.10. The early model by Lavaste et al. [79] was generated from global X-ray based measurements, including the width and height of the vertebral

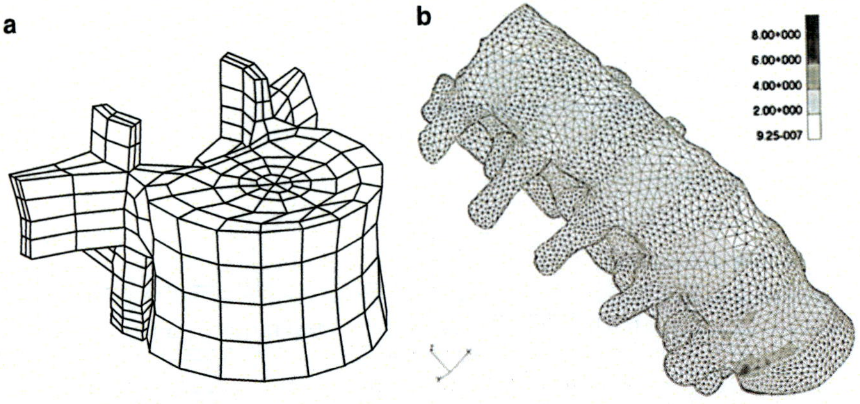

Fig. 9.10 Finite-element approximations of the spine. (**a**): Early parametric model of a vertebral body (adapted from [79]). The vertebral body is generated from X-ray measurements, such as height, width, and curvature. (**b**): Comprehensive finite-element model of the lumbar spine and the computed stress magnitude (adapted from [80] with permission to reprint through the Creative Commons License)

body, the diameter of its waist, and the length of the vertebral processes. As such, it is a semi-individual model that reflects gross measurements combined with a generalized shape model. Individual vertebral bodies can be combined to form a semi-individualized model of the spine or segments of the spine [79]. The recent model by Kuo et al. was segmented from volumetric high-resolution CT images (0.35 mm pixel size), and different material properties were assigned to spongy bone, the cortical shell, the endplates, and the intervertebral discs. Such a model may easily contain 20,000 elements per vertebra, and it accurately reflects the geometry of an individual spine.

The assignment of material properties is an ongoing question. In many models, including the two examples presented above, the spongiosa is modeled as a homogeneous material – in the example of Kuo et al., spongy bone is assigned a Young's modulus of 100 MPa compared to cortical bone with 12,000 MPa. However, spongy bone may have a larger local variation of its mechanical strength than the model allows. General models for spongy bone include hexagonal or cubic stick models [81, 82]. Specific models have been developed for spongy bone, often based on micro-CT or micro-MRI images that can resolve individual trabeculae [83,84]. Once again, the model strongly depends on the accurate representation of the geometry and the local material properties. To determine the load-bearing capacity of an entire vertebra, the spongiosa model needs to be appropriately connected to the cortical shell and the endplates. Furthermore, detailed models of the spongiosa cannot presently be used in a clinical setting, because whole-body scanners do not provide the necessary microscopic resolution to build a detailed model of the trabeculae.

These considerations nonwithstanding, finite-element models have become an important component in the toolkit of image analysis for bone. Whereas the analysis of texture may provide information on the bone structure, which may complement average bone density in an empirical fashion, finite-element models allow a more rigorous approach to examine the load distribution in bone.

9.4 Trends in Imaging of Bone

A focus of this chapter lies on the use of texture analysis methods to obtain density-independent information on bone microarchitecture. The underlying idea, that the combination of bone density and microarchitecture leads to an improved assessment of the individual fracture risk, has been validated in many studies. However, none of the texture analysis methods has found its way into clinical practice. In fact, several studies found a low predictive value of structure analysis, and most of the fracture risk was explained by bone density. The key reasons are:

- Bone density and microarchitectural/structural information cannot be truly orthogonal, because reduced bone density appears as a consequence of trabecular thinning.
- Unless microscopic methods are used that can resolve individual trabeculae, images of trabecular bone are subject to the point-spread function and the noise contribution from the imaging device. These artifacts can influence the metrics obtained.
- The same artifact prevents most metrics to be comparable between modalities or even between different scanners. No single universal (or even widely applicable) method has emerged.
- Texture methods usually do not take into account the load-bearing capacity of the cortical shell.
- Additional factors have a strong impact on the fracture risk. These include muscle mass and overall strength, age, body mass index, current medication, dementia, and ancillary diseases.

On the other hand, it is indisputable that bone density, whose measurement involves averaging over a relatively large volume, is associated with loss of information. In a frequency-domain interpretation, bone density contains only the very low-frequency components of the bone, and the high-frequency information is discarded. The latter component is related to the complexity of the trabecular network, and the complexity and interconnectedness of trabeculae have been linked to load-bearing capacity. Furthermore, bone density overlaps between patients with and without fractures, a fact that further supports the idea that more factors than bone density should be included to assess the fracture risk. In this respect, the inclusion of structural or textural parameters is certainly a step in the right direction.

These observations define the future trends in imaging of bone. First, existing methods for obtaining more complete information on a specific bone can be refined and new methods developed. Second, those methods must become more universal to allow adoption independent of imaging parameters, devices, and modalities. Third, ancillary factors need to be identified and included in the fracture risk assessment.

Progress is in part driven by the availability of higher-resolution modalities. Over the last 10–15 years, mainstream clinical scanners have increased their spatial resolution by an order of magnitude. Three-dimensional constructs can be extracted from CT and – more recently – MRI [85] that strongly resemble actual trabecular structure. With this development, biomechanical modeling of the fracture load of individual peripheral bones comes within reach. For the thoracic and lumbar spine, this extreme resolution is not readily available. Furthermore, present trends in healthcare policies could prevent relatively expensive three-dimensional imaging modalities from becoming more widely adopted. It is more reasonable to assume that the diagnosis will continue to be based on relatively inexpensive bone density estimation with ultrasound or DEXA. For this reason, continued research on texture-based methods gains importance as a means to obtain information that complements gross bone density. These methods would particularly focus on low-resolution modalities, such as DEXA and possibly ultrasound imaging. In fact, the prediction of bone strength with ultrasonic techniques has recently received increased attention [86–88].

An area where image-based assessment of bone strength becomes more attractive is the evaluation of anti-osteoporosis drugs [89]. A case in point is the controversial treatment of bone loss with fluoride, which leads to a rapid gain in bone density [90], but not to a matching gain in bone strength [13]. In fact, Riggs et al. found an increased fracture rate after fluoride treatment [13]. Grynpas [91] presumes that fluoride leads to the formation of larger bone mineral crystals, which make bone more brittle. This example highlights the importance of the microstructure particularly well. The methods to assess bone microstructure can therefore aid in drug development and evaluation: by providing information on the microstructure during treatment, and by providing the tools to noninvasively estimate or even compute bone strength. One recent example is a micro-CT study by Jiang et al. [92] on the effects of hormone therapy on bone microstructure.

In conclusion, there is wide agreement that bone density alone may be sufficient to diagnose early bone loss, but is insufficient to accurately predict the individual fracture risk. Bone microarchitecture holds complementary information. Microstructural information can be obtained from biopsies or, noninvasively, by suitable high-resolution imaging techniques. Depending on the resolution, the trabecular structure and its interface with the cortical shell can be directly reconstructed, or indirect quantitative metrics can be obtained that reflect the microarchitecture only to some extent. When the bone structure can be fully reconstructed, direct modeling of bone strength is possible, for example, with finite-element methods. Conversely, indirect metrics that are obtained from lower-resolution modalities or projection imaging can be combined with mineral density in an empirical fashion. The combined metrics often correlate better with age and existing fractures than density alone. Although it would be desirable to have assessment methods for the bone microarchitecture in routine, low-resolution modalities (e.g., DEXA), no single method has emerged as a routine complement for bone densitometry. Due to their costs, high-resolution modalities are rarely used in clinical practice, but could turn out to be promising in drug development.

References

1. Klibanski, A., Adams-Campbell, L., Bassford, T., Blair, S.N., Boden, S.D., Dickersin, K., et al.: Osteoporosis prevention, diagnosis, and therapy. J. Am. Med. Assoc **285**(6), 785–795 (2001)
2. Hernandez, C.J., Keaveny, T.M.: A biomechanical perspective on bone quality. Bone **39**(6), 1173–1181 (2006)
3. Holroyd, C., Cooper, C., Dennison, E.: Epidemiology of osteoporosis. Best Pract. Res. Clin. Endocrinol. Metabol. **22**(5), 671–685 (2008)
4. Ritchie, R.O.: How does human bone resist fracture? Ann. New York Acad. Sci. **1192**, 72–80 (2010)
5. Small, R.E.: Uses and limitations of bone mineral density measurements in the management of osteoporosis. Medsc. Gen. Med. **7**(2), 3 (2005)
6. Gennari, C.: Calcium and vitamin D nutrition and bone disease of the elderly. Publ. Health Nutr. **4**, 547–559 (2001)
7. Rittweger, J.: Can exercise prevent osteoporosis? J. Musculosceletal Neuronal Interact. **6**(2), 162 (2006)
8. Felsenberg, D., Boonen, S.: The bone quality framework: Determinants of bone strength and their interrelationships, and implications for osteoporosis management. Clin. Therapeut. **27**(1), 1–11 (2005)
9. Frost, H.M.: Dynamics of bone remodeling. Bone Biodynamics 315 (1964)
10. Wolff, I., Van Croonenborg, J.J., Kemper, H.C.G., Kostense, P.J., Twisk, J.W.R.: The effect of exercise training programs on bone mass: a meta-analysis of published controlled trials in pre-and postmenopausal women. Osteoporos. Int. **9**(1), 1–12 (1999)
11. Karlsson, M.K., Nordqvist, A., Karlsson, C.: Physical activity, muscle function, falls and fractures. Food Nutr. Res. 52 (2008)
12. Meunier, P.J., Sebert, J.L., Reginster, J.Y., Briancon, D., Appelboom, T., Netter, P., et al.: Fluoride salts are no better at preventing new vertebral fractures than calcium-vitamin D in postmenopausal osteoporosis: the FAVOStudy. Osteoporos. Int. **8**(1), 4–12 (1998)
13. Riggs, B.L., Hodgson, S.F., O'Fallon, W.M., Chao, E., Wahner, H.W., Muhs, J.M., et al.: Effect of fluoride treatment on the fracture rate in postmenopausal women with osteoporosis. Obstet. Gynecol. Surv. **45**(8), 542 (1990)
14. McCreadie, B.R., Goldstein, S.A.: Biomechanics of fracture: Is bone mineral density sufficient to assess risk? J. Bone Miner. Res. **15**(12), 2305–2308 (2000)
15. Rockoff, S.D., Sweet, E., Bleustein, J.: The relative contribution of trabecular and cortical bone to the strength of human lumbar vertebrae. Calcif. Tissue Int. **3**(1), 163–175 (1969)
16. Fields, A.J., Eswaran, S.K., Jekir, M.G., Keaveny, T.M.: Role of trabecular microarchitecture in whole-vertebral body biomechanical behavior. J. Bone Miner. Res. **24**(9), 1523–1530 (2009)
17. Keaveny, T.M., Morgan, E.F., Niebur, G.L., Yeh, O.C.: Biomechanics of trabecular bone. Annu. Rev. Biomed. Eng. **3**(1), 307–333 (2001)
18. Hernandez, C.J.: How can bone turnover modify bone strength independent of bone mass? Bone **42**(6), 1014–1020 (2008)
19. Ammann, P., Rizzoli, R.: Bone strength and its determinants. Osteoporos. Int. **14**(S3), 13–18 (2003)

20. Chappard, D., Baslé, M.F., Legrand, E., Audran, M.: Trabecular bone microarchitecture: A review. Morphologie **92**(299), 162–170 (2008)
21. Svendsen, O.L., Haarbo, J., Hassager, C., Christiansen, C.: Accuracy of measurements of body composition by dual-energy x-ray absorptiometry in vivo. Am. J. Clin. Nutr. **57**(5), 605 (1993)
22. Lang, T.F.: Quantitative computed tomography. Radiol. Clin. N. Am. **48**(3), 589–600 (2010)
23. Bushberg, J., Seibert, J., Leidholdt, Jr. E.M., Boone, J.M.: The essential Physics of medical imaging. Lippincott Williams & Wilkins, New York (2002)
24. Njeh, C.F., Boivin, C.M., Langton, C.M.: The role of ultrasound in the assessment of osteoporosis: a review. Osteoporos. Int. 7(1), 7–22 (1997)
25. Liu, X.S., Sajda, P., Saha, P.K., Wehrli, F.W., Bevill, G., Keaveny, T.M., et al.: Complete volumetric decomposition of individual trabecular plates and rods and its morphological correlations with anisotropic elastic moduli in human trabecular bone. J. Bone Miner. Res. **23**(2), 223–235 (2008)
26. Parfitt, A.M.: Bone histomorphometry: standardization of nomenclature, symbols and units (summary of proposed system). Bone **9**(1), 67–69 (1988)
27. Hildebrand, T., Laib, A., Müller, R., Dequeker, J., Rüegsegger, P.: Direct three dimensional morphometric analysis of human cancellous bone: microstructural data from Spine, Femur, Iliac Crest, and Calcaneus. J. Bone Miner. Res. **14**(7), 1167–1174 (1999)
28. Hernandez, C.J., Beaupre, G.S., Keller, T.S., Carter, D.R.: The influence of bone volume fraction and ash fraction on bone strength and modulus. Bone **29**(1), 74–78 (2001)
29. Hildebrand, T., Rüegsegger, P.: A new method for the model-independent assessment of thickness in three-dimensional images. J. Microsc. **185**(1), 67–75 (1997)
30. Cortet, B., Bourel, P., Dubois, P., Boutry, N., Cotten, A., Marchandise, X.: CT scan texture analysis of the distal radius: influence of age and menopausal status. Rev. Rhum. (English edn.) **65**(2), 109 (1998)
31. Ito, M., Ohki, M., Hayashi, K., Yamada, M., Uetani, M., Nakamura, T.: Trabecular texture analysis of CT images in the relationship with spinal fracture. Radiology **194**(1), 55 (1995)
32. Thomsen, J.S., Ebbesen, E.N., Mosekilde, L.: Relationships between static histomorphometry and bone strength measurements in human iliac crest bone biopsies. Bone **22**(2), 153–163 (1998)
33. Saha, P.K., Gomberg, B.R., Wehrli, F.W.: Three-dimensional digital topological characterization of cancellous bone architecture. Int. J. Imag. Syst. Tech. **11**(1), 81–90 (2000)
34. Le, H.M., Holmes, R.E., Shors, E.C., Rosenstein, D.A.: Computerized quantitative analysis of the interconnectivity of porous biomaterials. Acta. Stereologica. **11**, 267–267 (1992)
35. Vesterby, A., Gundersen, H.J.G., Melsen, F.: Star volume of marrow space and trabeculae of the first lumbar vertebra: sampling efficiency and biological variation. Bone **10**(1), 7–13 (1989)
36. Hahn, M., Vogel, M., Pompesius-Kempa, M., Delling, G.: Trabecular bone pattern factor–a new parameter for simple quantification of bone microarchitecture. Bone **13**(4), 327–330 (1992)
37. Laib, A., Hildebrand, T., Häuselmann, H.J., Rüegsegger, P.: Ridge number density: a new parameter for in vivo bone structure analysis. Bone **21**(6), 541–546 (1997)
38. Hildebrand, T., Rüegsegger, P.: Quantification of bone microarchitecture with the structure model index. Comput. Meth. Biomech. Biomed. Eng. **1**(1), 15–23 (1997)
39. Haidekker, M.A.: Advanced Biomedical Image Analysis. Wiley, Hoboken, NJ (2011)
40. Dougherty, G.: Image enhancement in the spatial domain. In: Digital image processing for medical applications, p. 170–188. Cambridge University Press, New York (2009)
41. Caldwell, C.B., Willett, K., Cuncins, A.V., Hearn, T.C.: Characterization of vertebral strength using digital radiographic analysis of bone structure. Med. Phys. **22**, 611 (1995)
42. Lespessailles, E., Gadois, C., Kousignian, I., Neveu, J.P., Fardellone, P., Kolta, S., et al.: Clinical interest of bone texture analysis in osteoporosis: a case control multicenter study. Osteoporos. Int. **19**(7), 1019–1028 (2008)
43. Haidekker, M.A., Andresen, R., Evertsz, C.J., Banzer, D., Peitgen, H.O.: Issues of threshold selection when determining the fractal dimension in HRCT slices of lumbar vertebrae. Br. J. Radiol. **73**(865), 69 (2000)

44. Haralick, R.M., Shanmugam, K., Dinstein, I.H.: Textural features for image classification. IEEE Trans. Syst. Man. Cybern. Syst. Hum. **3**(6), 610–621 (1973)
45. Laws, K.I.: Texture energy measures. Proc DARPA Image Unerstanding Workshop, pp. 47–51 (1979)
46. Lee, R.L., Dacre, J.E., Hart, D.J., Spector, T.D.: Femoral neck trabecular patterns predict osteoporotic fractures. Med. Phys. **29**, 1391 (2002)
47. Lespessailles, E., Gadois, C., Lemineur, G., Do-Huu, J.P., Benhamou, L.: Bone texture analysis on direct digital radiographic images: precision study and relationship with bone mineral density at the os calcis. Calcif. Tissue Int. **80**(2), 97–102 (2007)
48. Rachidi, M., Marchadier, A., Gadois, C., Lespessailles, E., Chappard, C., Benhamou, C.L.: Laws' masks descriptors applied to bone texture analysis: an innovative and discriminant tool in osteoporosis. Skeletal. Radiol. **37**(6), 541–548 (2008)
49. Vokes, T., Lauderdale, D., Ma, S.L., Chinander, M., Childs, K., Giger, M.: Radiographic texture analysis of densitometric calcaneal images: Relationship to clinical characteristics and to bone fragility. J. Bone Miner. Res. **25**(1), 56–63 (2010)
50. Wilkie, J.R., Giger, M.L., Engh, Sr. C.A., Hopper, Jr. R.H., Martell, J.M.: Radiographic texture analysis in the characterization of trabecular patterns in periprosthetic osteolysis1. Acad. Radiol. **15**(2), 176–185 (2008)
51. Chappard, C., Brunet-Imbault, B., Lemineur, G., Giraudeau, B., Basillais, A., Harba, R., et al.: Anisotropy changes in post-menopausal osteoporosis: characterization by a new index applied to trabecular bone radiographic images. Osteoporos. Int. **16**(10), 1193–1202 (2005)
52. Brunet-Imbault, B., Lemineur, G., Chappard, C., Harba, R., Benhamou, C.L.: A new anisotropy index on trabecular bone radiographic images using the fast Fourier transform. BMC Med. Imag. **5**(1), 4 (2005)
53. Peitgen, H.O., Jürgens, H., Saupe, D.: Chaos and fractals: new frontiers of science. Springer, New York (2004)
54. Mandelbrot, B.B.: The Fractal Geometry of Nature. Freeman, USA (1982)
55. Martínez-Lopez, F., Cabrerizo-Vílchez, M., Hidalgo-Alvarez, R.: A study of the different methods usually employed to compute the fractal dimension1. Phys. Stat. Mech. Appl. **311**, 411–428 (2002)
56. Saupe, D.: Algorithms for random fractals. In: Peitgen, H.-O., and Saupe, D. (eds.), The Science of Fractal Images, pp. 71–136. Springer, New York (1988)
57. Stein, M.C.: Nonparametric estimation of fractal dimension, vol. 1001 of Society of Photo-Optical Instrumentation Engineers (SPIE) Conference Series. SPIE (1988)
58. Chung, H.W., Chu, C.C., Underweiser, M., Wehrli, F.W.: On the fractal nature of trabecular structure. Med. Phys. **21**, 1535 (1994)
59. Dubuc, B., Zucker, S., Tricot, C., Quiniou, J., Wehbi, D.: Evaluating the fractal dimension of surfaces. Proc. Roy. Soc. Lond. Math. Phys. Sci. **425**(1868), 113–127 (1989)
60. Huang, Q., Lorch, J.R., Dubes, R.C.: Can the fractal dimension of images be measured? Pattern Recogn. **27**(3), 339–349 (1994)
61. Geraets, W.G., Van Der Stelt, P.F.: Fractal properties of bone. Dentomaxillofacial Radiology **29**(3), 144 (2000)
62. Lopes, R., Betrouni, N.: Fractal and multifractal analysis: A review. Med. Image. Anal. **13**(4), 634–649 (2009)
63. Lundahl, T., Ohley, W., Kuklinski, W.: Analysis and interpolation of angiographic images by use of fractals. Computers in Cardiology, p. 355. Linkoping, Sweden (1985)
64. Ruttimann, U.E., Webber, R.L., Hazelrig, J.B.: Fractal dimension from radiographs of peridental alveolar bone:: A possible diagnostic indicator of osteoporosis. Oral. Surg. Oral. Med. Oral. Pathol. **74**(1), 98–110 (1992)
65. Webber, R., Underhill, T., Horton, R., Dixon, R., Pope, Jr. T.: Predicting osseous changes in ankle fractures. IEEE Eng. Med. Biol. Mag. **12**(1), 103–110 (2002)
66. Majumdar, S., Weinstein, R.S., Prasad, R.R.: Application of fractal geometry techniques to the study of trabecular bone. Med. Phys. **20**, 1611 (1993)

67. Southard, T.E., Southard, K.A.: Detection of simulated osteoporosis in maxillae using radiographic texture analysis. IEEE Trans. Biomed. Eng. **43**(2), 123–132 (2002)
68. Veenland, J., Grashuis, J., Van der Meer, F., Beckers, A., Gelsema, E.: Estimation of fractal dimension in radiographs. Med. Phys. **23**, 585 (1996)
69. Fortin, C., Kumaresan, R., Ohley, W., Hoefer, S.: Fractal dimension in the analysis of medical images. IEEE Eng. Med. Biol. Mag. **11**(2), 65–71 (2002)
70. Messent, E., Buckland-Wright, J., Blake, G.: Fractal analysis of trabecular bone in knee osteoarthritis (OA) is a more sensitive marker of disease status than bone mineral density (BMD). Calcif. Tissue Int. **76**(6), 419–425 (2005)
71. Dougherty, G., Henebry, G.M.: Fractal signature and lacunarity in the measurement of the texture of trabecular bone in clinical CT images. Med. Eng. Phys. **23**(6), 369–380 (2001)
72. Dong, P.: Test of a new lacunarity estimation method for image texture analysis. Int. J. Rem. Sens. **21**(17), 3369–3373 (2000)
73. Plotnick, R.E., Gardner, R.H., Hargrove, W.W., Prestegaard, K., Perlmutter, M.: Lacunarity analysis: a general technique for the analysis of spatial patterns. Phys. Rev. E **53**(5), 5461–5468 (1996)
74. Dougherty, G., Henebry, G.M.: Lacunarity analysis of spatial pattern in CT images of vertebral trabecular bone for assessing osteoporosis. Med. Eng. Phys. **24**(2), 129–138 (2002)
75. Zaia, A., Eleonori, R., Maponi, P., Rossi, R., Murri, R.: MR imaging and osteoporosis: Fractal lacunarity analysis of trabecular bone. IEEE Trans. Inform. Tech. Biomed. **10**(3), 484–489 (2006)
76. Panagiotopoulou, O.: Finite element analysis (FEA): applying an engineering method to functional morphology in anthropology and human biology. Ann. Hum. Biol. **36**(5), 609–623 (2009)
77. Vesterby, A., Mosekilde, L., Gundersen, H.J.G., Melsen, F., Holme, K., Sørensen, S.: Biologically meaningful determinants of the in vitro strength of lumbar vertebrae. Bone **12**(3), 219–224 (1991)
78. Jones, A.C., Wilcox, R.K.: Finite element analysis of the spine: Towards a framework of verification, validation and sensitivity analysis. Med. Eng. Phys. **30**(10), 1287–1304 (2008)
79. Lavaste, F., Skalli, W., Robin, S., Roy-Camille, R., Mazel, C.: Three-dimensional geometrical and mechanical modelling of the lumbar spine. J. Biomech. **25**(10), 1153–1164 (1992)
80. Kuo, C.S., Hu, H.T., Lin, R.M., Huang, K.Y., Lin, P.C., Zhong, Z.C., et al.: Biomechanical analysis of the lumbar spine on facet joint force and intradiscal pressure-a finite element study. BMC Muscoskel. Disord. **11**, 151 (2010)
81. Gibson, L.J.: The mechanical behaviour of cancellous bone. J. Biomech. **18**(5), 317–328 (1985)
82. Jensen, K.S., Mosekilde, L.: A model of vertebral trabecular bone architecture and its mechanical properties. Bone **11**(6), 417–423 (1990)
83. Hollister, S.J., Brennan, J.M., Kikuchi, N.: A homogenization sampling procedure for calculating trabecular bone effective stiffness and tissue level stress. J. Biomech. **27**(4), 433–444 (1994)
84. Müller, R., Rüegsegger, P.: Three-dimensional finite element modelling of non-invasively assessed trabecular bone structures. Med. Eng. Phys. **17**(2), 126–133 (1995)
85. Magland, J., Vasilic, B., Wehrli, F.W.: Fast Low Angle Dual Spin Echo (FLADE): A new robust pulse sequence for structural imaging of trabecular bone. Magn. Reson. Med. **55**(3), 465–471 (2006)
86. Karjalainen, J.P., Toyras, J., Riekkinen, O., Hakulinen, M., Jurvelin, P.S.: Ultrasound backscatter imaging provides frequency-dependent information on structure, composition and mechanical properties of human trabecular bone. Ultrasound Med. Biol. **35**(8), 1376–1384 (2009)
87. Haïat, G., Padilla, F., Svrcekova, M., Chevalier, Y., Pahr, D., Peyrin, F., et al.: Relationship between ultrasonic parameters and apparent trabecular bone elastic modulus: A numerical approach. J. Biomech. **42**(13), 2033–2039 (2009)

88. Hosokawa, A.: Effect of porosity distribution in the propagation direction on ultrasound waves through cancellous bone. IEEE Trans. Ultrason. Ferroelectrics Freq. Contr. **57**(6), 1320–1328 (2010)
89. Davison, K.S., Kendler, D.L., Ammann, P., Bauer, D.C., Dempster, D.W., Dian, L., et al.: Assessing fracture risk and effects of osteoporosis drugs: bone mineral density and beyond. Am. J. Med. **122**(11), 992–997 (2009)
90. Resch, H., Libanati, C., Farley, S., Bettica, P., Schulz, E., Baylink, D.J.: Evidence that fluoride therapy increases trabecular bone density in a peripheral skeletal site. J. Clin. Endocrinol. Metabol. **76**(6), 1622 (1993)
91. Grynpas, M.D.: Fluoride effects on bone crystals. J. Bone Miner. Res. **5**(S1), S169–S175 (1990)
92. Jiang, Y., Zhao, J., Liao, E.Y., Dai, R.C., Wu, X.P., Genant, H.K.: Application of micro-CT assessment of 3-D bone microstructure in preclinical and clinical studies. J. Bone Miner. Metabol. **23**, 122–131 (2005)

Chapter 10
Applications of Medical Image Processing in the Diagnosis and Treatment of Spinal Deformity

Clayton Adam and Geoff Dougherty

10.1 Introduction

Spinal deformities are a group of disorders characterized by abnormal curvature of the spine. In the healthy spine, natural curves occur in the sagittal plane, with a lordosis (concave curvature) in the lower back (lumbar) region and kyphosis (convex curvature) in the upper back (thoracic) region. In some spinal deformities, these natural curves can be either suppressed or amplified, as in the case of hypokyphosis (flatback) or hyperkyphosis (exaggerated thoracic curvature or 'hunchback'). However, the most common type of deformity is scoliosis, which is defined as abnormal lateral (side to side) curvature of the spine accompanied by axial rotation. Because of the combined sagittal curvature, abnormal lateral curvature, and axial rotation, scoliosis is a complex three-dimensional deformity which cannot be visualised in any single viewing plane (Fig. 10.1).

This chapter describes the application of image processing to the assessment and treatment of spinal deformity, with a focus on the most common deformity, adolescent idiopathic scoliosis (AIS). We will briefly describe the natural history of spinal deformity and current approaches to surgical and non-surgical treatment to give some background to the problem, and present an overview of current clinically used imaging modalities. We will introduce the key metric currently used to assess the severity and progression of spinal deformities from medical images, the Cobb angle, and discuss the uncertainties involved in manual Cobb angle measurements. An alternative metric, the Ferguson angle, will also be discussed. This provides the context for an examination of semi-automated image processing approaches for improved measurement of spinal curve shape and severity, including the development of discrete and continuum representations of the thoracolumbar

C. Adam (✉)
Queensland University of Technology, Brisbane, Australia
e-mail: c.adam@qut.edu.au

G. Dougherty (ed.), *Medical Image Processing: Techniques and Applications*, Biological and Medical Physics, Biomedical Engineering, DOI 10.1007/978-1-4419-9779-1_10,

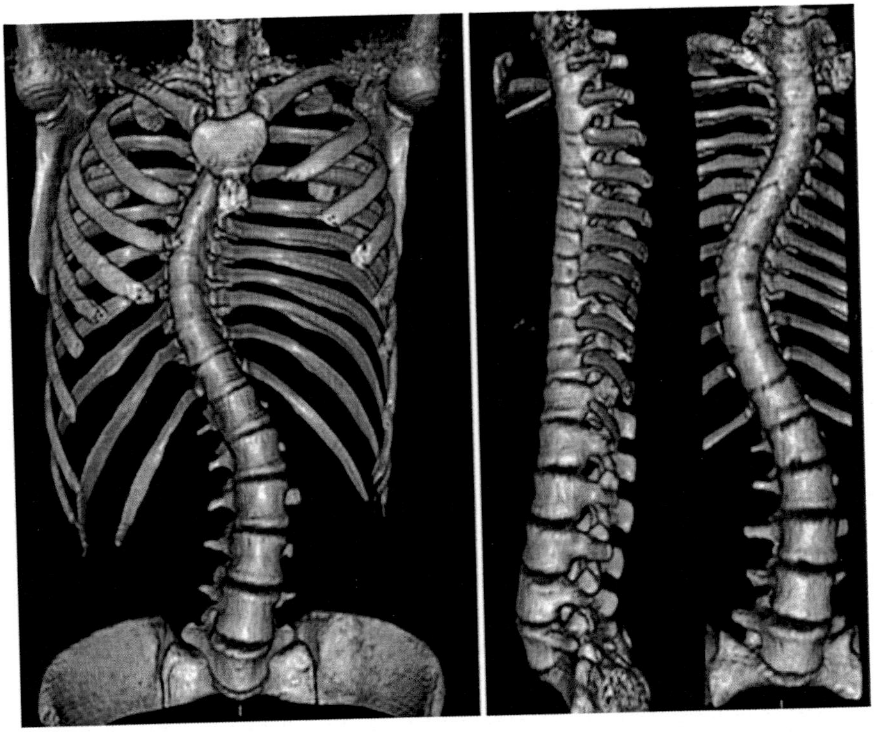

Fig. 10.1 Three dimensional CT reconstructions of a scoliotic spine showing overall spine and ribcage shape (*left*), sagittal, and coronal close-up views of the vertebral column (*right*)

spine and tortuosity measures. The newly defined metrics are applied to a dataset of idiopathic scoliosis patients and assessed by comparison with clinical Cobb angle measurements for the same patient group. Finally, areas for future image processing research applied to spinal deformity assessment and treatment are discussed.

10.1.1 Adolescent Idiopathic Scoliosis

While scoliosis can occur as a secondary consequence of a primary pathology (such as a leg length inequality or congenital malformation), most (70–80%) scoliosis cases occur during the adolescent growth spurt without known cause. These deformities are classified as AIS, and affect 2–4% of the population.

People with scoliosis often have no symptoms beyond a slight increase in pain and reduced lung capacity. However, progressive scoliosis leads to an increasingly severe cosmetic deformity and can compromise the function of internal organs in severe cases. For these reasons, both conservative (non-surgical) and surgical treatments for spinal deformities have been developed. Bracing is a common

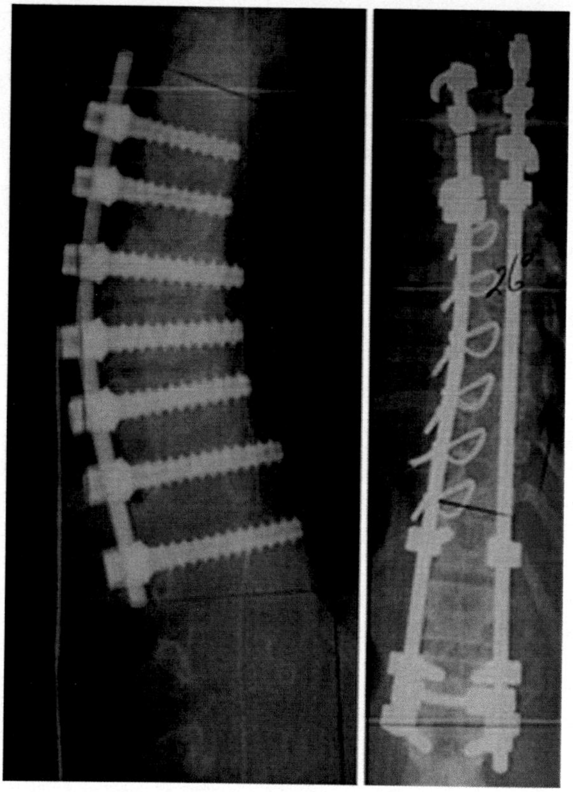

Fig. 10.2 Post-operative X-rays single rod showing anterior (*left*) and dual rod posterior (*right*) implants for scoliosis correction

conservative treatment in which the patient is asked to wear an orthotic brace which attempts to exert corrective forces on the deformed spine. Surgical approaches to scoliosis correction (Fig. 10.2) involve attachment of implants to the spine to restore a more normal curvature. Successful surgical approaches typically achieve a 60% reduction of the deformity and prevent further progression; however, there are risks of complication and further deformity progression after surgery which mandate ongoing imaging to monitor the corrected spine.

10.2 Imaging Modalities Used for Spinal Deformity Assessment

Many spinal deformities are visible just by looking at a patient. However, due to differences in body fat levels and bone structure between patients, they cannot be accurately assessed by visual examination of the patient's appearance. Medical imaging, therefore, plays a key role both in the monitoring of spinal deformity progression before treatment, and in assessing treatment outcomes. There are four medical imaging modalities relevant to the assessment of spinal deformities.

Planar and biplanar radiography Plane radiographs (X-rays) are the gold standard for imaging spinal deformities, and are used in spine clinics worldwide. A relatively new technology developed in France, the EOS system (EOS imaging, Paris, France), uses bi-planar radiography to simultaneously obtain coronal and sagittal radiographs of standing scoliosis patients with low radiation dose.

Computed Tomography (CT) CT is currently not used routinely for clinical assessment of scoliosis due to its greater cost and higher radiation dose than plane X-rays [1, 2]. However, low dose pre-operative CT is clinically indicated in endoscopic or keyhole scoliosis surgery for planning implant placement [3], and advances in scanner technology now allow CT with much lower radiation doses than previously possible [4]. In the research context, a number of groups have used CT to assess spinal deformities (in particular vertebral rotation in the axial plane) due to the 3D information provided [5–10].

Magnetic Resonance (MR) MR imaging is sometimes used clinically to detect soft tissue abnormalities (particularly the presence of syringomyelia in scoliosis patients), but clinical use of MR for spinal deformities is generally limited. Several research studies have used MR to investigate scoliosis [11–15]. MR is a useful research tool because of the absence of ionizing radiation, although the ability of MR to define bony anatomy is limited and thus image processing of MR datasets is often labour-intensive.

Back surface topography Various optical surface topography systems have been used to directly visualise the cosmetic effects of spinal deformity by assessing back shape in 3D [16–24]. Back surface topography does not involve ionising radiation, and so is useful for functional evaluations involving repeated assessments [25]. However, the relationship between spinal deformity and back shape is complicated by factors such as body positioning, trunk rotation, body build and fat folds [26].

10.2.1 Current Clinical Practice: The Cobb Angle

Currently, the accepted measure for clinical assessment of scoliosis is the *Cobb angle* [27]. The Cobb angle is measured on plane radiographs by drawing a line through the superior endplate of the superior end vertebra of a scoliotic curve, and another line through the inferior endplate of the inferior-most vertebra of the same scoliotic curve, and then measuring the angle between these lines (Fig. 10.3). Clinically, many Cobb measurements are still performed manually using pencil and ruler on hardcopy X-ray films, but PACS systems (viz. computer networks) are increasingly used which allow manual Cobb measurements to be performed digitally by clinicians on the computer screen. As well as being used to assess scoliosis in the coronal plane, the Cobb angle is used on sagittal plane radiographs to assess thoracic kyphosis and lumbar lordosis.

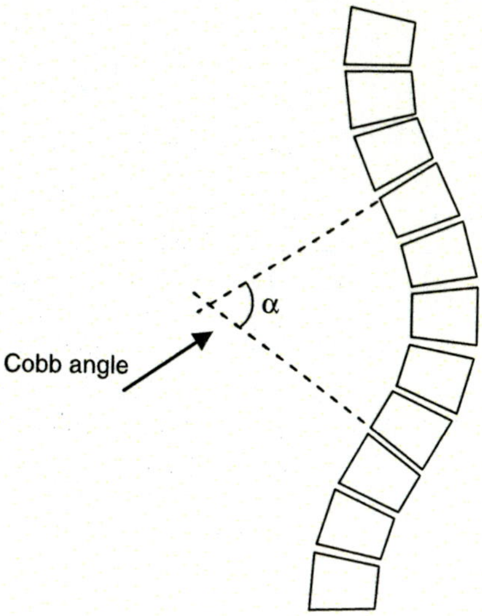

Fig. 10.3 Schematic view of a scoliotic deformity showing measurement of the Cobb angle

Although widely used for scoliosis assessment for its simplicity, the Cobb angle has several shortcomings. First, numerous studies have shown that the inter- and intra-observer measurement variabilities associated with the Cobb angle are high. If the vertebral endplates on a plane X-ray appear blurred due to a forward or backward tilt, considerable inter-observer errors can be introduced in the Cobb method as a result of the difficulty in selecting the endplate orientation. Such errors are even more pronounced in the presence of contour changes resulting from osteoporosis [28, 29]. Figure 10.4 shows a summary of Cobb variability studies performed between 1982 and 2005, indicating that the 95% confidence interval for the difference between two measurements by the same observer is around 5–7°, and for two measurements by different observers is around 6–8°. These measurement errors are large enough to make the difference between a diagnosis of progression (requiring treatment) or stability, and so introduce uncertainty into the assessment and treatment process.

Second, because of its simplicity, the Cobb angle omits potentially useful information about the shape of a scoliotic spine. Specifically, the Cobb angle cannot differentiate between a large scoliotic curve which may span 8 or 9 vertebral levels, and a small scoliotic curve which may only span 2 or 3 vertebral levels but that has the same endplate angulation due to its severity (Fig. 10.5).

10.2.2 An Alternative: The Ferguson Angle

Prior to the adoption of the Cobb angle as standard practice by the Scoliosis Research Society in 1966, a number of other scoliosis measurement techniques

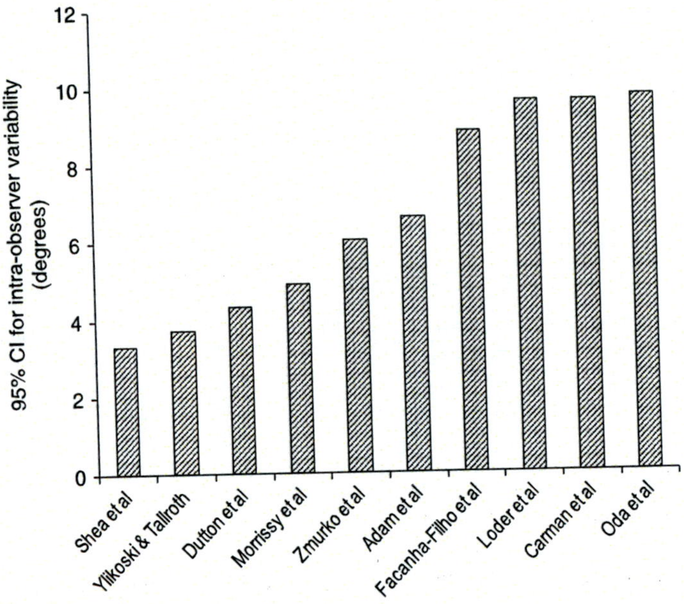

Fig. 10.4 Ninety five percent confidence intervals for intra-observer measurement variability using the Cobb angle from previous studies (figure based on [30])

Fig. 10.5 Simplified representation of three different scoliotic curves with increasing radius and a greater number of vertebral levels in the curve, but the same Cobb angle

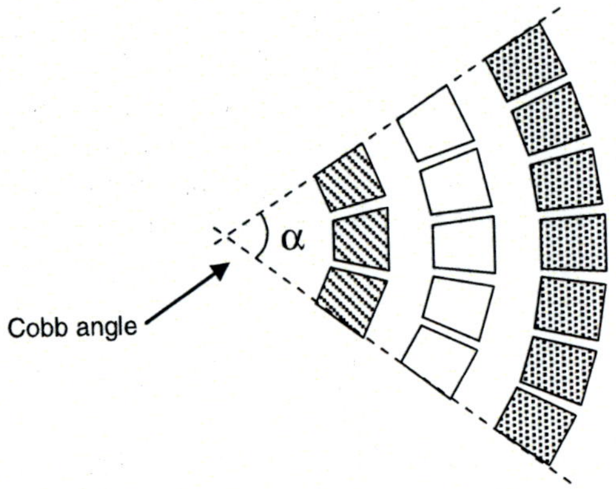

had been proposed, including the Ferguson angle [31]. The Ferguson angle requires identification of three landmark points, the geometric centres of the upper, apical (i.e. the most laterally deviated) and lower vertebrae in a scoliotic curve (Fig. 10.6). Not only does the Ferguson angle take into account the position of the apical vertebra, which the Cobb angle does not, it is less influenced by changes in the shape of the vertebrae [32].

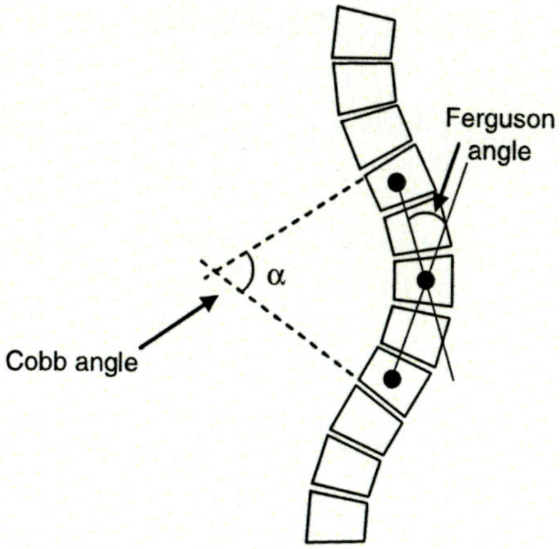

Fig. 10.6 Schematic view of a scoliotic deformity showing measurement of the Ferguson angle (Cobb angle shown for comparison)

The Ferguson angle is slightly more complicated to measure than Cobb due to the three landmark points (rather than two endplates in the Cobb technique). There can be difficulties in identifying the centres of the vertebrae, depending on their shape. The diagonals of the vertebra, as used in the standard Ferguson method, do not intersect at its geometric centre. The intersection of perpendicular lines drawn through the midpoints of the upper and lower endplates of a vertebra, proposed as a more accurate method of obtaining the geometric centre especially with prominently wedge-shaped vertebrae [32], is also flawed since it assumes that the four corners of the vertebral body lie on a circle. An alternative, more reliable method of finding the geometric centres has been proposed [33]. It has been shown that the Ferguson angle is closely related to the Cobb angle, with the ratio between Cobb and Ferguson angles for the same curve being ~1.35 [34]. Measurement variability for the Ferguson angle appears to be comparable with, or slightly higher than that of the Cobb angle [32, 34, 35].

10.3 Image Processing Methods

The increasing adoption of digital imaging provides a growing opportunity to (a) develop more detailed metrics for spinal deformity assessment which consider all vertebral levels in a scoliotic curve rather than just the two end vertebrae, (b) implement these metrics quickly and accurately using semi- and fully-automated image processing tools, and (c) measure scoliosis curvature using 3D datasets from modalities such as biplanar radiography, CT and MR. The remainder of this chapter presents details of image processing applications which are being developed by the authors in an attempt to avoid the manual measurement variability mentioned above, and also to allow development of new metrics which may in the longer term improve surgical planning and treatment decision making for spinal deformity patients.

10.3.1 Previous Studies

Development of medical image processing algorithms for assessment of scoliosis has been sparse to date. Several studies have used algorithms to perform computer-aided measurement of Cobb angle and other standard measures based on digitized anatomical landmarks manually selected by a user on a computer screen [36–39]. Gerard et al. [40] developed a semi-automated algorithm for scoliotic deformity measurement using dynamic programming optimisation. An automated algorithm for measuring vertebral rotation from a CT slices using the inherent symmetry of the vertebral cross-section was developed by Adam et al. [10]. Algorithms have also been developed to process back surface topography images (see Sect. 10.2), including a recent non-rigid registration algorithm [41]. However, as mentioned previously, back shape approaches are limited in their ability to assess the underlying spinal deformity and so are not widely used clinically.

10.3.2 Discrete and Continuum Functions for Spinal Curvature

A key aspect of developing more detailed metrics for spinal deformity assessment is measuring all vertebrae in a scoliotic curve, not just the end or apical vertebrae (as in the Cobb and Ferguson methods). Measuring all vertebrae provides a fuller understanding of how deformities affect each vertebral level before treatment, and also captures changes in spinal configuration after surgical treatment which may occur over only one or two vertebral levels (such as decompensation at the top of an implant construct). However, the challenges in measuring all vertebral levels are (a) defining appropriate anatomical landmarks for reliable, semi- or fully-automated detection from radiographic images, and (b) processing these landmark coordinates to obtain meaningful measures of deformity shape which will assist in diagnosis and treatment.

One approach being developed by the authors for use with low dose CT datasets of AIS patients is to use the edge of the vertebral canal as a robust anatomical landmark suitable for semi-automated detection. Figure 10.7 shows a transverse CT slice of a scoliosis patient, which demonstrates the clearly defined, high contrast, enclosed inner boundary of the vertebral canal.

Even though individual vertebrae may be substantially tilted relative to the transverse CT slices, even in large scoliosis curves (Cobb angle $\sim$60°), one or more CT slices will contain an enclosed vertebral canal such as that shown in Fig. 10.7. From these images, a spinal canal tracking algorithm has been developed using the ImageJ software (version 1.43u, National Institutes of Health, USA), which uses ImageJ's automatic edge tracing command to locate the inner boundary of the vertebral canal, and determine the canal centroid coordinates. By applying this algorithm for each vertebral level in a scoliosis curve, a series of 17 x,y,z datapoints for the entire thoracolumbar spine can be generated, and pairs of x,z and y,z data points define spinal curvature in the coronal and sagittal planes respectively.

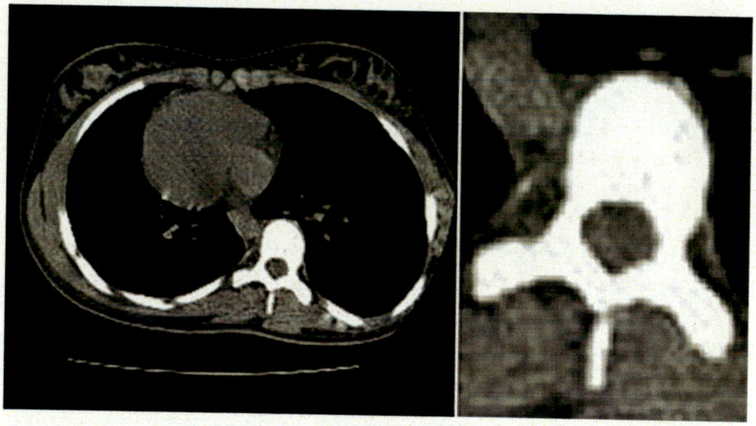

Fig. 10.7 *Left*: Axial CT slice showing the enclosed vertebral canal which is used as a landmark detection point by the semi-automated 'canal tracker' algorithm. Note that the anterior vertebral column is more rotated and deviated away from the mid-sagittal plane than the posterior part of the vertebra. *Right*: Close-up view of vertebral canal showing outline traced by the ImageJ edge detection algorithm and geometric centre of the detected outline which is used as the vertebral canal landmark

Having measured vertebral canal landmark coordinates for each vertebral level, these coordinates (which represent a 'discrete' representation of the scoliotic spine curvature) can then be used to generate a 'continuum' representation of scoliotic spinal curvature by fitting a mathematical function to the 17 sets of landmark coordinates. The approach used here is to fit two sixth order polynomial functions to pairs of x, z (coronal plane) and y, z (sagittal plane) coordinates respectively,

$$x = c_0 + c_1 z + c_2 z^2 + c_3 z^3 + c_4 z^4 + c_5 z^5 + c_6 z^6 \quad (10.1)$$

$$y = S_0 + S_1 z + S_2 z^2 + S_3 z^3 + S_4 z^4 + S_5 z^5 + S_6 z^6, \quad (10.2)$$

where c_n and s_n are the coronal and sagittal polynomial coefficients, respectively. Of course, other mathematical functions could be used. Figure 10.8 shows examples of two scoliosis patients where vertebral canal landmarks have been measured and continuum representations of the spinal curvature generated using this approach.

To show the relationship of the vertebral canal landmarks (which are located posteriorly in the vertebral canal) to the anterior vertebral column (where the Cobb and Ferguson angles are measured), Fig. 10.8 overlays the vertebral canal landmark points on CT reconstructions of the anterior vertebral columns. From these images it is apparent that first, while the vertebral canal landmarks and continuum curves closely follow the anterior column contours, the vertebral canal curvature tends to be less severe than that of the anterior column. This is to be expected because the anterior column is more rotated and displaced relative to the mid-sagittal plane than the vertebral canal (Fig. 10.7). Second, the vertebral canal landmark points do not always lie at mid-height relative to the anterior vertebral bodies.

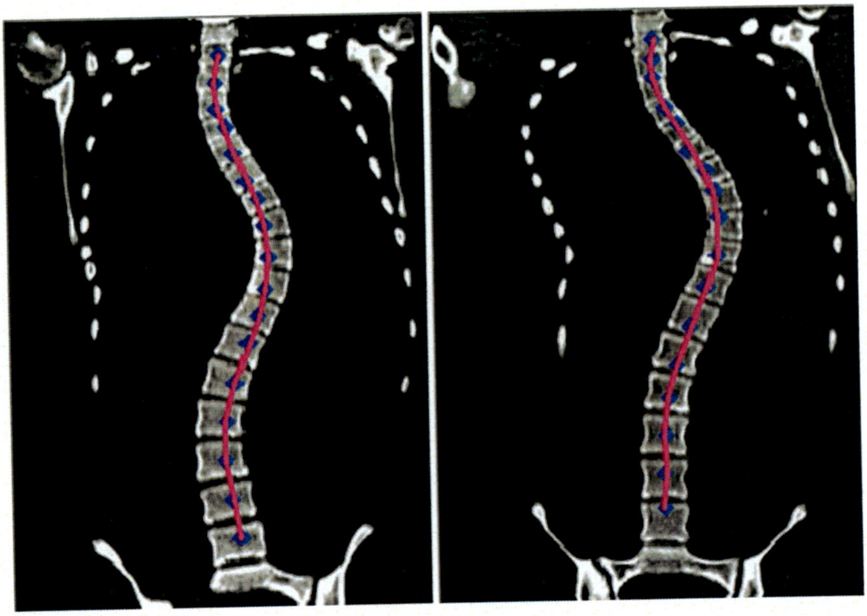

Fig. 10.8 Coronal CT reconstructions of two scoliosis patients showing the vertebral canal landmark points (*dark blue diamonds*) and polynomial fits (*pink lines*) to the vertebral canal landmark points which provide a continuum mathematical description of the spinal curvature

This can occur firstly because the vertebra may be tilted in the sagittal plane (with the anterior vertebral body higher than the posterior vertebral canal – see the sagittal CT view in Fig. 10.1), and also because the current implementation of the algorithm is semi-automated, so there may be a choice of several axial CT slices in which an enclosed vertebral canal can be clearly seen by the user, leading to a user variability in slice selection. We note, however, that the resulting continuum mathematical representation is insensitive to variations in the CT slice chosen within a particular vertebrae for spinal canal landmark detection.

Having obtained discrete landmark points and a continuum representation of the spinal curvature, various curve metrics can be extracted. In particular, the inflection points of the coronal plane polynomial can be readily located by finding the zeros of the second derivative of the polynomial (i.e. the roots of a fourth order polynomial),

$$\frac{d^2x}{dz^2} = 2c_2 + 6c_3z + 12c_4z^2 + 20c_5z^3 + 30c_6z^4 = 0, \qquad (10.3)$$

and the angle between the normals to the coronal curve at two neighboring inflection points can then be determined to provide a 'Cobb-equivalent angle'. This approach is analogous with the clinical definition of the coronal Cobb angle, which is based on locating the 'most tilted' endplates in a scoliotic curve.

In some cases, the coronal polynomial curve has only one inflection point, and in these cases an alternative approach must be used to generate the 'Cobb-equivalent' metric. Here, we propose that the z-coordinates of the upper and lower end vertebrae of the scoliotic curve *as determined clinically* are used in this situation. The angle between the normals to the coronal curve at these vertebral levels is again a Cobb-equivalent measure, with the disadvantage that a manual selection of levels was required.

The two approaches just presented; (1) Cobb-equivalent angle determined as the angle between inflection points of coronal polynomial, and (2) Cobb-equivalent angle determined as the angle between manually selected vertebral locations of coronal polynomial are denoted as the 'Cobb-equivalent 1' and 'Cobb-equivalent 2' angles, respectively.

10.3.3 Tortuosity

The concept of tortuosity, the accumulation of curvature along a curve, has been used to characterise blood vessels and their risk of aneurysm formation or rupture [33, 42–47]. A variety of possible metrics for tortuosity have been proposed, such as the distance factor (the relative length increase from a straight line) [48–50] or sinuosity [51], the number of inflection points along the curve [52], the angle change along segments [53, 54], and various line integrals of local curvature values [42, 44], which can be computed from second differences of the curve [43].

We have developed two tortuosity metrics [45] which are amenable to automation and can be used as putative scoliosis metrics for measuring the severity of the condition. Both are inherently three-dimensional, although they can be applied to two-dimensional projections. (A third possible metric, the integral of the square of the derivative of curvature of a spline-fit smoothest path, was found not to be scale-invariant, and subsequently abandoned [55]).

The first metric delivers a scoliotic angle, which can be considered an extension of the Ferguson angle. It is the accumulated angle turned along the length of the section of spine of interest, calculated as the sum of the magnitudes of the angles between straight line segments connecting the consecutive centres of all the vertebrae under consideration (Fig. 10.9). We have previously designated it as M. It can be applied to the whole spine, or a designated curve within it.

Figure 10.9 shows that M can be considered an extension of the Ferguson angle (referred to here as the *segmental Ferguson angle*). For a given scoliotic curve, the segmental Ferguson angle will be larger than the conventional Ferguson angle and smaller than the Cobb angle, although in the special (theoretical) case of a circular arc comprising many short segments its value approaches that of the Cobb angle. This special case is shown schematically in Fig. 10.10.

The second metric is based on a continuum function for the spinal curve, based on a unit speed parameterization of the vertebral centres. A piece-wise spline is used to produce a continuous function, which is the 'smoothest path' connecting the vertebral centres (Fig. 10.11), and it is used to compute a normalized

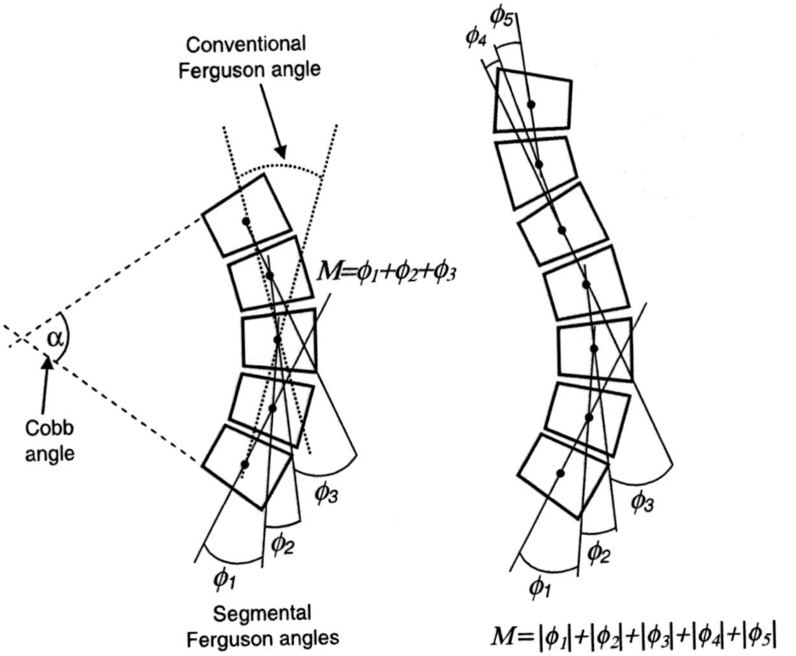

Fig. 10.9 *Left*: Portion of a scoliotic curve showing conventional Cobb and Ferguson angles as well as the segmental Ferguson angles which are summed to give the coronal tortuosity metric M. *Right*: Absolute segmental angles are summed in the case of a spinal curve containing both positive and negative angles

Fig. 10.10 For the special case of a circular arc the Cobb angle α is twice the Ferguson angle ϕ. As the arc is divided into successively more segments, the coronal tortuosity (sum of the segmental Ferguson angles $M = \phi_1 + \phi_2 + \phi_3 + \ldots$) approaches the Cobb angle

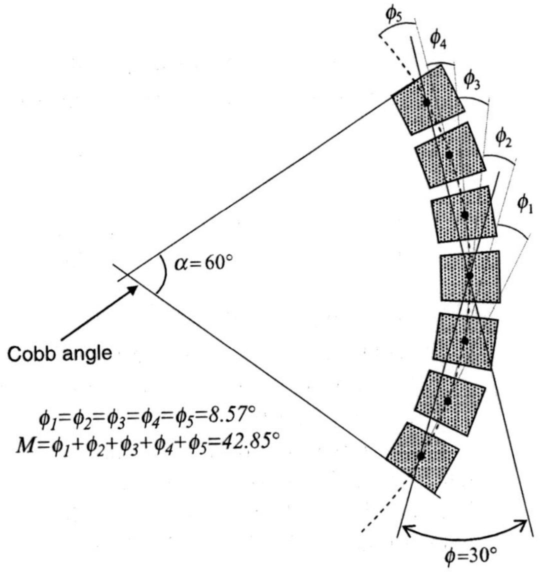

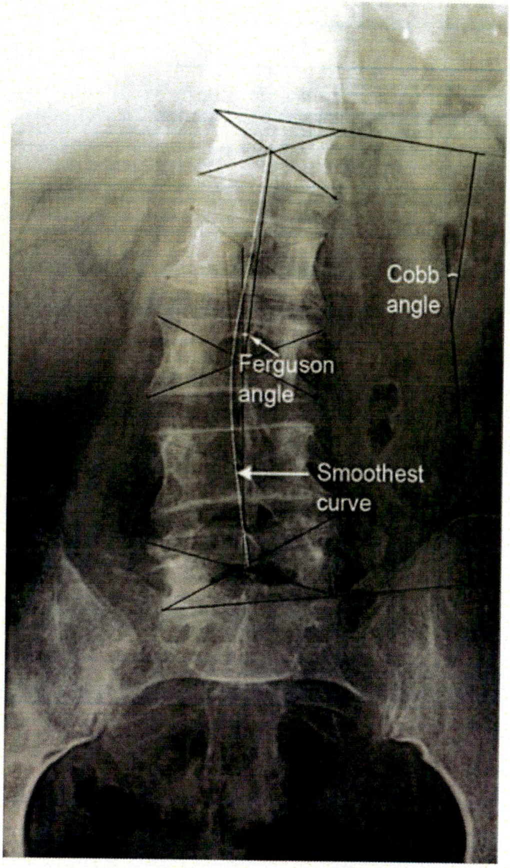

Fig. 10.11 Anterior–posterior (AP) radiograph, illustrating the measurement of the conventional Cobb and Ferguson angles, and showing the smoothest-path piece-wise spline iteratively fitted to the geometric centres of the vertebrae

root-mean-square (rms) curvature, which we designated K. It is defined in terms of the root-mean-square curvature, J, of the smoothest path by

$$K = \sqrt{J.L}, \tag{10.4}$$

where L is the length of the smoothed curve. The 'normalization' by $\sqrt{L}$ ensures that K is dimensionless (viz. an angle). While M is an accumulated angle using straight line segments between the vertebral centres, K is an accumulated angle using the smoothest path connecting the centres. With K, the accumulation is not democratic; rather contributions of higher curvature are given more weight than contributions of lower curvature. (If the curvature is constant, then K is forced to accumulate democratically and $K = M$.)

Both metrics have been shown to be scale invariant and additive, and K is essentially insensitive to digitization errors [45] Their usefulness has been demonstrated in discriminating between arteries of different tortuosities in assessing the relative utility of the arteries for endoluminal repair of aneurysms [33].

10.4 Assessment of Image Processing Methods

Before adopting an automated method for the assessment of spinal deformities, there is a need to compare the proposed method with the results from current practice, in particular the Cobb angle which is the clinically accepted measure of scoliosis severity. Here, we apply the image processing techniques described above to derive Cobb-equivalent and tortuosity metrics in a series of AIS patients. We then compare the new metrics with existing, clinically measured Cobb angles for the patients in the series.

10.4.1 Patient Dataset and Image Processing

The patient group comprised 79 AIS patients from the Mater Children's Hospital in Brisbane, Australia. Each of the patients in this series underwent thoracoscopic (keyhole) anterior surgery for correction of their deformity, and prior to surgery each patient received a single, low-dose, pre-operative CT scan for surgical planning purposes. Pre-operative CT allows safer screw sizing and positioning in keyhole scoliosis surgery procedures [56]. The estimated CT radiation dose for the scanning protocol used was 3.7 mSv [57].

Following the procedure described in Sect. 10.3.2, vertebral canal landmark co-ordinates were measured for each thoracolumbar vertebrae in each patient in the series. We note that measurement of spinal deformities from supine CT scans yields lower values of the Cobb angle than measurements on standing patients. Torell et al. [58] reported a 9° reduction in Cobb angle for supine compared to standing patients, and Yazici et al. [59] showed a reduction in average Cobb angle from 56 to 39° between standing and supine positions.

10.4.2 Results and Discussion

The patient group comprised of 74 females and 5 males with a mean age of 15.6 years (range 9.9–41.2) at the time the CT scan was performed. The mean height was 161.5 cm (range 139.5–175) and mean weight was 53.4 kg (range 30.6–84.7). All 79 patients had right-sided major scoliotic curves.[1] The mean clinically measured major Cobb angle for the group was 51.9° (range 38–68°). The clinical Cobb measurements were performed manually at the Mater Children's Hospital spinal clinic by experienced clinicians, using standing radiographs. Figure 10.12 shows a

[1] The major curve is defined as the curve with the largest Cobb angle in a scoliotic spine. Typically, adolescent idiopathic scoliosis major curves are convex to the right in the mid-thoracic spine, with smaller (minor) curves above and below, convex to the left.

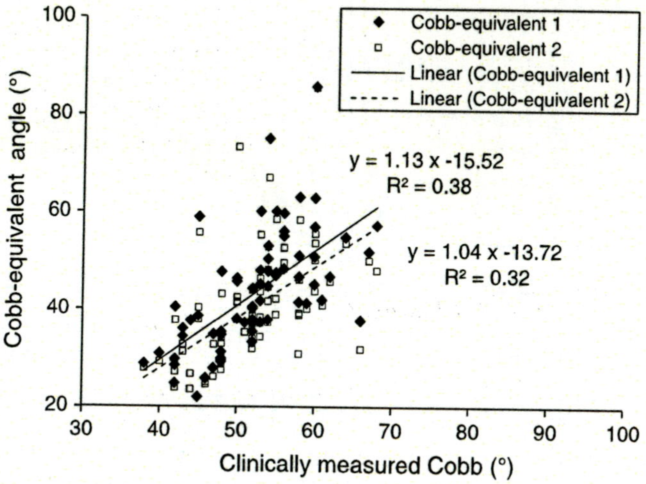

Fig. 10.12 Plot of Cobb-equivalent 1 and Cobb-equivalent 2 angles against clinically measured coronal Cobb angle for each patient in the series. Note that for 10 of the 79 patients, there was only one inflection point on the polynomial curve so Cobb-equivalent1 could not be determined for these ten patients

comparison between the clinically measured (standing) Cobb angle and the Cobb-equivalent 1 and Cobb-equivalent 2 metrics derived from the supine CT scans for the patient group.

With respect to Fig. 10.12, the relatively low R^2 values of 0.38 and 0.32 (Cobb-equivalent 1 and Cobb-equivalent 2, respectively) show there are substantial variations between individual clinical Cobb measurements from standing radiographs, and the Cobb-equivalent angles derived from continuum representations of spine shape on supine CT scans. The 13.7 and 15.5° offsets in the linear regression equations for the two Cobb-equivalent angles are consistent with the magnitude of change in Cobb angle between standing and supine postures of 9° and 17° reported by Torell et al. [58] and Yazici et al. [59]. The gradients of the regression lines for Cobb-equivalent 1 (1.13) and Cobb-equivalent 2 (1.04) in Fig. 10.13 are close to unity as would be expected, but the slightly greater than unity values suggest that either (1) bigger scoliotic curves are more affected by gravity (i.e. the difference between standing and supine Cobb increases with increasing Cobb angle), or (2) there is greater rotation of the anterior vertebral column (where clinical Cobb angles are measured from endplates) compared to the posterior vertebral canal (where the Cobb-equivalent landmarks are measured) in patients with larger deformities. Note that although not shown in Fig. 10.12, Cobb-equivalent 1 and Cobb-equivalent 2 are highly correlated with each other ($R^2 = 0.96$).

Figure 10.13 shows the close correlation between two of the new metrics, the coronal tortuosity (segmental Ferguson angle) of the major curve and the Cobb-equivalent 2. The coronal tortuosity, M (or segmental Ferguson angle), is strongly correlated ($R^2 = 0.906$, $p < 0.0000001$) with the Cobb-equivalent 2 angle.

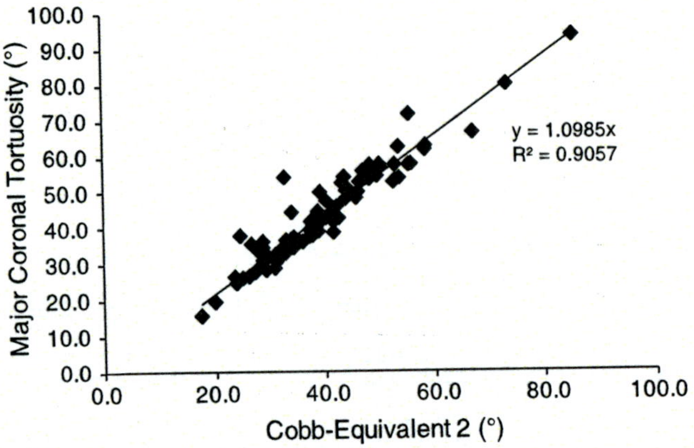

Fig. 10.13 Major coronal tortuosity (segmental Ferguson angle) vs. Cobb-equivalent 2 for the patient group, showing the close correlation between these two metrics

It is almost 10% larger for all angles, as expected from a metric based on the Ferguson angle, and there is no offset angle. This represents a strong internal consistency for these two semi-automated metrics. *M* follows the Cobb-equivalent 2 angle in being larger than the measured Cobb angle and having a significant correlation ($R^2 = 0.30$, $p < 0.007$) to it.

To remove the influence of (1) patient positioning (supine vs. standing) and (2) measurement error associated with a single clinical Cobb measurement, we performed a separate sub-study on 12 of the 79 patients in the main patient group.[2] For each patient in the sub-study, repeated manual Cobb measurements (six clinicians measured each Cobb angle on three separate occasions at least a week apart) were made using 2D coronal reconstructions from the supine CT scans. This allowed direct comparison of manual and Cobb-equivalent metrics using the same supine CT datasets, and reduced manual measurement variability by using repeated measures by multiple observers. Figure 10.14 shows the result of this comparison.

The R^2 value of 0.88 in Fig. 10.14 suggests that when postural differences are accounted for, the Cobb-equivalent metric is strongly correlated to manual Cobb measurements, but has the advantage of not being prone to the substantial manual measurement variability which can occur with a single Cobb measurement by a single observer. Note that the intercept and gradient of the regression equation in Fig. 10.14 suggest that although the metric is strongly correlated with supine manual measures, there is still a difference between the magnitudes of the two Cobb measures, perhaps due to the difference in anatomical location of the landmarks used in each case (manual Cobb uses endplates in the anterior column, whereas

[2] Note that the measurements described in this sub-study were performed before the main study, so there was no bias in the selection of the 12 patients based on the results from the entire patient group.

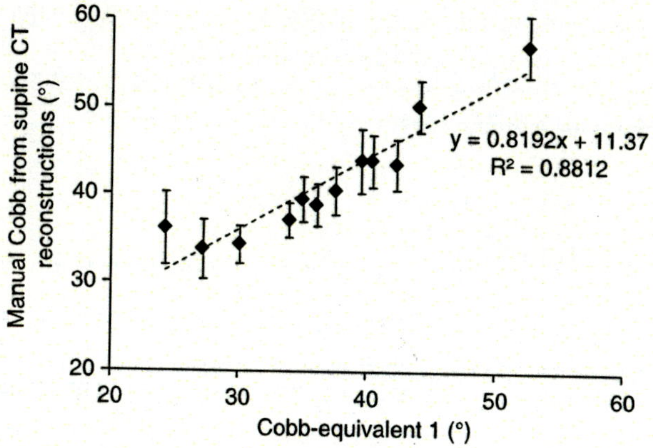

Fig. 10.14 Comparison of manually measured Cobb and Cobb-equivalent 1 for a subgroup of 12 patients, where manual Cobb angles were measured from 2D coronal supine CT reconstructions. Each data point represents the mean of 18 manual measurements (six observers on three occasions each). *Error bars* are the standard deviation of the manual measurements

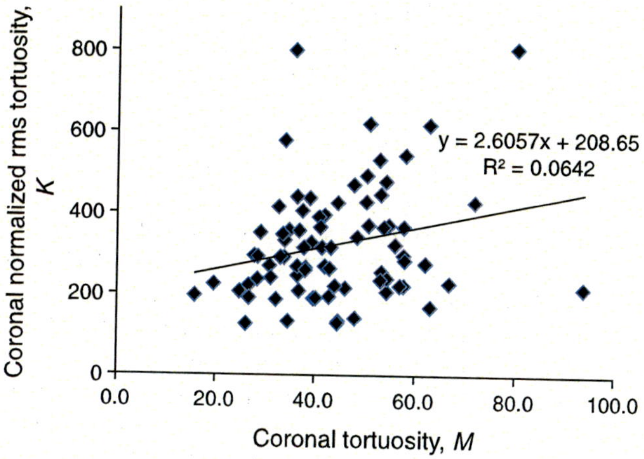

Fig. 10.15 Plot of normalized root-mean-square tortuosity, *K*, against tortuosity, *M*, both measured in the coronal (AP) plane, for each patient in the series

Cobb-equivalent metric uses vertebral canal). Also the number of patients used in this sub-study was relatively small, due to the constraints associated with obtaining multiple repeat Cobb measurements by a group of clinicians.

Figure 10.15 shows the relationship between the two tortuosity-based metrics, *K* and *M* (Sect. 10.3.3). Clearly, the correlation is very poor, although both metrics have been shown to correlate well with the ranking of an expert panel when used with retinal vessels [47]. However, in this application, the data is very sparse and there are no 'data balls' which can be used to constrain the spline-fitting [45].

Under these circumstances, the utility of the K metric is questionable. Since it computes tortuosity differently, by emphasizing contributions of high curvature, it is not directly comparable to any of the other methods and seems to have limited applicability to scoliotic angles.

10.5 Summary

Current clinical approaches to spinal deformity assessment and treatment are based on manual (printed film or computer screen) measurement of plane radiographs, along with limited use of other modalities such as CT/MRI or back shape analysis. The Cobb angle is currently the standard clinical metric for assessing the severity of a scoliotic curve. It reduces the 3D curvature to a single angle, measured at the upper and lower vertebral endplates of the curve. The Cobb angle is a key parameter used in surgical decision-making, yet measurement variability studies have demonstrated that it is a relatively 'noisy' measure (Sect. 10.2.1). The alternative, the Ferguson angle, includes lateral deviation at the apex of the deformity but the geometric centres of the vertebrae are difficult to establish from a plane radiograph (Sect. 10.2.2), especially when the vertebrae are wedge-shaped [32].

Given these uncertainties in manual measurement and the increasing availability of digitized medical images, there are emerging opportunities for the development of medical image processing techniques to assess spinal deformities. Both discrete and continuum representations of spinal curvature on a vertebral level-by-level basis offer the potential for better reproducibility and sensitivity so that the progression of disease can be followed using automated or semi-automated selection of anatomical landmarks such as the vertebral canal landmark detection approach demonstrated here. Image processing approaches also offer the potential to develop new metrics which use data from all of the vertebrae in a scoliotic curve rather than only two or three manually selected vertebrae.

One practical issue around the development of new spinal deformity assessment techniques is how they compare with existing clinical measures, and for this reason we included a comparison of several new metrics (Cobb equivalent 1, Cobb equivalent 2 and tortuosity metrics) with manual Cobb measurements for a group of AIS patients. This comparison showed that a single manual Cobb measurement by a single observer is subject to significant measurement variability, which results in scatter when comparing manual and Cobb-equivalent measures (Fig. 10.12). However, when a group of manual measurements of the same image are averaged, there is much closer agreement between manual Cobb and Cobb-equivalent metrics (Fig. 10.14). Further, the Cobb-equivalent 1, Cobb-equivalent 2 and coronal tortuosity metrics are all closely correlated. These initial results show that continuum and discrete representations of entire thoracolumbar spinal curves can be interrogated to yield simple clinical measures which agree closely with current manual measurements, but more work is required to extend the comparison to 3D (sagittal and axial planes), and to other clinical measures than the Cobb angle.

The image processing metrics which we presented here were based on semi-automated landmark detection in the vertebral canal, which is a high-contrast landmark on transverse CT slices; however, semi-automated detection of the anterior vertebral column would be a valuable direction for future study, as the anterior column in scoliosis tends to be more deformed than the posterior region.

We note again that although CT is not current clinical practice for scoliosis assessment (except in the case of keyhole surgery planning), advances in CT scanner technology have dramatically reduced radiation dose compared to earlier scanners [6], and CT or biplanar radiography (with their associated advantages of 3D reconstruction with good bony resolution) may become more common. One issue with CT is the relatively large difference in deformity magnitude between supine and standing postures (which in itself is a potentially valuable indicator of spine flexibility). A move toward 3D imaging modalities is likely considering the increasing realisation of the need to consider scoliosis as a 3D deformity [60].

There is much potential for future development of image processing algorithms based on 3D imaging modalities for improved assessment and treatment of spinal deformities. New metrics can assist in surgical planning by highlighting 3D aspects of the deformity, by feeding into biomechanical analysis tools (such as finite element simulations of scoliosis [61], and by interfacing with existing classification systems [39, 62, 63] to provide automated classification.

References

1. Nash, C.L. Jr., Gregg, E.C., Brown, R.H., et al.: Risks of exposure to X-rays in patients undergoing long-term treatment for scoliosis. J. Bone Joint Surg Am. **61**, 371–374 (1979)
2. Levy, A.R., Goldberg, M.S., Mayo, N.E., et al.: Reducing the lifetime risk of cancer from spinal radiographs among people with adolescent idiopathic scoliosis. Spine (Phila. Pa. 1976) **21**, 1540–1547 (1996); discussion 1548
3. Kamimura, M., Kinoshita, T., Itoh, H., et al.: Preoperative CT examination for accurate and safe anterior spinal instrumentation surgery with endoscopic approach. J. Spinal Disord. Tech. **15**, 47–51 (2002); discussion 51–42
4. Abul-Kasim, K., Overgaard, A., Maly, P., et al.: Low-dose helical computed tomography (CT) in the perioperative workup of adolescent idiopathic scoliosis. Eur. Radiol. **19**, 610–618 (2009)
5. Aaro, S., Dahlborn, M.: Estimation of vertebral rotation and the spinal and rib cage deformity in scoliosis by computer tomography. Spine **6**, 460–467 (1981)
6. Ho, E.K., Upadhyay, S.S., Ferris, L., et al.: A comparative study of computed tomographic and plain radiographic methods to measure vertebral rotation in adolescent idiopathic scoliosis. Spine **17**, 771–774 (1992)
7. Krismer, M., Sterzinger, W., Haid, C., et al.: Axial rotation measurement of scoliotic vertebrae by means of computed tomography scans. Spine **21**, 576–581 (1996)
8. Krismer, M., Chen, A.M., Steinlechner, M., et al.: Measurement of vertebral rotation: a comparison of two methods based on CT scans. J. Spinal Disord. **12**, 126–130 (1999)
9. Gocen, S., Havitcioglu, H., Alici, E.: A new method to measure vertebral rotation from CT scans. Eur. Spine J. **8**, 261–265 (1999)
10. Adam, C.J., Askin, G.N.: Automatic measurement of vertebral rotation in idiopathic scoliosis. Spine (Phila. Pa. 1976) **31**, E80–E83 (2006)

11. Perie, D., Sales de Gauzy, J., Curnier, D., et al.: Intervertebral disc modeling using a MRI method: migration of the nucleus zone within scoliotic intervertebral discs. Magn. Reson. Imag. **19**, 1245–1248 (2001)
12. Perie, D., Curnier, D., de Gauzy, J.S.: Correlation between nucleus zone migration within scoliotic intervertebral discs and mechanical properties distribution within scoliotic vertebrae. Magn. Reson. Imag. **21**, 949–953 (2003)
13. Violas, P., Estivalezes, E., Briot, J., et al.: Objective quantification of intervertebral disc volume properties using MRI in idiopathic scoliosis surgery. Magn. Reson. Imag. **25**, 386–391 (2007)
14. Wessberg, P., Danielson, B.I., Willen, J.: Comparison of Cobb angles in idiopathic scoliosis on standing radiographs and supine axially loaded MRI. Spine (Phila. Pa. 1976) **31**, 3039–3044 (2006)
15. Adam, C., Izatt, M., Askin, G.: Design and evaluation of an MRI compatible axial compression device for 3D assessment of spinal deformity and flexibility in AIS. Stud. Health Technol. Inform. **158**, 38–43 (2010)
16. Willner, S.: Moiré topography for the diagnosis and documentation of scoliosis. Acta Orthop. Scand. **50**, 295–302 (1979)
17. Stokes, I.A., Moreland, M.S.: Measurement of the shape of the surface of the back in patients with scoliosis. The standing and forward-bending positions. J. Bone Joint Surg. Am. **69**, 203–211 (1987)
18. Turner-Smith, A.R., Harris, J.D., Houghton, G.R., et al.: A method for analysis of back shape in scoliosis. J. Biomech. **21**, 497–509 (1988)
19. Weisz, I., Jefferson, R.J., Turner-Smith, A.R., et al.: ISIS scanning: a useful assessment technique in the management of scoliosis. Spine (Phila. Pa. 1976) **13**, 405–408 (1988)
20. Theologis, T.N., Fairbank, J.C., Turner-Smith, A.R., et al.: Early detection of progression in adolescent idiopathic scoliosis by measurement of changes in back shape with the integrated shape imaging system scanner. Spine **22**, 1223–1227 (1997); discussion 1228
21. Hackenberg, L., Hierholzer, E., Potzl, W., et al.: Rasterstereographic back shape analysis in idiopathic scoliosis after posterior correction and fusion. Clin. Biomech. (Bristol, Avon) **18**, 883–889 (2003)
22. Berryman, F., Pynsent, P., Fairbank, J., et al.: A new system for measuring three-dimensional back shape in scoliosis. Eur. Spine. J. **17**, 663–672 (2008)
23. Shannon, T.M.: Development of an apparatus to evaluate Adolescent Idiopathic Scoliosis by dynamic surface topography. Stud. Health Technol. Inform. **140**, 121–127 (2008)
24. Zubovic, A., Davies, N., Berryman, F., et al.: New method of scoliosis deformity assessment: ISIS2 System. Stud. Health Technol. Inform. **140**, 157–160 (2008)
25. Drerup, B., Ellger, B., Meyer zu Bentrup, F.M., et al.: Functional raster stereographic images: A new method for biomechanical analysis of skeletal geometry. Orthopade **30**, 242–250 (2001)
26. Wong, H.K., Balasubramaniam, P., Rajan, U., et al.: Direct spinal curvature digitization in scoliosis screening – a comparative study with Moiré contourgraphy. J. Spinal Disord. **10**, 185–192 (1997)
27. Cobb, J.R.: Outline for the study of scoliosis. American Academy of Orthopedic Surgeons Instructional Course Lectures (1948)
28. Genant, H.K., Wu, C.Y., van Kuijk, C., et al.: Vertebral fracture assessment using a semiquantitative technique. J. Bone Miner Res. **8**, 1137–1148 (1993)
29. Polly, D.W., Jr., Kilkelly, F.X., McHale, K.A., et al.: Measurement of lumbar lordosis. Evaluation of intraobserver, interobserver, and technique variability. Spine (Phila. Pa. 1976) **21**, 1530–1535 (1996); discussion 1535–1536
30. Adam, C.J., Izatt, M.T., Harvey, J.R., et al.: Variability in Cobb angle measurements using reformatted computerized tomography scans. Spine **30**, 1664–1669 (2005)
31. Ferguson, A.B.: Roentgen diagnosis of the extremities and spine, pp. 414–415. Hoeber, New York (1949)
32. Diab, K.M., Sevastik, J.A., Hedlund, R., et al.: Accuracy and applicability of measurement of the scoliotic angle at the frontal plane by Cobb's method, by Ferguson's method and by a new method. Eur. Spine J. **4**, 291–295 (1995)

33. Dougherty, G., Johnson, M.J.: Assessment of scoliosis by direct measurement of the curvature of the spine. Proc. SPIE **7260**, 72603Q (2009). doi:10.1117/12.806655
34. Stokes, I.A., Aronson, D.D., Ronchetti, P.J., et al.: Reexamination of the Cobb and Ferguson angles: Bigger is not always better. J. Spinal Disord. **6**, 333–338 (1993)
35. Gupta, M.C., Wijesekera, S., Sossan, A., et al.: Reliability of radiographic parameters in neuromuscular scoliosis. Spine **32**, 691–695 (2007)
36. Chockalingam, N., Dangerfield, P.H., Giakas, G., et al.: Computer-assisted Cobb measurement of scoliosis. Eur. Spine J. **11**, 353–357 (2002)
37. Cheung, J., Wever, D.J., Veldhuizen, A.G., et al.: The reliability of quantitative analysis on digital images of the scoliotic spine. Eur. Spine J. **11**, 535–542 (2002)
38. Zhang, J., Lou, E., Hill, D.L., et al.: Computer-aided assessment of scoliosis on posteroanterior radiographs. Med. Biol. Eng. Comput. **48**, 185–195 (2010)
39. Stokes, I.A., Aronsson, D.D.: Computer-assisted algorithms improve reliability of King classification and Cobb angle measurement of scoliosis. Spine **31**, 665–670 (2006)
40. Gerard, O., Lelong, P., Planells-Rodriguez, M., et al.: Semi-automatic landmark detection in digital X-ray images of the spine. Stud. Health Technol. Inform. **88**, 132–135 (2002)
41. Mitchell, H.L., Ang, K.S.: Non-rigid surface shape registration to monitor change in back surface topography. Stud. Health Technol. Inform. **158**, 29–33 (2010)
42. Hart, W.E., Goldbaum, M., Cote, B., et al.: Measurement and classification of retinal vascular tortuosity. Int. J. Med. Inform. **53**, 239–252 (1999)
43. Dougherty, G., Varro, J.: A quantitative index for the measurement of the tortuosity of blood vessels. Med. Eng. Phys. **22**, 567–574 (2000)
44. Bullitt, E., Gerig, G., Pizer, S.M., et al.: Measuring tortuosity of the intracerebral vasculature from MRA images. IEEE Trans. Med. Imag. **22**, 1163–1171 (2003)
45. Johnson, M.J., Dougherty, G.: Robust measures of three-dimensional vascular tortuosity based on the minimum curvature of approximating polynomial spline fits to the vessel mid-line. Med. Eng. Phys. **29**, 677–690 (2007)
46. Dougherty, G., Johnson, M.J.: Clinical validation of three-dimensional tortuosity metrics based on the minimum curvature of approximating polynomial splines. Med. Eng. Phys. **30**, 190–198 (2008)
47. Dougherty, G., Johnson, M.J., Wiers, M.D.: Measurement of retinal vascular tortuosity and its application to retinal pathologies. Med. Biol. Eng. Comput. **48**, 87–95 (2010)
48. Capowski, J.J., Kylstra, J.A., Freedman, S.F.: A numeric index based on spatial frequency for the tortuosity of retinal vessels and its application to plus disease in retinopathy of prematurity. Retina **15**, 490–500 (1995)
49. Grisan, E., Foracchia, M., Ruggeri, A.: A novel method for the automatic grading of retinal vessel tortuosity. IEEE Trans. Med. Imag. **27**, 310–319 (2008)
50. Wallace, D.K.: Computer-assisted quantification of vascular tortuosity in retinopathy of prematurity. Trans. Am. Ophthalmol. Soc. **105**, 594–615 (2007)
51. Benhamou, S.: How to reliably estimate the tortuosity of an animal's path: straightness, sinuosity, or fractal dimension? J. Theor. Biol. **229**, 209–220 (2004)
52. Smedby, O., Hogman, N., Nilsson, S., et al.: Two-dimensional tortuosity of the superficial femoral artery in early atherosclerosis. J. Vasc. Res. **30**, 181–191 (1993)
53. Kimball, B.P., Bui, S., Dafopoulos, N.: Angiographic features associated with acute coronary artery occlusion during elective angioplasty. Can. J. Cardiol. **6**, 327–332 (1990)
54. Brinkman, A.M., Baker, P.B., Newman, W.P., et al.: Variability of human coronary artery geometry: an angiographic study of the left anterior descending arteries of 30 autopsy hearts. Ann. Biomed. Eng. **22**, 34–44 (1994)
55. Patasius, M., Marozas, V., Lukosevicius, A., et al.: Model based investigation of retinal vessel tortuosity as a function of blood pressure: Preliminary results. Conf. Proc. IEEE Eng. Med. Biol. Soc. **2007**, 6460–6463 (2007)
56. Kamimura, M., Kinoshita, T., Itoh, H., et al.: Preoperative CT examination for accurate and safe anterior spinal instrumentation surgery with endoscopic approach. J. Spinal Disord. Tech. **15**, 47–51 (2002)

57. Schick, D.: Computed tomography radiation doses for paediatric scoliosis scans. Internal report commissioned by QUT/Mater Health Services Paediatric Spine Research Group from Queensland Health Biomedical Technology Services (2004)
58. Torell, G., Nachemson, A., Haderspeck-Grib, K., et al.: Standing and supine Cobb measures in girls with idiopathic scoliosis. Spine **10**, 425–427 (1985)
59. Yazici, M., Acaroglu, E.R., Alanay, A., et al.: Measurement of vertebral rotation in standing versus supine position in adolescent idiopathic scoliosis. J. Pediatr. Orthop. **21**, 252–256 (2001)
60. Krawczynski, A., Kotwicki, T., Szulc, A., et al.: Clinical and radiological assessment of vertebral rotation in idiopathic scoliosis. Ortop. Traumatol. Rehabil. **8**, 602–607 (2006)
61. Little, J.P., Adam, C.J.: The effect of soft tissue properties on spinal flexibility in scoliosis: biomechanical simulation of fulcrum bending. Spine **34**, E76–82 (2009)
62. King, H.A., Moe, J.H., Bradford, D.S., et al..: The selection of fusion levels in thoracic idiopathic scoliosis. J. Bone Joint Surg. Am. **65**, 1302–1313 (1983)
63. Lenke, L.G., Betz, R.R., Harms, J., et al.: Adolescent idiopathic scoliosis: a new classification to determine extent of spinal arthrodesis. J. Bone Joint Surg. Am. **83-A**, 1169–1181 (2001)

Chapter 11
Image Analysis of Retinal Images

Michael J. Cree and Herbert F. Jelinek

11.1 Introduction

The eye is sometimes said to provide a window into the health of a person for it is only in the eye that one can actually see the exposed flesh of the subject without using invasive procedures. That 'exposed flesh' is, of course, the retina, the light sensitive layer at the back of the eye. There are a number of diseases, particularly vascular disease, that leave tell-tale markers in the retina. The retina can be photographed relatively straightforwardly with a fundus camera and now with direct digital imaging there is much interest in computer analysis of retinal images for identifying and quantifying the effects of diseases such as diabetes.

It is a particularly exciting and interesting field for the image analysis expert because of the richness and depth of detail in retinal images and the challenges presented for analysis. There are many distinctive lesions and features for segmentation and quantification ranging from those requiring straightforward implementations to those presenting formidable challenges that remain largely unsolved. Finding solutions to these problems present enormous opportunity to positively impact on the health care of millions of people.

In this chapter, we present a tutorial introduction to some of the image processing techniques used in analysis of retinal images. Some space is given to the simpler approaches to image preprocessing and the detection of two major features. The first, the blood vessel network, is ubiquitous to all retinal images and can provide a wealth of health and disease information. The second, microaneurysms, is a lesion particularly associated with diabetic retinopathy – a disease of the retina resulting from diabetes. This is polished off with some more recent and sophisticated techniques in wavelet and fractal analysis of the vessel network.

M.J. Cree (✉)
University of Waikato, Hamilton, New Zealand
e-mail: cree@waikato.ac.nz

G. Dougherty (ed.), *Medical Image Processing: Techniques and Applications*, Biological and Medical Physics, Biomedical Engineering, DOI 10.1007/978-1-4419-9779-1_11,

But first, there is some notation and jargon that is necessary for talking about retinal images. We turn to that first followed by a brief expansion upon the motivation for automated computer analysis of retinal images, and an introduction to the technologies used to capture retinal images.

11.2 Retinal Imaging

11.2.1 Features of a Retinal Image

The *retina* is the light sensitive layer at the back of the eye that is visualisable with specialist equipment when imaging through the pupil. The features of a typical view of the retina (see Fig. 11.1) include the *optic disc* where the blood vessels and nerves enter from the back of the eye into the retina. The blood vessels emerge from the optic disc and branch out to cover most of the retina. The *macula* is the central region of the retina about which the blood vessels circle and partially penetrate (the view shown in Fig. 11.1 has the optic disc on the left and the macula towards the centre-right) and is the most important for vision.

There are a number of diseases of the retina of which *diabetic retinopathy* (pathology of the retina due to diabetes) has generated the most interest for automated computer detection. Diabetic retinopathy (DR) is a progressive disease that results in eye-sight loss or even blindness if not treated. Pre-proliferative diabetic retinopathy (loosely DR that is not immediately threatening eye-sight loss) is characterized by a number of clinical symptoms, including microaneurysms (small round outgrowths from capillaries that appear as small round red dots less

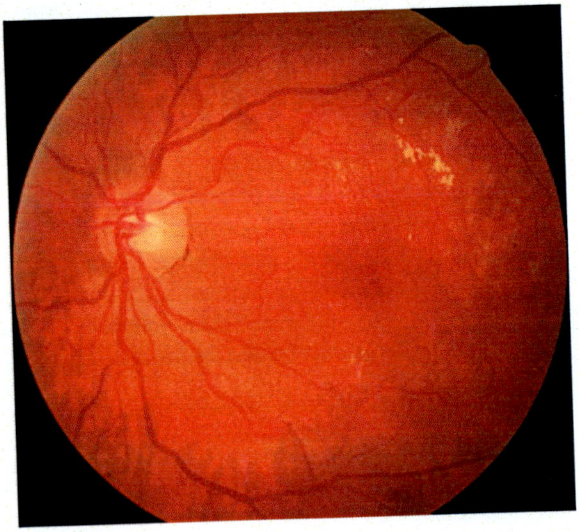

Fig. 11.1 Color retinal image showing features of diabetic retinopathy including microaneurysms and exudate

than 125 μm in diameter in color retinal images), dot-haemorrhages (which are often indistinguishable from microaneurysms), exudate (fatty lipid deposits that appear as yellow irregular patches with sharp edges often organized in clusters) and haemorrhage (clotting of leaked blood into the retinal tissue). These symptoms are more serious if located near the centre of the macular.

Proliferative diabetic retinopathy (PDR) is the more advanced form that poses significant risk of eye-sight loss. Features that lead to a diagnosis of PDR include leakage of blood or extensive exudate near the macular, ischaemia, new vessel growth and changes in vessel diameter such as narrowing of the arterioles and venous beading (venules alternately pinching and dilating that look like a string of sausages). Indeed, there is a reconfiguring of the blood vessel network that we have more to say about later.

11.2.2 The Reason for Automated Retinal Analysis

Recent data suggest that there are 37 million blind people and 124 million with low vision, excluding those with uncorrected refractive errors. The main causes of global blindness are cataract, glaucoma, corneal scaring, age-related macular degeneration, and diabetic retinopathy. The global Vision 2020 initiative is having an impact to reduce avoidable blindness particularly from ocular infections, but more needs to be done to address cataract, glaucoma, and diabetic retinopathy [14]. Screening is generally considered effective if a number of criteria are met including identification of disease at an early, preferably preclinical, stage and that the disease in its early or late stage is amenable to treatment. Screening for diabetic retinopathy, for example, and monitoring progression, especially in the early asymptomatic stage has been shown to be effective in the prevention of vision loss and cost.

Automated screening (for example, by computer analysis of retinal images) allows a greater number of people to be assessed, is more economical and accessible in rural and remote areas where there is a lack of eye specialists. Automated assessment of eye disease as an ophthalmological equivalent to the haematology point-of care testing such as blood glucose levels has been subject to intense research over the past 40 years by many groups with algorithms being proposed for identification of the optic disc, retinal lesions such as microaneurysms, haemorrhage, cotton wool spots and hard exudates, and retinal blood vessel changes.

Retinal morphology and associated blood vessel pattern can give an indication of risk of hypertension (high blood pressure), cardiovascular and cerebrovascular disease as well as diabetes [23, 28, 36]. With the increase in cardiovascular disease, diabetes and an aging population, a greater number of people will need to be screened yet screening a large number of people is difficult with limited resources necessitating a review of health care services.

Early identification of people at risk of morbidity and mortality due to diverse disease processes allows preventative measures to be commenced with the greatest efficacy. However in many instances preclinical signs are not easily recognized and often appear as signs or symptoms that are not specific for a particular disease. The retina and its blood vessel characteristics however have been shown to be a window into several disease processes. The identification of increased risk of disease progression is based on several markers in the eye including venous dilatation, vessel tortuosity and the change in the ratio of the arteriolar to venular vessel diameter especially in proximity to the optic disc. This morphological characteristic allows the application of image analysis and automated classification [23, 28, 36].

11.2.3 Acquisition of Retinal Images

Digital images of the human retina are typically acquired with a digital fundus camera, which is a specialized camera that images the retina via the pupil of the eye. The camera contains an illumination system to illuminate the retina and optics to focus the image to a 35 mm SLR camera. Modern systems image at high-resolution and in color with Nikon or Canon digital SLR camera backends. The field of view (FOV) of the retina that is imaged can usually be adjusted from 25° to 60° (as determined from the pupil) in two or three small steps. The smaller FOV has better detail but this is at the expense of a reduced view of the retina.

When monochromatic film was commonplace a blue-green filter was sometimes placed in the optical path of the fundus camera as the greatest contrast in retinal images occurs in the green wavelengths of light. An image acquired in such a manner is referred to as a red-free image. With color digital imaging it is common practice to take the green field of a RGB image as a close approximation to the red-free image.

In fluorescein angiographic imaging, the patient is injected with a sodium fluorescein drug which is transported in the blood supply to the retina. The fundus camera uses a flash filtered to the blue spectrum (465–490 nm) to activate the fluorescein, which fluoresces back in the green part of the spectrum (520–630 nm). The collected light is filtered with a barrier filter so that only the fluorescence from the retina is photographed. Images obtained with fluorescein angiography are monochromatic and highlight the blood flow in the retina. Since the fluorescein, when injected into the blood stream, takes a few seconds to completely fill the retinal vessels, images taken early in the angiographic sequence show the vasculature filling with fluorescein. First the fluorescein streams into the arterioles, and then a couple or so seconds later fills the venules. Over time (minutes) the images fade as fluorescein is flushed out of the retinal blood supply.

Angiographic retinal images better highlight vascular lesions such as microaneurysms, ischaemia (absence of blood flow) and oedema (leakage of blood into the surrounding tissues). The downside of the use of fluorescein is the inherent risk to the patient with about 1 in 200,000 patients suffering anaphylactic shock. Indocyanine green (ICG) is sometimes used for imaging the choroidal vasculature but requires a specially designed fundus camera due to the low intensity fluorescence.

Other imaging technologies such as the scanning laser ophthalmoscope (SLO) and optical coherence tomography (OCT) may be encountered. The SLO scans a laser point on to the retina and simultaneously collects light reflected back from the retina with a photodiode. The photodiode has no pixels; an image is formed by the scanning of the light source in a raster fashion over the retina in time.

The raster scanning of the SLO has been exploited in some interesting applications, such as forming images on the retina in a form of heads-up display, and projecting images on to the retina to measure retinal distortion. In another application, *in vivo* study of cell movement in retinal vessels is made with a single scan only by fluorescent labelling of the blood cells [18,38]. This is possible because the SLO used scans the retina with interlacing, namely it scans the odd lines of the raster first (the odd field) and then scans the intervening even lines (the even field). In between scanning the two fields the leucocyte moves so it appears in two locations in one image. Matching the two appearances of the leucocyte together enables a calculation of leucocyte speed in the blood stream.

11.3 Preprocessing of Retinal Images

As the photographer does not have complete control over the patient's eye which forms a part of the imaging optical system, retinal images often contain artifacts and/or are of poorer quality than desirable. Patients often have tears covering the eye and, particularly the elderly, may have cataract that obscures and blurs the view of the retina. In addition, patients often do not or cannot hold their eye still during the imaging process hence retinal images are often unevenly illuminated with parts of the retinal image brighter or darker than the rest of the image, or, in worst cases, washed out with a substantial or complete loss of contrast.

Not much attention has been given to the blurring effect of cataract and tears, maybe because one can sometimes choose another image out of a sequence that is better and, in any case, it has a position dependent blurring function that varies from image to image, making restoration difficult. Much more problematic is the uneven illumination of the retina, partly because it occurs more often, but also in part because in its extreme form can obliterate almost all the detail in a substantial part of the retinal image.

It should be recognized that in addition to the uneven illumination due to failures in the imaging system, the retina varies in intensity due to its own natural appearance. This distinction, though, is not too important for general image processing and pattern recognition of the features of the retina, but must be considered in quantitative analysis of illumination such as that occurs, for example, in fluorescence change over time in angiographic sequences [10].

Minor unevenness in illumination occurs in most retinal images and it is usually advantageous to correct for during preprocessing for successful automated detection of retinal lesions. For example, Chaudhuri et al. [3] described one of the very first attempts at vessel detection in retinal images (the algorithm is described in more detail in Sect. 11.4.1 below). Recent proposals for vessel detection are often compared to the algorithm of Chaudhuri et al. [3] and if they are any good they show substantial improvement. But it has not always been realized that the algorithm proposed by Chaudhuri et al. does not include shade- (or illumination-) correction as a preprocessing step. Preprocessing with shade-correction (as described below) can substantially improve the [3] algorithm on certain images.

The shading effect due to uneven illumination is a slowly changing function of the spatial coordinates, that is, it consists of low frequency content only, thus can be isolated by a low-pass filter. While it is possible to do this with a filter in the Fourier domain, it is much more common to isolate the illumination changes using a gross mean or median filtering of the image. In the past, the mean filter was sometimes preferred because it was much quicker to compute than the median filter, however, the median filter has better edge preserving properties and typically gives better results on retinal images. Now that a very efficient implementation of the median filter whose computational efficiency is almost independent of the kernel size is widely available [27], the median filter should be preferred.

How large the median filter kernel should be is determined by the resolution of the retinal image and the size of the objects/lesions that one wishes to segment. If the lesions are small (for example microaneurysms and dot-haemorrhages) then it would not matter if one underestimated the size of the kernel as long as it is much bigger than the largest lesion to be detected. On the hand if large lesions such as extensive haemorrhage that cover a large part of the retinal image are to be detected then determining the size of the kernel becomes a tough proposition as the lesion size is on the same scale as the illumination changes. Methods more sophisticated than those described here are then needed for shade-correcting the image.

If we take the illumination of the retina by the camera to be $L(x,y)$ where (x,y) labels the pixels in the image, and $f(x,y)$ to be the perfect image of the retina, then the camera measures $g(x,y)$ given by

$$g = Lf, \tag{11.1}$$

where the multiplication is pixel-wise. It is clear that we should divide the captured retinal image g by the illumination estimated by gross median filtering to give a reasonable approximation f^* to the true image f. Of course, the goodness of the approximation depends on our ability to estimate the illumination and the goodness of the model expressed by (11.1).

The above, as illustrated in Fig. 11.2, generally works well for preparing color fundus images for the segmentation of lesions/objects that are smaller than the scale of illumination change. It may be seen in Fig. 11.2 that the technique is correcting for more than unevenness in illumination, but also for intrinsic background intensity changes of the retina, and that is advantageous in many applications.

The illumination expressed in (11.1) is, however, not an appropriate model for the shade-correction of fluorescein angiographic images. This is because the capillary bed of the retina (the capillaries themselves are not resolvable in the typical retinal image) contributes a background glow due to the fluorescein in the blood. Where the retina is densely populated with capillaries the background glow is substantial, and where the retina is absent of capillaries, i.e. the foveal avascular zone, there is little or no background glow.

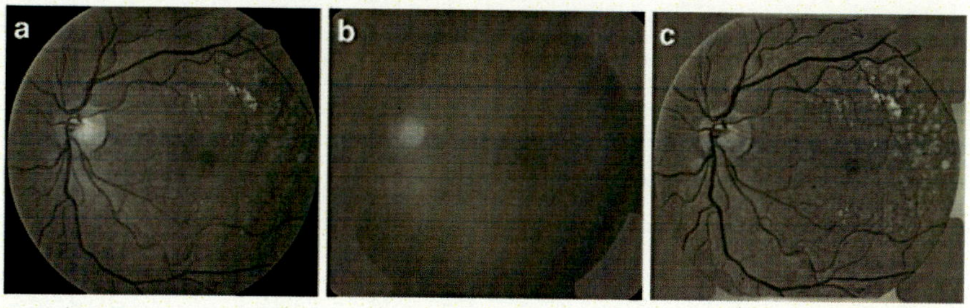

Fig. 11.2 Shade-correction of a retinal image. (**a**) Green plane, (**b**) background illumination estimated by median filtering, and (**c**) shade-correction by dividing green plane by estimated illumination (contrast adjusted by linear stretch for display purposes)

Applying a gross median filter to an angiographic image does determine the illumination change across the image, however, it is not solely due to the illumination function L but includes the background fluorescence B due to the capillary bed. In this case the illumination model is better described as

$$g = L(f + B). \tag{11.2}$$

Since the contribution due to B is substantial, reducing (11.2) to $g \approx f + I$, where I is the illumination estimated with a gross median filtering of f, thus estimating f by subtracting I from g usually produces sufficiently good results for detecting small lesions. Indeed, this is the approach used by early microaneurysm detection algorithms that were designed for use with angiographic images [9, 31].

For the segmentation of large lesions such as extensive haemorrhage or quantifying changes in brightness over time (for example, quantifying blood leakage into surrounding tissues during an angiographic sequence) then more sophisticated physics inspired preprocessing approaches are required [10, 13]. Some authors have noted that the global approach to shade-correction described above still leaves some room for improvement when segmenting microaneurysms in regions of poor contrast. Huang and Yan [19], Fleming et al. [12] and Walter et al. [35] all resort to some form of locally adaptive shade-correction with contrast enhancement to eke out slight improvements in microaneurysm detection.

11.4 Lesion Based Detection

We now turn attention to the segmentation of features of interest and lesions in retinal images. In the following pages three general techniques in image processing, namely linear filtering in image space, morphological processing, and wavelets are illustrated by way of application to the detection of retinal blood vessels and microaneurysms. Analysis of blood vessels is of particular interest as vascular

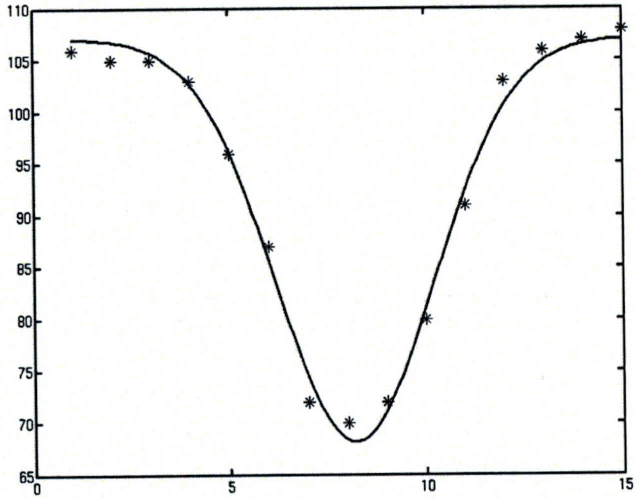

Fig. 11.3 Cross-section of a blood vessel. The asterisks show the intensity at pixel locations across the vessel and the solid line is the fitted Gaussian function

disease such as diabetes cause visible and measurable changes to the blood vessel network. Detecting (i.e. segmenting) the blood vessels and measuring blood vessel parameters provides information on the severity and likely progression of a variety of diseases. Microaneurysms are a particular vascular disorder in which small pouches grow out of the side of capillaries. They appear in color fundus images as small round red dots and in fluorescein angiographic images as small hyperfluorescent round dots. The detected number of microaneurysms is known to correlate with the severity and likely progression of diabetic retinopathy.

11.4.1 Matched Filtering for Blood Vessel Segmentation

One of the earliest and reasonably effective proposals for the segmentation of blood vessels in retinal images [3] is the use of oriented matched-filters for the detection of long linear structures.

Blood vessels often have a Gaussian like cross-section (see Fig. 11.3) that is fairly consistent along the length of vessel segments. Provided the vessels are not too tortuous then they can be approximated as elongated cylinders of Gaussian cross-section between the vessel branch points. Thus, the two-dimensional model consisting of an elongated cylinder of Gaussian cross-section should correlate well with a vessel segment provided they have both the same orientation. The model is moved to each possible position in the image and the correlation of the local patch of image to the model is calculated to form a correlation image. Peaks in the correlation image occur at the locations of the blood vessels.

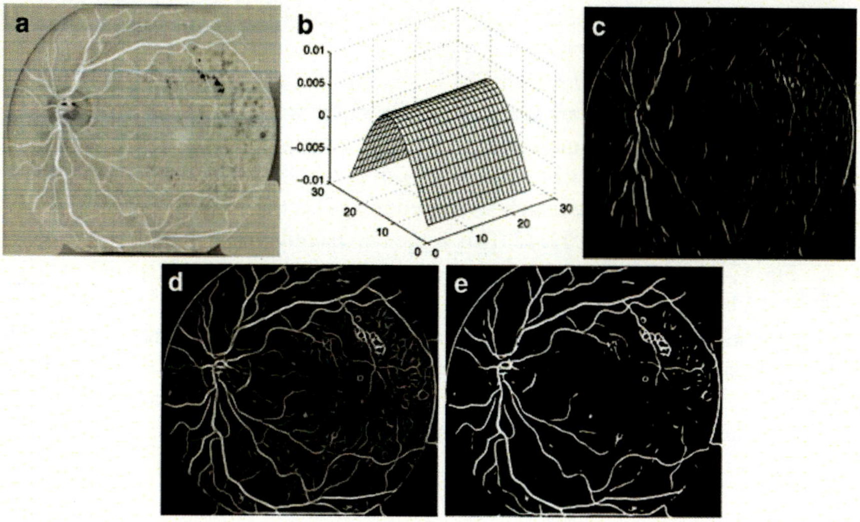

Fig. 11.4 Segmentation of blood vessels by matched-filtering: (**a**) Inverted shade-corrected green component of the retinal image of Fig. 11.1, (**b**) vessel model used for matched-filtering, (**c**) the result match-filtering with the vertical orientation model, (**d**) combined matched-filters applied in all orientations, and (**e**) thresholded to give the blood vessels

A better theoretical formulation can be given to support the argument [3]. Take $f(x)$ to be a signal and $F(f)$ be its Fourier transform (that is, the spectrum of the signal f). Consider $f(x)$ contaminated by additive Gaussian white noise with spectrum $N(f)$. The optimal linear filter in the sense of maximising the signal to noise ratio that recovers f in the presence of the noise N is F^*, the complex conjugate of F. That is, if we calculate

$$f_0(x) = \int H(f)\,(F(f) + N(f))\,e^{2\pi i f x}\,dx \qquad (11.3)$$

then $H(f) = F^*(f)$ gives the best approximation of $f_0(x)$ as $f(x)$. Equation (11.3) is the correlation of f with itself.

Now if f is a localized patch of retinal image (the generalisation to 2D does not change the argument) then correlation of f with the blood vessel signal is the optimal linear filter for detecting the blood vessel. Of course, this assumes that everything else in the localized patch of retinal image is Gaussian noise, which is certainly not true. Let us proceed anyway despite the flawed assumption.

The blood vessel model described above is correlated with a small patch of image to isolate the blood vessel section. This is repeated over every local region of the image to form a correlation image. The peaks in the correlation image correspond to the locations of the model in the image. This process is commonly referred to as a *matched-filter* or as *matched-filtering*.

Figure 11.4a shows the shade-corrected image of a color retinal image. It is inverted to make the vessels appear bright and a median filter with a kernel size of 5×5 pixels has been applied to reduce pixel noise. The vessel model in the vertical orientation is shown in Fig. 11.4b. It is correlated with the retinal image to produce the image shown in Fig. 11.4c. All vertical sections of blood vessels are strongly emphasized and everything else in the image is suppressed. The model is rotated by a small amount and the matched-filter is applied again to the image. This is repeated for all possible orientations of the model, and the maximum response at each pixel over all the matched-filtered images is taken as the final response for the pixel as shown in Fig. 11.4d. This is then thresholded to give the vessel network, see Fig. 11.4e. A failing of this approach is evident in the example given, namely that it has false-detected on the exudate at the top-right of the image.

The amount to rotate the model for each application of the matched-filter should be small enough so that all vessel segments are segmented in at least one of the matched-filtered images but not so small that processing time becomes excessive. Chaudhuri et al. used 15° angular increments with a kernel size of 32×32 pixels on images of size 512×480.

11.4.2 Morphological Operators in Retinal Imaging

Morphological operators are based on mathematical set theory and provide a natural way of analysing images for geometrical structure. The basic operators are the dilation and the erosion. The dilation has the effect of dilating objects so that closely located structures become joined and small holes are filled. The erosion has the effect of eroding objects with sufficiently thin structures eliminated. But it is better than that; the direction, size and even shape, of the erosion or dilation can be controlled with a mask called the *structuring element*.

The downside of the basic operators is that while they have extremely useful properties they nevertheless do not retain the object size. Dilation, not unsurprisingly, causes objects to grow in size and erosion causes them to shrink. Better is a combination of the two operators to form the opening and the closing.

The opening is an erosion followed by a dilation. It has the effect of returning objects to their near original size (since the dilation reverses somewhat the effect of the erosion) with the destruction of very small objects and thin joins between objects that are smaller than the structuring element (since the dilation cannot dilate structures that have been entirely eliminated by the opening). The closing is a dilation followed by an erosion and has the effect of returning objects to near original size but with small holes filled and objects that are very close to each other joined together.

The general description of gray-scale morphology treats both the image and the structuring element as gray-scale [17, 30] but in many implementations the structuring element is taken to be a set of connected pixels. The erosion of image $f(x,y)$ by structuring element b, written $f \ominus b$ is

$$(f \ominus b)(x,y) = \min_{(x',y') \in b} f(x+x', y+y'), \tag{11.4}$$

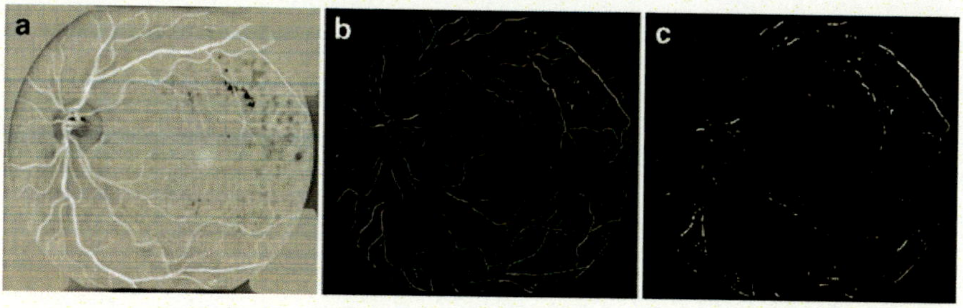

Fig. 11.5 The tophat operator used to detect microaneurysms. (**a**) median filtered inverted green plane of retinal image, (**b**) the result of the tophat operator with a 6-pixel radius disc structuring element, and (**c**) thresholded to segment microaneurysms and, unfortunately, many other spurious features (contrasts adjusted by linear stretch for display purposes)

and the dilation, written as $f \oplus b$, is[1]

$$(f \oplus b)(x,y) = \max_{(x',y') \in b} f(x-x', y-y'). \tag{11.5}$$

The closing of f by mask b, written $f \bullet b$, is

$$f \bullet b = (f \oplus b) \ominus b, \tag{11.6}$$

and the opening is

$$f \circ b = (f \ominus b) \oplus b. \tag{11.7}$$

The opening and closing operators are idempotent; repeated application of the operator does not change the result further.

An opening removes all objects that cannot be enclosed by the structuring element from the image. Subtracting the opening off the original image, namely calculating,

$$m = f - f \circ b \tag{11.8}$$

eliminates most structure in the image except for objects smaller than the structuring element. This procedure is referred to as the *tophat* transform.

As described in Sect. 11.2.1 microaneurysms appear in retinal images as little round dots (see Fig. 11.5). If the structuring element is chosen to be round and just bigger than the largest microaneurysm and f is the inverted green plane of a retinal image, then m is an image that contains the microaneurysms. This, with a final thresholding of m to give a binary image, is the basis of some of the earliest proposals for microaneurysm detection in retinal images [2,24], but, it is not specific enough. Other small objects, bits of background texture, and vessel bits are all

[1]The reflection of the structuring element in the dilation that is not present in the erosion is intended as it simplifies the definitions of the opening and closing following.

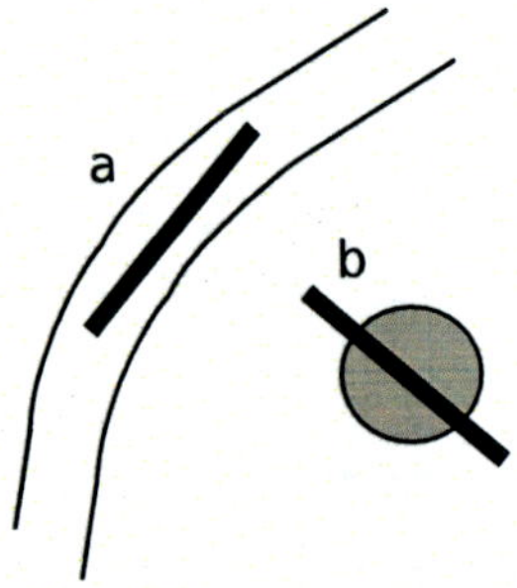

Fig. 11.6 The structuring element to segment vessels is long and narrow and segments all objects that it can fit into. Vessels (**a**) are segmented because the structuring element fits in them when correctly orientated and small objects (**b**) are eliminated because they cannot enclose the structuring element at any orientation

picked up in the result. In Fig. 11.5 the inverted green plane of the retinal image was median filtered with a 5 × 5 pixel kernel first because of pixel noise in the image. Then the procedure described above with a round disc structuring element of radius 6 pixels was applied to highlight the microaneurysms. It should be obvious that a suitable threshold to segment the microaneurysms cannot be chosen.

A very similar procedure can be used to detect the blood vessels. If the structuring element is a long thin linear structure it can be use to segment sections of the blood vessel. The structuring element is normally taken to be one pixel wide and enough pixels long that it is wider than any one vessel but not so long that it cannot fit into any vessel segment provided it is orientated correctly (see Fig. 11.6). The opening of the inverted green plane of the retinal image is taken with the structuring element at a number of orientations. The maximal response of all openings at each pixel location is calculated. The resultant image contains the blood vessels, but not small structures such as the microaneurysms and background texture. This approach for detecting the vessels is not specific enough as it also segments large structures such as extensive haemorrhage as vessels.

Even though this "vessel detection" algorithm is not brilliant at solely detecting vessels, it does have the feature that it does not detect microaneurysms, thus it can be used as a vessel removal procedure in microaneurysm detection to reduce the number of false detections of microaneurysms. One approach is to detect microaneurysms (say as described above) and remove the ones that are on vessels detected with the tophat transform. Since the line structuring element used to detect blood vessels also removes all larger objects an extra tophat operation with a round structuring element is not needed and it is best to apply an opening with a small round structuring element that is smaller than the smallest microaneurysm. That opening removes all small spurious objects due to noise.

To put this on a more formal setting, take f to be the inverted green plane of the retinal image, b_m to be a round structuring element that is just small enough that no microaneurysm can fully enclose it, and b_{v,θ_i} to be a one-pixel wide structuring element of suitable length to detect vessels when correctly orientated (θ_i) along the length of the vessel. The image with the vessels (and other large structures) removed is constructed by tophat transform, viz

$$f_1 = f - \max_i f \circ b_{v,\theta_i}. \tag{11.9}$$

This image contains the microaneurysms and other small detections. They are removed with the opening,

$$m = f_1 \circ b_m, \tag{11.10}$$

then m is thresholded to those objects of enough depth to be a microaneurysm. This reduces false-detections of microaneurysms on vessels but false-detections of small bits of texture, retinal-pigment epithelium defects, and so on, still occur so improvements can yet be made.

Some prefer to use a matched-filter approach to detect the microaneurysms. A good model of microaneurysm intensity is a circularly symmetric Gaussian function. Applying such a model to the image f_1 of (11.9) (i.e. the image with vessels removed) with a matched-filter then thresholding to isolate the microaneurysms gives a reasonable result but, yet again, it is not good enough [32].

A small improvement to the above algorithms can be made by using morphological reconstruction [33]. The tophat transform to detect objects does not preserve the segmented objects' shape precisely; morphological reconstruction addresses this problem. First some operator is applied to identify objects of interest. It need not segment the objects in their entirety; just having one of the brightest pixels within the object is sufficient. This image is called the *marker* and the original image is the *mask*. The marker is dilated by one pixel with the limitation that it cannot dilate further than any objects in the mask. Note that this restricted dilation by one pixel, called the *geodesic dilation*, is different to the dilation described by (11.5) in that it is limited by structures in the mask. The geodesic dilation is repeatedly applied until the limit is reached when no more changes occur. This process is called *reconstruction by dilation*.

If the marker has a peak that is in the mask then the shape of the peak is returned from the mask in reconstruction by dilation. If the marker has no feature at a peak in the mask then the peak in the mask is eliminated. Morphological reconstruction by dilation better preserves the shapes of the segmented features in the image than does an opening or tophat transform. *Opening by reconstruction* is the process of reconstructing by dilation an image with its erosion with a structuring element as the marker. It has a similar effect as the morphological opening but with a much better preservation of the shape of the segmented objects.

As described above microaneurysm detection often proceeds by removing the blood vessels before segmenting the microaneurysms. Tests show that removing the blood vessels in a retinal image with tophat by reconstruction instead of the morphological tophat transform reduces the false detection of microaneurysms in blood vessels [16]. An extra opening with a disc structuring element smaller than

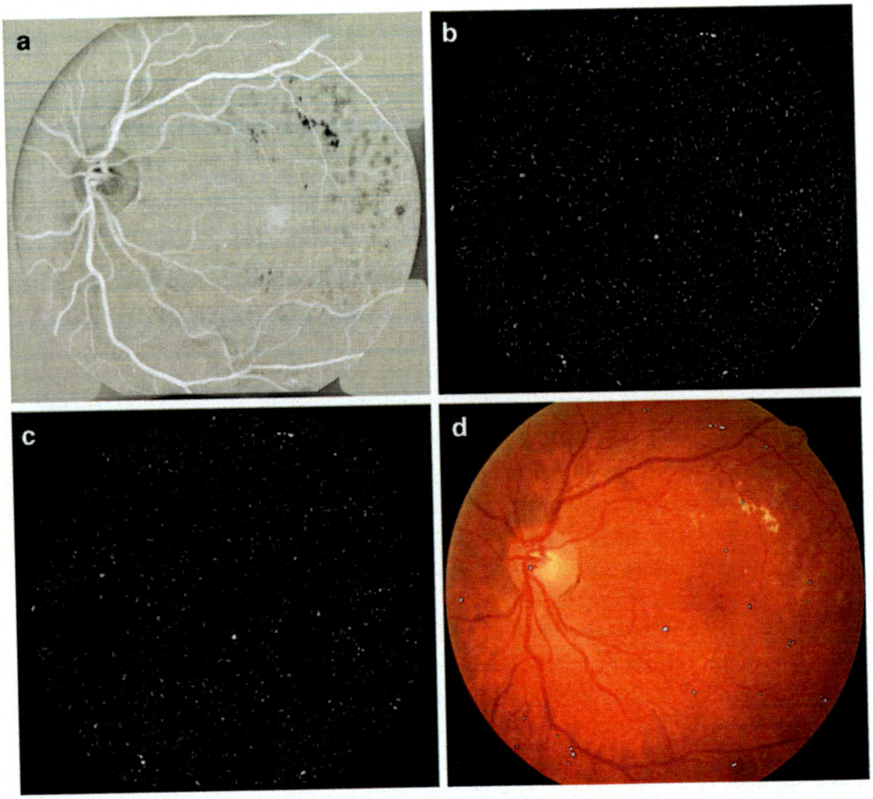

Fig. 11.7 Detection of microaneurysms by way of vessel removal (**a**) median filtered inverted green plane of retinal image, (**b**) tophat by reconstruction of the image to remove vessels, then (**c**) opened with a structuring element smaller than microaneurysms to remove small specks, and (**d**) thresholded to isolate the microaneurysms and overlaid original image

any microaneurysm helps to remove spurious noise. See Fig. 11.7 for an illustration of the process. The detection of microaneurysms shown in Fig. 11.7c is a pretty good result but there are a couple of false detections, one on the optic disc and one amongst the patch of exudate, thus there is still room for improvement. There are also other non-trivial issues, such as how to automatically adapt the threshold for each image.

Walter and Klein [34] and Walter et al. [35] argue that the diameter opening (and closing) is a better morphological operator to detect microaneurysms. The diameter opening is the maximum of all openings with structuring elements with a diameter greater than or equal to a required diameter. Here, the diameter of a connected group of pixels is the maximum distance from any one pixel in the object to a pixel on the other side of the object. Nevertheless, to improve specificity most authors that use morphology or matched filtering or some combination of both to segment microaneurysms, measure features such as size, shape, and intensity on the segmented candidate microaneurysms, and apply machine learning techniques to refine the detection of the microaneurysms [8, 9, 26, 31, 35].

The above illustrates an important point in processing complicated medical images rich in detail such as retinal images. With basic and well understood image processing operators it is easy to do reasonably well in detecting a certain lesion or feature. But reasonably well is not good enough. One missed lesion that has serious clinical implications (such as missing indications of proliferative retinopathy in retinal images when it can lead to significant eye-sight loss within a few days or weeks) will not generate trust from the medical profession. On the other hand, false detections on far too many images is useless. Getting the first 80% sensitivity and specificity is easy; improving that to better than 90% sensitivity at 90% specificity is the hard problem.

11.5 Global Analysis of Retinal Vessel Patterns

So far we have discussed segmenting specific lesions (in particular microaneurysms) and features (vasculature) in retinal images. Typically, lesion specific measures such as number of microaneurysms or localized vessel parameters that are known to predict or correlate with disease are measured. The analysis of vessel properties, in particular, often involves sophisticated fitting of parametric models to the blood vessels in a local region from which features such as vessel diameter, tortuosity and branching ratios are measured [15, 37]. Vessel diameters of arterioles and venules, for example, can be used to calculate arterio-venous diameter ratios that predict hypertensive (high blood pressure) disorders [6].

In this section, we take a different tack and examine an approach to measure general disease condition with global vessel analysis. Proliferative diabetic retinopathy (PDR) – an advanced form of retinopathy due to diabetes with high risk of eye-sight loss if not treated [7] – is characterized by a reconfiguration of the blood vessels resulting from ischaemia in parts of the retina and new vessel growth that emerges from the area of the optic disc or from peripheral vessels [22]. Given that the vessel pattern is noticeably different in PDR than non-proliferative diabetic retinopathy it is tempting to ask whether a global operator applied to the vessel patterns is capable of characterising the changes in the vasculature and, therefore, detect PDR [21]. We, therefore, seek such operators.

Traditional and simpler shape analysis features often used in image processing, such as area (a), perimeter (p), and circularity (typically calculated as p^2/a) are not expected to be powerful enough as these features are dependent on a change of the amount of vasculature whereas the existing vessel network can reconfigure with minimal change in the amount of vasculature. Jelinek et al. [21] explore global vessel analysis with wavelet and fractal inspired features to characterise the vessel reconfiguration that occurs in PDR. The vessels are detected with a sophisticated wavelet based segmentation [29], then reduced by a morphological skeletonisation to a 1-pixel wide skeleton that represents the vessel tracks.

The vector gradient of the vessel skeleton image is calculated via analysing wavelets, namely the partial derivatives of the Gaussian, viz,

$$\psi_1(x,y) = \frac{\partial g(x,y)x}{\partial x}, \qquad \psi_2(x,y) = \frac{\partial g(x,y)y}{\partial y}, \tag{11.11}$$

where $g(x,y)$ denotes the two-dimensional Gaussian. Calculating the wavelet transform of the skeletonized vessel image f with the two wavelets at displacement $\mathbf{b}$ and scale a and forming the vector,

$$\mathbf{T}_\psi[f](\mathbf{b},a) = \begin{pmatrix} T_{\psi_1}[f](\mathbf{b},a) \\ T_{\psi_2}[f](\mathbf{b},a) \end{pmatrix} \tag{11.12}$$

where the entries are the two wavelet transforms is an estimate of the vector gradient. This can be efficiently implemented with the fast Fourier transform.

The vector gradient image is analysed on the boundary of the vessels. It has two orthogonal components, namely orientation and magnitude, which can be analysed separately. Let us first consider calculating the entropy of the orientations as entropy is a quantitative measure of manifest disorder. If the vessels all tend to be aligned in the same direction (i.e. order) then the orientation entropy will be low, whereas if the vessels are randomly orientated throughout the image (disorder) then the entropy will be high. The suspicion is that the new vessel growth in PDR may affect the distribution of vessel orientations and, hence, the orientation of the gradient field. The orientation entropy also has the advantage that it is invariant against rotations and reflections of the image.

The orientation entropy s is straightforward to calculate and involves forming a histogram of orientations and calculating

$$s = -\sum_i p_i \ln p_i \tag{11.13}$$

where i indexes the histogram bins, that is orientations, and p_i is the frequency of occurrence of orientation i.

One can also analyse the magnitudes of the gradient vector field. A useful measure is the second moment which when applied to $\mathbf{T}_\psi$ indicates bias in the gradient vector field. A histogram can be formed from the magnitudes then the CWT second moment is

$$m_2 = \sum_i i^2 q_i \tag{11.14}$$

where the i are the centers of the histogram bins and q_i is the frequency of occurrence of the magnitudes in the bin.

There are other features that can be measured on the vessels with a global approach. The vessels are shapes and an important parameter of a shape is the curvature which characterises how the direction of a unit tangent vector varies along the shape contour. What is interesting is that the curvature can be measured with a two-dimensional global Fourier analysis of the image without having to perform sophisticated one-dimensional parameterisations of the shape contours [11]. Jelinek et al. [21] apply this technique to the skeletonized vessel patterns to estimate a global measure of vessel curvature.

The blood vessels in the retina branch a number of times each time forming a vessel tree that is similar in characteristics to the branch it came from. Systems that have self-similarity at multiple scales are known as fractals. There are some reports of measuring fractal properties of blood vessel patterns of the retina; most have involved manual segmentation of the blood vessel patterns [4,5,25]. Now with reliable automated vessel segmentation, attention is turning to analysing the retinal vasculature as a fractal. As the retinal tree branches it fills the two-dimensional space of the image. One way to measure the space-filling is with various fractal dimensions of which the correlation dimension $D_c(\varepsilon)$ is a common choice [1]. It is defined as

$$D_c = \lim_{r \to 0} \frac{\log C(r)}{\log r}, \tag{11.15}$$

where $C(r)$ is the correlation integral given by

$$C(r) = \frac{\text{number of distances less than } r}{\text{total number of distances}}. \tag{11.16}$$

The limit is usually approximated by evaluating the slope of the straight line segments of a plot of $\log C(r)$ against $\log r$. In practice the first and last line segments should be disregarded since at these scales of r little fractal information is brought from the shape. Jelinek et al. [21] make two calculations of correlation dimension over the skeletonized retinal vessel shapes. The first is the median of the ranked line segment slopes and the second is the global correlation dimension calculated by adding all slopes except those from the first and last line segments.

Jelinek et al. [21] test the features described on a database of 27 retinal images (16 with PDR; 11 with pathology but not PDR) for the ability to predict PDR. The traditional measures (area, perimeter, and circularity) showed no predictive power whatsoever when used separately and failed statistical significance at 95% confidence when used in combination to predict PDR. All the wavelet based features when used separately showed predictive power but only the curvature achieved 95% confidence in predicting PDR. Combining features and using linear discriminant analysis as the classifier correctly classified all images except three.

11.6 Conclusion

Automated analysis of retinal images is an ongoing active field of research. In the above some of the simpler and some of the more recent analytical tools bought to analyse retinal images have been discussed. We have seen that standard image processing techniques that may be found in any good text on general image processing can go a long way to detecting certain features/lesions in retinal images and produce seemingly good results, but do not provide the quality of result required in clinical practice. Sophisticated image analysis techniques are now being explored for the quantification of previously difficult to assess features and to improve existing methods. The interested reader can find a useful summary and inspiration for further research in the text by Jelinek and Cree [20].

References

1. Asvestas, P., Matsopoulos, G.K., Nikita, K.S.: Estimation of fractal dimension of images using a fixed mass appoach. Patt. Recog. Lett. **20**, 347–354 (1999)
2. Baudoin, C., Maneschi, F., Quentel, G., Soubrane, G., Hayes, T., Jones, G., Coscas, G., Kohner, E.M.: Quantitative-evaluation of fluorescein angiograms – microaneurysm counts. Diabetes **32**, 8–13 (1983)
3. Chaudhuri, S., Chatterjee, S., Katz, N.P., Nelson, M., Goldbaum, M.H.: Detection of blood vessels in retinal images using two-dimensional matched filters. IEEE Trans. Med. Im. **8**, 263–269 (1989)
4. Cheng, S.C., Huang, Y.M.: A novel approach to diagnose diabetes based on the fractal characteristics of retinal images. IEEE Trans. Inf. Technol. Biomed. **7**(3), 163–170 (2003)
5. Cheung, N., Donaghue, K.C., Liew, G., Rogers, S.L., Wang, J.J., Lim, S.W., Jenkins, A.J., Hsu, W., Lee, M.L., Wong, T.Y.: Quantitative assessment of early diabetic retinopathy using fractal analysis. Diabetes Care **32**, 106–110 (2009)
6. Cheung, N., Wong, T.Y., Hodgson, L.: Retinal vascular changes as biomarkers of systemic cardiovasular diseases. In: Jelinek, H.F., Cree, M.J. (eds.) Automated Image Detection of Retinal Pathology, pp. 185–219. CRC Press, Boca Raton, FL (2010)
7. Coyne, K.S., Margolis, M.K., Kennedy-Matrin, T., Baker, T.M., Klein, R., Paul, M.D., Revicki, D.A.: The impact of diabetic retinopathy: Perspectives from patient focus groups. Fam. Pract. **21**, 447–453 (2004)
8. Cree, M.J.: Automated microaneurysm detection for screening. In: Jelinek, H.F., Cree, M.J. (eds.) Automated Image Detection of Retinal Pathology, pp. 155–184. CRC Press, Boca Raton, FL (2010)
9. Cree, M.J., Olson, J.A., McHardy, K.C., Sharp, P.F., Forrester, J.V.: A fully automated comparative microaneurysm digital detection system. Eye **11**, 622–628 (1997)
10. Cree, M.J., Olson, J.A., McHardy, K.C., Sharp, P.F., Forrester, J.V.: The preprocessing of retinal images for the detection of fluorescein leakage. Phys. Med. Biol. **44**, 293–308 (1999)
11. Estrozi, L.F., Rios, L.G., Campos, A.G., Cesar Jr, R.M., d. F. Costa, L.: 1D and 2D Fourier-based approaches to numeric curvature estimation and their comparative performance assessment. Digit. Signal Process. **13**, 172–197 (2003)
12. Fleming, A.D., Philip, S., Goatman, K.A., Olson, J.A., Sharp, P.F.: Automated microaneurysm detection using local contrast normalization and local vessel detection. IEEE Trans. Med. Im. **25**(9), 1223–1232 (2006)
13. Foracchia, M., Grisan, E., Ruggeri, A.: Luminosity and contrast normalization in retinal images. Med. Im. Anal. **9**, 179–190 (2005)
14. Foster, A., Resnikoff, S.: The impact of Vision 2020 on global blindness. Eye **19**(10), 1133–1135 (2005)
15. Fritzsche, K.H., Stewart, C.V., Roysam, B.: Determining retinal vessel widths and detection of width changes. In: Jelinek, H.F., Cree, M.J. (eds.) Automated Image Detection of Retinal Pathology, pp. 269–304. CRC Press, Boca Raton, FL (2010)

16. Gamble, E.: Microaneurysm detection in directly acquired colour digital fundus images. Master's thesis, University of Waikato, Hamilton, New Zealand (2005)
17. Gonzalez, R.C., Woods, R.E.: Digital Image Processing, 2nd edn. Prentice-Hall, Upper Saddle River, NJ (2002)
18. Hossain, P., Liversidge, J., Cree, M.J., Manivannan, A., Vieira, P., Sharp, P.F., Brown, G.C., Forrester, J.V.: In vivo cell tracking by scanning laser ophthalmoscopy: Quantification of leukocyte kinetics. Invest. Ophthalmol. Vis. Sci. **39**, 1879–1887 (1998)
19. Huang, K., Yan, M.: A local adaptive algorithm for microaneurysms detection in digital fundus images. In: Proceedings of Computer Vision for Biomedical Image Applications, *Lecture Notes in Computer Science*, vol. 3765, pp. 103–113 (2005)
20. Jelinek, H.F., Cree, M.J. (eds.): Automated Image Detection of Retinal Pathology. CRC Press, Boca Raton, FL (2010)
21. Jelinek, H.F., Cree, M.J., Leandro, J.J.G., Soares, J.V.B., Cesar Jr, R.M.: Automated segmentation of retinal blood vessels and identification of proliferative diabetic retinopathy. J. Opt. Soc. Am. A **24**, 1448–1456 (2007)
22. Kanski, J.: Clinical Ophthalmology: A Systematic Approach. Butterworth-Heinemann, Boston (1989)
23. Klein, R., Klein, B.E., Moss, S.E., Wong, T.Y., Hubbard, L., Cruickshanks, K.J., Palta, M.: The relation of retinal vessel caliber to the incidence and progression of diabetic retinopathy: XIX: The Wisconsin Epidemiologic Study of Diabetic Retinopathy. Arch. Ophthalmol. **122**(1), 76–83 (2004)
24. Laÿ, B., Baudoin, C., Klein, J.C.: Automatic detection of microaneurysms in retinopathy fluoro-angiogram. Proc. SPIE **432**, 165–173 (1983)
25. Masters, B.R.: Fractal analysis of the vascular tree in the human retina. Annu. Rev. Biomed. Eng. **6**, 427–452 (2004)
26. Niemeijer, M., van Ginneken, B., Cree, M.J., Mizutani, A., Quellec, G., Sanchez, C.I., Zhang, B., Hornero, R., Lamard, M., Muramatsu, C., Wu, X., Cazuguel, G., You, J., Mayo, A., Li, Q., Hatanaka, Y., Cochener, B., Roux, C., Karray, F., Garcia, M., Fujita, H., Abramoff, M.D.: Retinopathy online challenge: Automatic detection of microaneurysms in digital color fundus photographs. IEEE Trans. Med. Im. **29**, 185–195 (2010)
27. Perreault, S., Hébert, P.: Median filtering in constant time. IEEE Trans. Im. Proc. **16**(9), 2389–2394 (2007)
28. Sherry, L.M., Wang, J.J., Rochtchina, E., Wong, T., Klein, R., Hubbard, L.D., Mitchell, P.: Reliability of computer-assisted retinal vessel measurement in a population. Clin. Experiment. Ophthalmol. **30**, 179–82 (2002)
29. Soares, J.V.B., Leandro, J.J.G., Cesar Jr., R.M., Jelinek, H.F., Cree, M.J.: Retinal vessel segmentation using the 2-D Gabor wavelet and supervised classification. IEEE Trans. Med. Im. **25**, 1214–1222 (2006)
30. Soille, P.: Morphological Image Analysis, 2nd edn. Springer, Berlin, Germany (2004)
31. Spencer, T., Olson, J.A., McHardy, K.C., Sharp, P.F., Forrester, J.V.: An image-processing strategy for the segmentation and quantification of microaneurysms in fluorescein angiograms of the ocular fundus. Comp. Biomed. Res. **29**, 284–302 (1996)
32. Spencer, T., Phillips, R.P., Sharp, P.F., Forrester, J.V.: Automated detection and quantification of microaneurysms in fluorescein angiograms. Graefes Arch. Clin. Exp. Ophthalmol. **230**, 36–41 (1992)
33. Vincent, L.: Morphological grayscale reconstruction in image analysis: Applications and efficient algorithms. IEEE Trans. Im. Proc. **2**, 176–201 (1993)
34. Walter, T., Klein, J.C.: Automatic detection of microaneurysms in color fundus images of the human retina by means of the bounding box closing. In: Proceedings Medical Data Analysis, *Lecture Notes in Computer Science*, vol. 2526, pp. 210–220 (2002)
35. Walter, T., Massin, P., Erginay, A., Ordonez, R., Jeulin, C., Klein, J.C.: Automatic detection of microaneurysms in color fundus images. Med. Im. Anal. **11**, 555–566 (2007)

36. Wang, J.J., Mitchell, P., Sherry, L.M., Smith, W., Wong, T.Y., Klein, R., Hubbard, L.D., Leeder, S.R.: Generalized retinal arteriolar narrowing predicts 5-year cardiovascular and cerebrovascular mortality: Findings from the Blue Mountains Eye Study. Invest. Ophthalmol. Vis. Sci. p. 43 (2002)
37. Witt, N.W., Martinez-Pérez, M.E., Parker, K.H., Thom, S.A.M., Hughes, A.D.: Geometrical and topological analysis of vascular branches from fundus retinal images. In: Jelinek, H.F., Cree, M.J. (eds.) Automated Image Detection of Retinal Pathology, pp. 305–338. CRC Press, Boca Raton, FL (2010)
38. Xu, H., Manivannan, A., Goatman, K.A., Liversidge, J., Sharp, P.F., Forrester, J.V., Crane, I.J.: Improced leukocyte tracking in mouse retinal and choroidal circulation. Exp. Eye Res. **74**, 403–410 (2002)

Chapter 12
Tortuosity as an Indicator of the Severity of Diabetic Retinopathy

Michael Iorga and Geoff Dougherty

12.1 Introduction

The retinal vasculature can be viewed directly and noninvasively, offering a unique and accessible window to study the health of the human microvasculature in vivo. The appearance of the retinal blood vessels is an important diagnostic indicator for much systemic pathology, including diabetes mellitus, hypertension, cardiovascular and cerebrovascular disease, and atherosclerosis [1–3]. There is mounting evidence supporting the notion that the retinal vasculature may provide a lifetime summary measure of genetic and environmental exposure, and may therefore act as a valuable risk marker for future systemic diseases [4]. Using its characteristics may provide early identification of people at risk due to diverse disease processes [5].

12.1.1 The Progression of Diabetic Retinopathy

Diabetic retinopathy is the most frequent cause of new cases of blindness among adults aged 20–74 years [6]. It is a progressive disease, beginning with mild nonproliferative abnormalities, characterized by increased vascular permeability, and progressing through moderate and severe nonproliferative diabetic retinopathy (NPDR) characterized by vascular closure, to proliferative diabetic retinopathy (PDR) with the growth of new blood vessels on the retina and posterior surface of the vitreous. Macular edema, marked by retinal thickening from leaky blood vessels, can develop at all stages of retinopathy.

In the early state of the disease, symptoms are mild or nonexistent. The curvature of a blood vessel influences its local flow hemodynamics and may result in

G. Dougherty (✉)
California State University Channel Islands, Camarillo, CA 91320, USA
e-mail: geoff.dougherty@csuci.edu

G. Dougherty (ed.), *Medical Image Processing: Techniques and Applications*, Biological and Medical Physics, Biomedical Engineering, DOI 10.1007/978-1-4419-9779-1_12,

unfavorable clinical consequences [7–9]. The tortuosity of intracranial arteries, for example, has been implicated in the risk of aneurysm formation due to the high shear stress weakening the outer walls of the arteries [10]. The process may be similar in diabetic retinopathy. As the disease develops, increased tortuosity of the retinal blood vessels may result in the weakening of the outer walls and precede the formation of microaneurysms, which can leak fluid into the retina and cause swelling of the macula.

Microaneurysms are often the first clinical sign of diabetic retinopathy, and are seen as intraretinal deep red spots 10–100 μm in diameter. The significance of microaneurysm counts and their close correlation with the severity of the disease are well documented [11,12]. Microaneurysm formation and regression are dynamic processes [13], where microaneurysms form and then later clot and regress. More than 50% of them either form or regress within a 12-month period [14]. There is evidence that turnover, as well as absolute counts, is an early indicator of retinopathy progression [12, 15]. Rupture of microaneurysms gives rise to small round dot hemorrhages, which are indistinguishable from microaneurysms in color fundus images. Hemorrhages present a range of size, color, and texture from the dot hemorrhages, through blotch (cluster) hemorrhages to larger boat-shaped or flame-shaped hemorrhages. A pattern recognition approach may well be required to reliably detect all the variants.

White lesions comprise exudates, cotton wool spots, and drusen. Hard exudates are caused when weakened blood vessels in the eye leak lipids onto the retina, which in turn block it from sensing light. This results in blurred or completely obstructed vision in some areas of the eye. Since exudates are made of lipids, they appear as light yellow in fundus images. Early automated detection systems used thresholding of red-free images [16, 17], while a more recent study used a multilayer neural network to detect exudates [18]. The appearance of microaneurysms and hard exudates in the macular area is more serious, and is identified as "Diabetic Maculopathy" so as to highlight the potential sight-threatening nature of this condition.

As the disease advances (preproliferative retinopathy), circulation problems cause the retina to become more ischemic and cotton wool spots to become more prevalent. In proliferative retinopathy (PDR), new fragile blood vessels can begin to grow in the retina in an attempt to restore the malnourished area and prevent it from dying. These new blood vessels may leak blood into the vitreous, clouding vision. Other complications of PDR include detachment of the retina due to scar tissue formation and the development of glaucoma, an eye disease resulting in progressive damage to the optic nerve.

12.2 Automated Detection of Diabetic Retinopathy

Many studies have been initiated worldwide to develop advanced systems for the automated detection and monitoring of diabetic retinopathy. They comprise image analysis tools for detecting and measuring common lesions (such as microaneurysms, hemorrhages, and exudates) and may include indexing and automated retrieval techniques applied to image databases. A major issue is how to accurately and objectively assess results from different studies.

Microaneurysms are among the first signs of the presence of diabetic retinopathy and their numbers correlate well with the severity of the disease in its early stages [12, 13]. Since microaneurysms are extremely small and similarly colored to the background, they are very tedious to measure manually. Most publications on automated microaneurysm detection use monochromatic data from one of the color planes of a fundus image, even though full color information is available. (Microaneurysms are better visualized with fluorescein angiography but it is more invasive and therefore less practical for screening purposes). The green plane normally contains the best detail; the red plane, while brighter and sometimes saturated, has poorer contrast; and the blue plane is dark and of least use. Hemoglobin has an absorption peak in the green region of the spectrum, so that features containing hemoglobin (e.g., microaneurysms) absorb more green light than surrounding tissues and appear dark, giving the green plane a higher contrast. Red light penetrates deeper into the layers of the retina and is primarily reflected in the choroid, explaining the reddish appearance of fundus images. Because red light has a lower absorption than green in the tissues of the eye, the red plane has less contrast. Blue light is mostly absorbed by the lens, and then by melanin and hemoglobin, and is the most scattered light, so the blue plane shows very little contrast.

However, the appearance of color retinal images can vary considerably, especially between people of different race [16, 19], so that it may be prudent to use all the color information and to employ some form of color normalization. For example, divisive shade correction can be applied to each of the color planes, and then the individual contrasts normalized to a specific mean and standard deviation [20]. This process retains the overall shape of the color image histogram, but shifts the hue (which is often dominated by the ratio of green to red in retinal images) to be consistent between images.

12.2.1 Automated Detection of Microaneurysms

Over the years, a variety of algorithms have been used to automatically detect microaneurysms [21]. For example, a morphological top-hat transformation with a linear structuring element at different orientations was used to distinguish connected, elongated structures (i.e., the vessels) from unconnected circular objects (i.e., the microaneurysms) [22]. A shade-correction preprocessing step and a matched filtering postprocessing step were then added to the basic detection technique [23–32].[1,2,3]

---

[1] http:www.ces.clemson.edu/ahoover/stare.

[2] http://www.isi.uu.nl/Research/Databases/DRIVE/.

[3] http://messidor.crihan.fr.

The Waikato Microaneurysm Detector [26, 27] modified this procedure to work on full-color retinal images. The green plane of the retinal images was used to find all candidate objects. After normalization by subtracting a median filtered version of the image, noise was removed by median filtering using a small kernel. A top-hat transform was performed by morphological reconstruction [28], using an elongated structuring element at different orientations to detect the vasculature. Following the removal of the vasculature and a microaneurysm matched filtering step, an adaptive threshold isolated microaneurysm candidates and region-growing on the shade-corrected green plane at the positions of candidates was used to isolate the morphology of the underlying candidate. A number of features based on the color, intensity, and shape of the candidates were extracted [27], and a naïve Bayesian classifier was used to assign a likelihood to each of the found candidate objects that it is a true microaneurysm. Sensitivity can be traded against specificity by varying a threshold probability that determines whether or not a candidate should be considered a microaneurysm.

A recent study [29], using a variation of this processing method and a *k*-nearest neighbor (*k*-NN) classifier, reported a sensitivity of 88.5% for detecting microaneurysms. An alternative method [30] used template matching in the wavelet domain to find the microaneurysm candidates. It assumed that microaneurysms at a particular scale can be modeled with two-dimensional, rotation-symmetric generalized Gaussian functions.

12.3 Image Databases

The STARE (see footnote 1) and DRIVE (see footnote 2) [31, 32] databases of retinal images have been widely used to compare various vessel segmentation algorithms. A subset of the DRIVE database is being used for an online challenge (the Retinopathy Online Challenge [21]) to compare algorithms used to detect microaneurysms.

The Messidor project database (see footnote 3) is the largest database of retinal images currently available on the Internet. It was established to facilitate studies on computer-assisted diagnoses of diabetic retinopathy, and comprises 1,200 color fundus images of the posterior pole acquired using a Topcon TRC NW6 camera [Topcon Medical Systems, Inc. (TMS), of Paramus, NJ] with a 45° field of view. The 24-bit RGB color images are of various sizes: 1,440 × 960, 2,240 × 1,488 or 2,304 × 1,536 pixels. Eight hundred of the images were acquired with pupil dilation and 400 without dilation.

Two diagnoses are provided by expert ophthalmologists for each image: the retinopathy grade and the risk of macular edema. The retinopathy grade [0 (normal), 1, 2 or 3] is based on the numbers of microaneurysms (MA) and hemorrhages (H), and whether there is neovascularization (NV = 1) or not (NV = 0), using

0: $(\mathrm{MA} = 0)$ AND $(\mathrm{H} = 0)$
1: $(0 < \mathrm{MA} \leq 5)$ AND $(\mathrm{H} = 0)$

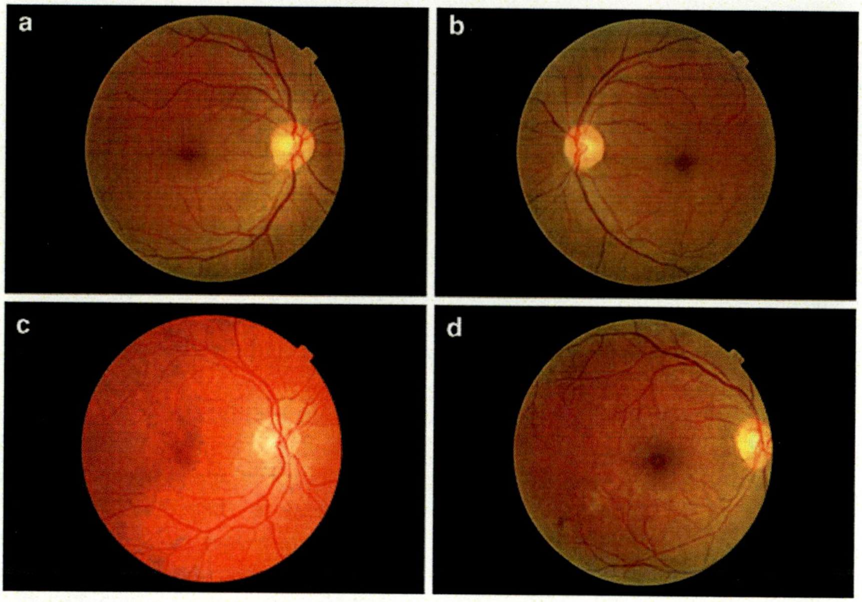

Fig. 12.1 Example images from the publicly available Messidor database: (**a**) grade 0, (**b**) grade 1, (**c**) grade 2, and (**d**) grade 3

2: $\left((5 < \text{MA} < 15) \text{ OR } (0 < \text{H} < 5)\right) \text{ AND } (\text{NV} = 0)$
3: $(\text{MA} \geq 15) \text{ OR } (\text{H} \geq 5) \text{ OR } (\text{NV} = 1)$

Typical images, for grades 0 through to 3, are shown in Fig. 12.1.

This reflects the order in which these pathologies appear, viz. first microaneurysms, then hemorrhages, and then neovascularization (which results in PDR). Grade 1 corresponds to the R1 (minimal) and R2 (mild) grades, and grade 2 to the R3 (moderate) and R4 (severe) grades, the main categories of the early treatment diabetic retinopathy study (ETDRS) classification system [33] used in clinical trials. Hard exudates were used to grade the risk of macular edema. We looked at example images from all grades, with a view to exploring whether tortuosity might be a useful indicator of the severity of the retinopathy.

In addition to the Messidor project database images, each assigned a retinopathy grade, we were supplied with a second database of 82 different images (with names MA_Originaux_XX, where XX is the Image ID starting from 01: courtesy of Dr. Jean-Claude Klein, Center of Mathematical Morphology of MINES ParisTech). Each image is 1,440 × 960 pixels (and 118 pixels/cm). The microaneurysms in these images were particularly clear, and were manually identified by three expert ophthalmologists. We will refer to this database as the "marked database."

12.4 Tortuosity

Normal retinal blood vessels are straight or gently curved, but they become dilated and tortuous in a number of disease classes, including high blood flow, angiogenesis, and blood vessel congestion [34]. It has been suggested that the severity of many retinal diseases, and the progression of retinopathy, could be inferred from the tortuosity of the blood vessel network if a consistent quantitative measure of tortuosity could be demonstrated [34]. In clinical practice, ophthalmologists commonly grade tortuosity using a qualitative scale (e.g., mild, moderate, severe, and extreme) [35], but a reliable quantitative measure would enable the automated measurement of retinal vascular tortuosity and its progression to be more easily discerned.

A multiplicity of tortuosity measures are in use, including the relative length increase over a straight vessel [36] or a smoothed curve through the vessel [37], the relative length increase for vessels in a range of spatial frequencies [36, 38], and various measures of integral curvature along the vessels [39–43]. Those based on relative length increase only measure vessel elongation and have no value in measuring morphology or hemodynamic consequences, while those using integrated curvature require arbitrary smoothing schemes to smooth the noise in the coordinates resulting from limited sampling.

12.4.1 *Tortuosity Metrics*

Two robust metrics have been proposed for quantifying vascular tortuosity in terms of three-dimensional (3-D) curvature [44]. They are additive and scale invariant, and largely independent of image noise (for signal-to-noise ratios greater than ~50 dB) and the resolution of the imaging system. The metrics were validated using both 2-D and 3-D clinical vascular systems [45], and are well suited to automated detection and measurement when used with a vessel tracking algorithm. In a preliminary application to retinal pathologies [46], they correlated strongly with the ranking of tortuosity by an expert panel of ophthalmologists, and were able to distinguish several pathologies from normal in a discretionary (i.e., referred) population.

One of these metrics, the mean tortuosity, is equivalent to the accumulating angle change along the length of a vessel considered to comprise straight-line segments between closely digitized points along its midline. Figure 12.2 shows how the tortuosity decreases as the length of these segments (viz., the sampling interval) increases. There are large digitization errors with small sampling intervals, which results in an artificially elevated tortuosity. Large sampling intervals miss high-frequency changes and underestimate the tortuosity of highly tortuous vessels. A sampling interval of five pixels minimized digitization errors and accurately traced the vessels in images corresponding to all retinopathy grades.

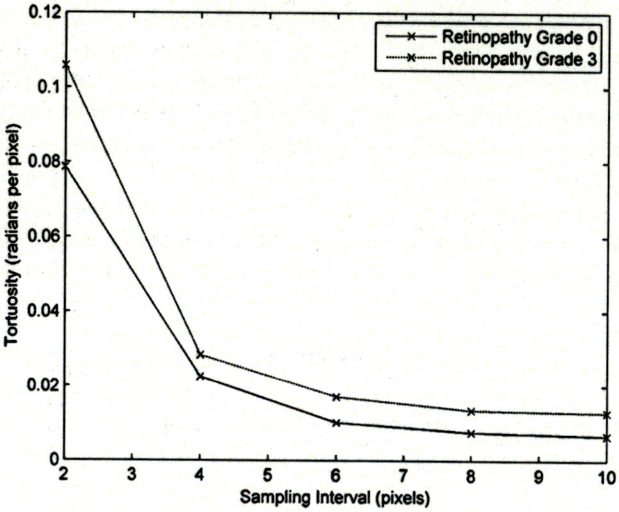

Fig. 12.2 Tortuosity measured using different sampling intervals along two typical vessels

Since tortuosity is additive, it is clear that it is the tortuosity per unit length, rather than tortuosity itself, that is the actual metric of interest. Therefore, Fig. 12.2 plots the tortuosity divided by the length of the vessel (in pixels).

12.5 Tracing Retinal Vessels

A common approach to tracing linear features (viz., thin objects across which the image presents an intensity maximum in the direction of the largest variance, gradient, or surface curvature (i.e., perpendicular to the linear feature)) in an image is to segment the image and perform skeletonization, after some initial preprocessing. Typically, an image will be hampered by noise (inevitable statistical fluctuations as well as other irrelevant structures), poor resolution, low contrast, and background gradients (nonuniform illumination). Although prevention is better than cure, to some extent these artifacts can be minimized by image processing operations such as (nonlinear) smoothing [47], deconvolution, shading correction, and morphological filtering [48]. Comparison of images often calls for histogram equalization or histogram matching.

Segmentation is a challenging process in all but the very simplest images. Methods for segmentation can be roughly categorized as region-based and boundary-based approaches. Region-based segmentation is usually implemented by some form of (adaptive) thresholding. However, intensity thresholding, while commonly used for its simplicity and efficiency, is generally known to be one of the most error-prone segmentation methods. Boundary-based methods include edge-detecting and subsequent linking, boundary tracking, active contours (see Chapter 4), and watershed segmentation. The next step would be to extract the centerlines, for which various skeletonization algorithms have been proposed [48]. The process is very sensitive to noise, and generally results in a number of errors such as spurious gaps and branches (spurs) and ambiguities (especially in 2-D) such as spurious loops and crossings. Various filling and pruning strategies must then be employed to try to rectify these retrospectively.

An alternative approach, which circumvents the problems inherent in segmentation and skeletonization, is to obtain the centerlines directly from the grayscale images by applying a Hessian [49–52] or Jacobian [53] based analysis of critical points, using matched or steerable filters [54] or by nonmaximum suppression [55, and Chapter 8].

The Hessian is a generalization of the Laplacian operator; it is a square matrix comprising second-order partial derivatives of the image, and can therefore be used to locate the center of a ridge-like structure. Specifically, the local principal ridge directions at any point in an image are given by the eigenvectors of the second-derivative matrix computed from the intensity values around that point.

The Hessian of an intensity image can be obtained at each point by computing

$$H(x,y) = \begin{bmatrix} \partial 2L/\partial x2 & \partial 2L/\partial x\partial y \\ \partial 2L/\partial x\partial y & \partial 2L/\partial y2 \end{bmatrix} = \begin{bmatrix} Lxx & Lxy \\ Lyx & Lyy \end{bmatrix}, \tag{12.1}$$

where

$$L(x,y;t) = g(x,y;t)^{*} f(x,y) \tag{12.2}$$

and $g(\cdot\,;t)$ is a Gaussian function with variance t, f is an image, (x,y) is a pixel location, and * represents the convolution operation. The partial derivatives can be computed by convolving the image f with a derivative-of-Gaussian kernel. Due to the symmetry of this matrix, the eigenvectors are orthogonal, with the eigenvector corresponding to the smaller absolute eigenvalue pointing in the longitudinal direction of the ridge. The scale-normalized determinant of the Hessian has better scale selection properties than the more commonly used Laplacian operator [56]. The Hessian values can be incorporated into a boundary tracking algorithm and linked using the so-called live-wire segmentation paradigm [57–59].

12.5.1 NeuronJ

This latter approach is the method of semiautomated tracing employed by NeuronJ [49], a plugin for ImageJ, which was developed to identify and trace neurons with limited user intervention but which we have used equally effectively to trace the centerlines of retinal blood vessels. The user selects a starting point and the search algorithm finds the optimal paths from that point to all other points in the image (on the basis of their Hessian "vesselness" values), where "optimal" means having a globally minimal cumulative cost according to a predefined function. The paths can

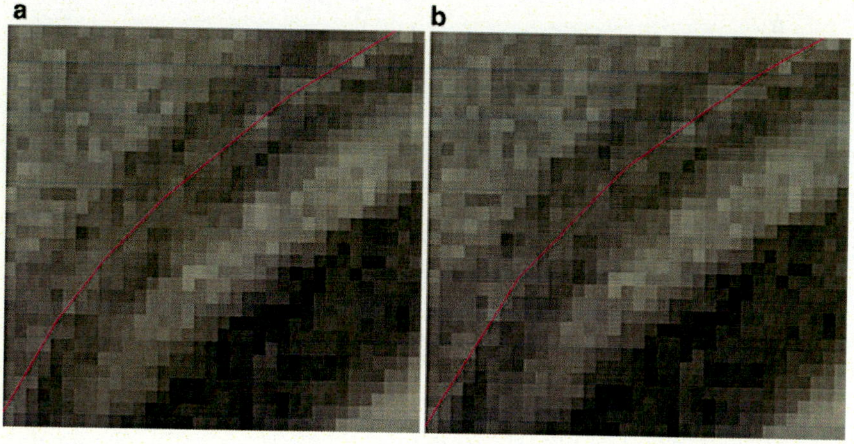

Fig. 12.3 Use of a tracing sampling interval of (**a**) 5 and (**b**) 10 in tracing a typical segment of a vessel

be displayed in real time as the user moves the cursor towards the end of the object of interest, until the presented path starts to deviate too much from what is considered the optimal tracing by the user. The tracing can then be anchored up to that point by a single mouse click, after which the algorithm proceeds by presenting optimal paths from that point. The process is iterated until the entire object has been traced. In cases where the user is not completely satisfied with the paths presented, which may sometimes occur in regions with very low contrast, it is possible to switch to manual delineation.

Before tracing can begin, several parameters must be selected. We chose parameter settings based on test tracings. For our images, we used a sampling interval of 5 (viz., 1 out of every 5 pixels along the tracings is used). Figure 12.3a shows a tracing with a sampling interval of 5, and Fig. 12.3b uses a sampling interval of 10. The former produces a smoother and better fit to the vessel, while the latter produces a more jagged centerline, which would artifactually result in a higher tortuosity. However for narrow vessel segments which are straight, or nearly so, a larger sampling interval performs better at finding the centerline. Figure 12.4a shows a nearly straight segment of a vessel using a sampling interval of 5. (Short lines are shown extending from each subsegment for clarity). In Fig. 12.4b, with a sampling interval of 10, the centerline tracing more closely follows the straight vessel. This is a consequence of NeuronJ using integer coordinates (taken as the centers of the pixels); representing lines other than those along the horizontal, vertical or at 45° to the grid can introduce subpixel errors which are more significant the smaller the subsegments. (We will return to this point later.)

The tracings are generally smoothed prior to sampling using a moving-average filter; we used a smoothing half-length of 5, namely a filter length of 11. We found that selecting a smoothing half-length lower than the sampling interval tends to produce more jagged lines, while a smoothing half-length larger than the sampling interval makes it difficult to follow sharply bending vessels.

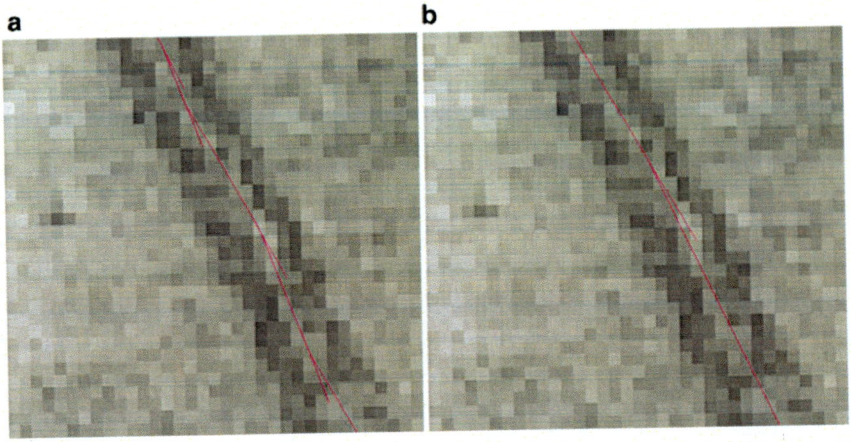

Fig. 12.4 Use of a tracing sampling interval of (**a**) 5 and (**b**) 10 in tracing a straight segment of a vessel. To highlight the error on straight lines, each segment has been extrapolated to show the angle it forms with the next

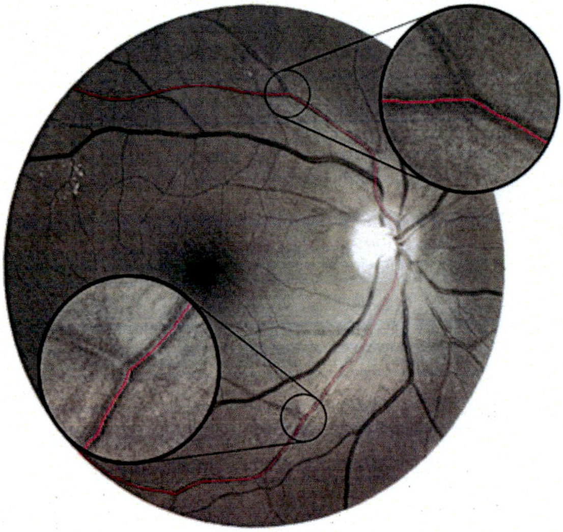

Fig. 12.5 Showing the choice of tracing at bifurcations (see magnified areas) in a particular image (MA_Originaux_33)

Typically, there are four long blood vessels which emerge from the optic disk in a retinal image – two arterioles and two venules. We identified the (oxygen-rich) arterioles as the redder vessels (they also tended to have smaller diameters and higher tortuosity than the venules) and traced two of them (an "upper" and a "lower" arteriole) from each image, starting where the vessels leave the optic disk. At a bifurcation, we selected the ongoing branch at the shallower branching angle, which generally corresponded to the thicker, longer branch (Fig. 12.5). The digitized coordinates of the vessel centerlines were then exported to an Excel file.

The quantization of the digitized coordinates to integers when tracing in NeuronJ is an unwanted limitation. It introduces an error that is particularly noticeable for line-like structures or segments. We mitigated this by applying a three-point moving average filter to the exported coordinates prior to calculating tortuosities.

NeuronJ is a semiautomated tracing method, as it requires the user to guide the vessel tracing. However, it has major advantages in that there is no need to preprocess the images (e.g., by histogram equalization) nor is segmentation required (i.e., the tracing can be performed on a grayscale image (usually the green plane of an RGB color image) directly).

12.5.2 Other Software Packages

Hessian-based detectors are computationally expensive. HCA-vision is a software platform for the automated detection and analysis of linear features based on multidirectional nonmaximum suppression (MDNMS). HCA-vision has been successfully applied to a number of applications including neurite tracing for drug discovery and functional genomics [60], quantifying astrocyte morphology [61], and separating adjacent bacteria under phase contrast microscopy [62]. It is discussed in detail in Chapter 8. The software can be downloaded after completing a request at http://www.hca-vision.com/product_download_hca_vision.html. The large noise content in our images precluded us from utilizing the program successfully.

Retinal vessel centerline extraction can also be achieved using multiscale matched filters, with a vessel confidence measure defined as a projection of a vector formed from a normalized pixel neighborhood on to a normalized ideal vessel profile [54]. Vessel boundary measures and associated confidences are computed at potential vessel boundaries. A training technique is used to develop a mapping of this vector to a likelihood ratio that measures the "vesselness" at each pixel. Results comparing this vesselness measure to matched filters alone and to measures based on the Hessian of intensities have shown substantial improvements both qualitatively and quantitatively. Binary executables of the code are available at http://www.sofka.com/LRV.html.

This represents a possible route for fully automated tracing of retinal vessels. A typical tracing is shown in Fig. 12.6. This is a promising result, although it was difficult to avoid breaks in the tracings without adding spurious vessels. It may be possible to achieve a more acceptable result with an optimal choice of tracer sensitivity, and judicious preprocessing of the image to remove background variation and reduce noise.

12.6 Experimental Results and Discussion

The grades associated with the Messidor project database (and all other databases) are rather coarse. To explore changes in the vessels with the severity of retinopathy, we need a finer grading scheme. The number of detected microaneurysms has been

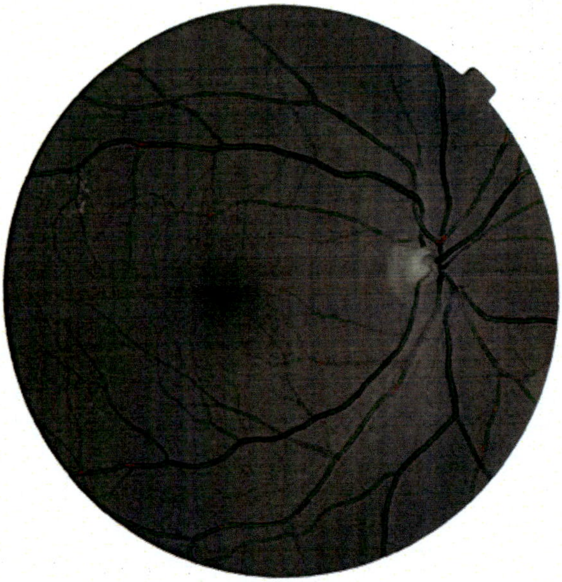

Fig. 12.6 The result of a fully automated tracing of retinal vessels using Sofka's program on the same image as in Fig. 12.5. The traced vessels are shown in *green*, with the bifurcations subsequently colored *red*

used as a surrogate for the severity of retinopathy in its early stages [12, 13]. This information was available in the marked database (of 82 unique images) supplied by MINES ParisTech.

Figure 12.7 shows that the tortuosity (per cm) increases steadily with the number of manually identified microaneurysms from images in the marked database. The correlation coefficient ($r = 0.4159$, $n = 46$) corresponds to a probability, p, for the null hypothesis of 0.004.[4] This suggests that tortuosity is related to microaneurysm count, and that the tortuosity increases with the severity of retinopathy.

The number of microaneurysms detected by the Waikato Automated Microaneurysm Detector did not match the numbers manually counted by experts for images from the marked database. Despite the normalization of the images prior to segmentation, the number of microaneurysms detected is still somewhat dependent on the threshold probability used in the Bayes classifier. Useful thresholds normally lie between 10^{-5} and 10^{-7} (Cree, private communication). The key is to select a threshold which is high enough to detect all the microaneurysms but small enough not to count noise as aneurysm, but this optimum threshold depends on the particular image. We investigated two methods for finding the optimum value of the threshold.

[4] http://faculty.vassar.edu/lowry/VassarStats.html.

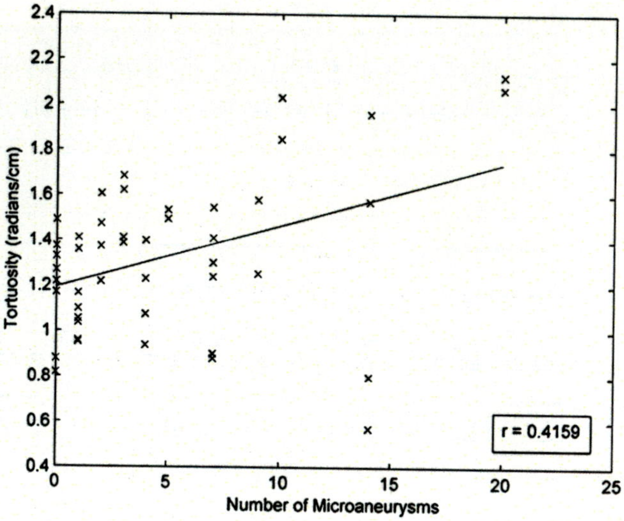

Fig. 12.7 The tortuosity/length for 46 vessels from the marked database. The best-fitted line is superimposed

For both methods, we plotted the number of microaneurysms detected in an image for a series of thresholds between 10^{-5} and 10^{-7}. The higher the threshold, the smaller the number of aneurysms detected. We considered the optimum threshold to be a balance between the microaneurysm count changing too fast with threshold and it hardly changing at all with threshold. In our first method (method 1), we considered this be the value where the plot of microaneurysm counts was furthest from a straight line connecting the two extreme thresholds (Fig. 12.8). In our second method (method 2), we calculated the local curvature of datapoints (by calculating the local tortuosity using five points centered on the datapoint of interest) in the plot of microaneurysm counts vs. threshold, and considered the largest value to indicate the position of the optimum threshold.

We tested both methods on the marked database, for which we know the microaneurysm counts (viz., the "gold standard" counts from manual counting by three experts). Table 12.1 shows the number of microaneurysms in the images, assessed by the various methods. Probability thresholds of 10^{-5}, 10^{-6}, and 10^{-7} are included, although microaneurysm counts for 100 different thresholds were computed. The correlation coefficient for these fixed values with the "gold standard" values varied between 0.4696 and 0.6805, depending on the value of the threshold. Although it is possible to achieve a correlation of 0.6805 with a fixed threshold, it would be highly unlikely that this particular value of the threshold would be chosen a priori. The correlation coefficient for the number of microaneurysms found by method 1 with the "gold standard" counts was 0.6458, while the correlation coefficient for the number of microaneurysms found by method 2 with the "gold standard" counts was higher at 0.7303. (This corresponds to $p < 0.001$ $(n = 82)$

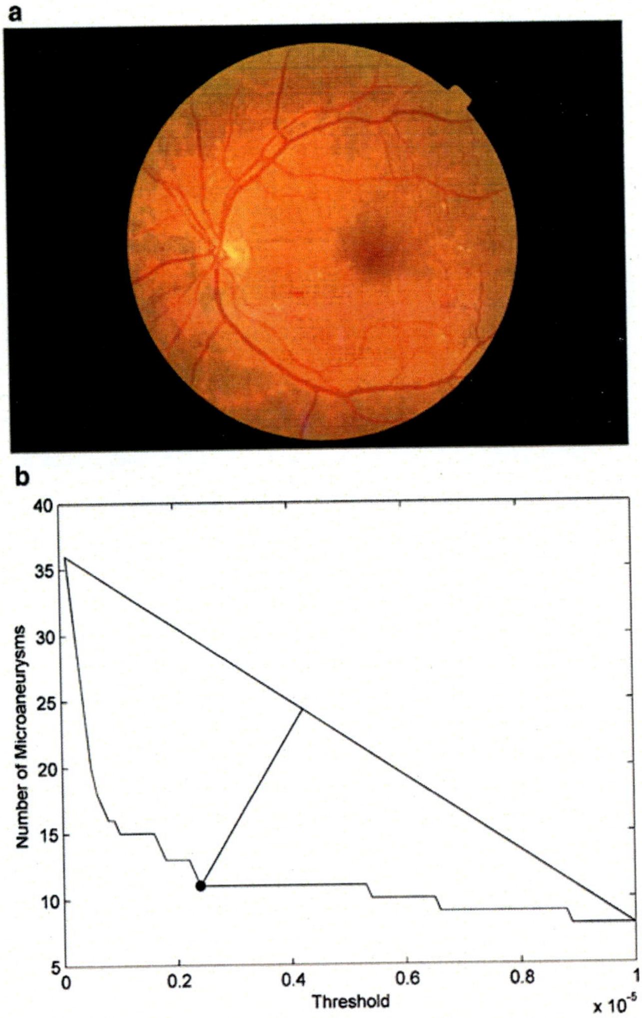

Fig. 12.8 (**a**) Image (MA_Originaux_09, from the marked database) and (**b**) a plot showing how the optimum threshold is obtained from this image using method 1

for the null hypothesis). In the light of these findings, we consider that method 2 is particularly well suited for finding the optimum threshold for use with the Waikato Automated Microaneurysm Detector.

The microaneurysms in the Messidor project database are not so clearly delineated. We detected and counted them with the Waikato Automated Microaneurysm Detector using method 2 (Fig. 12.9). Only images with a microaneurysm count of five or greater were considered, since we had greatest confidence that these corresponded to actual microaneurysms. The correlation

Table 12.1 The number of microaneurysms detected for the images of the marked database using various thresholds, and our methods 1 and 2, compared with the actual "gold standard" counts

Image ID	"Gold standard" counts	Threshold $= 10^{-5}$	Threshold $= 10^{-6}$	Threshold $= 10^{-7}$	Method 1	Method 2
1	2	0	1	11	1	2
2	3	2	3	11	3	4
3	2	1	1	4	1	2
4	1	3	4	16	3	4
5	4	2	4	19	4	4
6	16	5	17	34	11	16
7	3	1	6	11	2	7
8	8	3	7	12	4	7
9	9	8	15	36	11	11
10	19	7	20	39	16	16
11	14	4	15	41	11	9
12	3	2	3	11	3	4
13	3	2	4	11	4	5
14	4	0	4	7	1	4
15	1	3	4	16	3	4
16	5	2	4	19	4	4
17	11	3	10	22	10	11
18	16	7	17	29	13	15
19	7	2	8	21	4	4
20	0	1	1	7	1	3
21	2	0	1	4	1	1
22	3	1	6	11	2	7
23	6	2	7	11	4	5
24	7	3	13	25	11	10
25	0	1	1	3	1	3
26	0	0	1	6	0	1
27	6	3	10	39	8	6
28	10	8	17	48	14	14
29	11	0	3	12	2	3
30	5	5	13	20	5	15
31	2	1	7	21	5	6
32	8	4	13	36	9	7
33	7	0	6	13	3	6
34	5	4	7	33	6	8
35	3	1	3	15	1	3
36	0	1	2	43	2	2
37	1	0	6	34	2	2
38	18	5	12	24	10	10
39	18	1	13	34	8	14
40	22	5	17	42	11	17
41	7	3	9	25	9	9
42	17	3	8	31	5	8
43	10	5	18	57	10	10

(continued)

Table 12.1 (continued)

Image ID	"Gold standard" counts	Threshold $= 10^{-5}$	Threshold $= 10^{-6}$	Threshold $= 10^{-7}$	Method 1	Method 2
45[a]	4	2	6	11	7	4
46	4	1	4	22	3	3
47	1	0	1	10	5	2
48	18	4	9	25	1	9
49	2	1	2	14	9	2
50	13	5	15	42	2	12
51	10	4	9	30	12	8
52	7	1	8	23	9	4
53	1	0	1	4	2	1
54	0	1	4	11	1	4
55	3	1	3	6	1	3
56	12	3	5	36	2	5
57	14	3	17	52	5	11
58	8	2	5	8	11	6
59	0	0	2	5	3	2
60	2	2	3	12	2	3
61	4	1	2	9	3	3
62	11	3	11	22	2	9
63	21	4	11	25	9	12
64	5	5	12	27	6	14
65	14	8	21	36	12	13
66	5	1	2	17	16	4
67	2	0	2	17	2	2
68	0	1	2	22	0	2
69	0	0	0	3	2	3
70	2	3	7	24	0	7
71	1	1	2	5	6	2
72	1	1	1	14	1	2
73	13	9	26	61	1	14
74	12	15	22	93	14	20
75	6	3	15	50	22	13
76	0	0	1	4	11	3
77	0	0	1	3	0	1
78	4	10	37	111	1	20
79	2	1	3	7	25	3
80	1	0	1	4	3	1
81	1	0	8	39	1	8
82	15	5	15	32	3	16
83	6	5	15	8	9	7

[a]No ID 44 exists because the image is identical to ID 43

of microaneurysm count with tortuosity is not as strong ($r = 0.2236$ ($n = 34$), corresponding to a probability value for the null hypothesis of 0.1018) as with the marked database. Although we optimized the counting within the Waikato

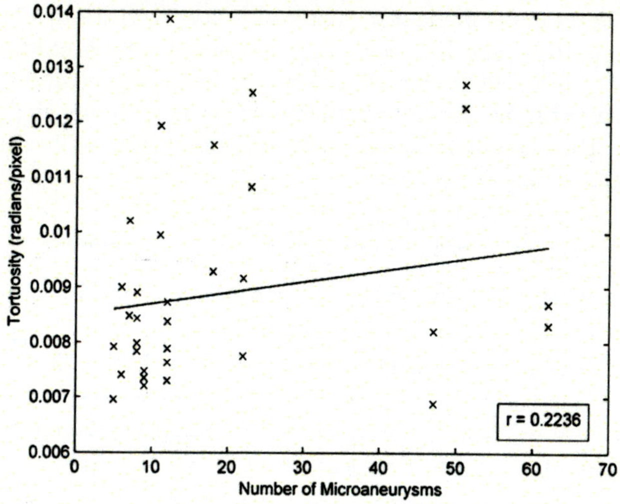

Fig. 12.9 The tortuosity/length for 34 vessels from the Messidor project database. The best-fitted line is superimposed

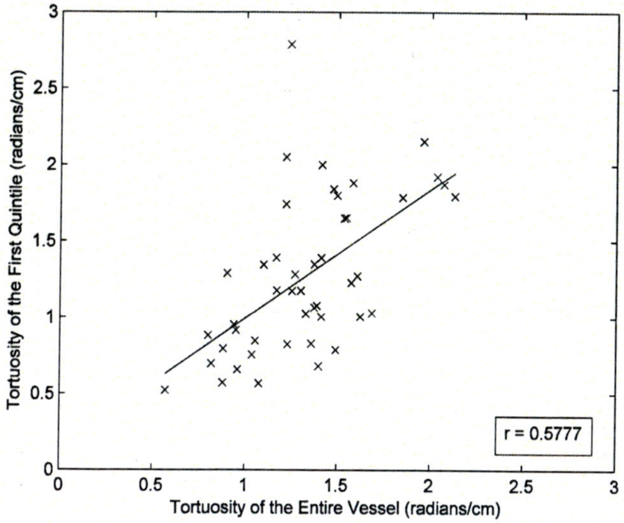

Fig. 12.10 The tortuosity of the first quintile length compared to the tortuosity of the entire vessel, for 46 vessels from the marked database

Automated Microaneurysm Detector, we do not expect the number of counts to be completely accurate. Despite this, the level of correlation supports our earlier finding with the marked database that tortuosity is related to microaneurysm count, and that tortuosity increases with severity of retinopathy.

A complicating factor in the use of vessel tortuosity as an indicator of the severity of retinopathy would be a change in tortuosity along an individual blood vessel. We measured the tortuosity of vessels along the quintiles of their length. Figure 12.10

plots the tortuosity of the first quintile of its length (closest to the optic disk) with the tortuosity of the entire vessel, for the 46 vessels of the marked database shown in Fig. 12.7. The correlation coefficient is 0.5777, corresponding to $p < 0.0001$. This indicates that there is little significant change in tortuosity along the length of a vessel, and therefore the tortuosity of the entire vessel can be confidently used to characterize each vessel.

12.7 Summary and Future Work

It has been recognized that the local flow hemodynamics of a curved vessel may dispose it to the formation of an aneurysm. The geometry of intracranial arteries, for example, has been implicated in the formation of aneurysms [10]. The branching patterns of the vessel network may also be useful in diagnosing and evaluating the severity of a disease such as diabetic retinopathy [34, 42].

We have shown that tortuosity is related to microaneurysm count, and we suggest that tortuosity increases with severity of diabetic retinopathy. It is too early to say with this limited data whether tortuosity could be used as an alternate predictor of the severity of such retinopathy. Longitudinal data would help to resolve the matter. Local flow hemodynamics will be affected by the tortuosity of the vessels, and it will affect the number of microaneurysms formed. Precisely which is cause and which is effect is difficult to ascertain, but as diabetic retinopathy becomes more severe it is likely that both tortuosity and microaneurysm count will increase, and our results confirm this trend. Blood pressure and the diameter of the vessels are also likely implicated in the changes. It may be that tortuosity is related to an integral effect of blood pressure, while microaneurysm responds more to local maxima.

Tortuosity can be measured easily from the digitized tracings of vessels in retinal images, and these tracings can be obtained using a semiautomated program such as NeuronJ. Fully automated tracing is an enticing prospect, although current algorithms would seem to require customized preprocessing of the image, which would then render the process semiautomatic again.

Fractal dimension (or the fractal signature [63]) may be an alternative method of measuring the bending within a blood vessel. Initial studies demonstrated that the blood vessels of the optic fundus are fractal, and that the fractal dimension can be used to identify PDR [64, 65]. Preliminary analysis of the skeletonized vascular patterns in the normal and NPDR macula suggested that vascular morphology had already changed by this relatively early stage of retinal disease [66].

A disadvantage of fractal dimension is that it is constrained within very tight limits (1–2 for a vessel tracing), and this limits its sensitivity. Another limitation is that different algorithms can result in different values [67]. Perhaps, its greatest potential is that it can deliver a quantitative summary value for the geometric complexity of a network, rather than a single vessel, and could therefore summarize the complete retinal vascular branching pattern in an image. Recent computerized studies [68, 69] suggest that increased fractal dimension of the retinal vasculature, reflecting increased geometric complexity of the retinal vascular branching pattern, is associated with early diabetic retinopathy microvascular damage. Although the differences in fractal dimension were small [69], the average fractal dimension was higher in participants with retinopathy than in those without retinopathy (median 1.46798 [interquartile range 1.45861–1.47626] compared with 1.46068 [1.44835–1.47062], respectively; $p < 0.001$). After adjustments for age and sex, greater fractal dimension was significantly associated with increased odds of retinopathy (odds ratio [OR] 4.04 [95%CI 2.21–7.49] comparing highest to lowest quartile of fractal dimension; OR 1.33 for each 0.01 increase in fractal dimension). This association remained with additional adjustments for diabetes duration, blood sugar, blood pressure, body mass index (BMI), and total cholesterol levels. Further adjustment for retinal arteriolar or venular caliber had minimal impact on the association (OR 3.92 [95% CI 1.98–7.75]).

Acknowledgments We acknowledge with thanks the help and assistance of Dr. Eric Meijering, Dr. Michal Sofka, and Dr. Michael Cree on issues regarding their respective software. A database of retinal images, with the microaneurysms manually identified by experts, was kindly supplied by Dr. Jean-Claude Klein of the Center of Mathematical Morphology of MINES, Paris Tech. One of us (G.D.) acknowledges the award of a Fulbright Senior Scholarship, which provided the opportunity to expand work in this area and others.

References

1. Hoover, A., Kouznetsova, V., Goldbaum, M.: Locating blood vessels in retinal images by piecewise threshold probing of a matched filter response. IEEE Trans. Med. Imaging **19**, 203–210 (2000)
2. Witt, N., Wong, T.Y., Hughes, A.D., et al.: Abnormalities of retinal vasculature structure and the risk of mortality from ischemic heart disease and stroke. Hypertension **47**, 975–981 (2006)
3. Wong, T.Y., Shankar, A., Klein, R., et al.: Retinal arteriolar narrowing, hypertension and subsequent risk of diabetes mellitus. Medicine **165**, 1060–1065 (2005)
4. Cheung, N., Wong, T.Y., Hodgson, L.: Retinal vascular changes as biomarkers of systemic cardiovascular diseases. In: Jelinek, H.F., Cree, M.J. (eds.) Automated Image Detection of Retinal Pathology, pp. 185–219, CRC Press, Boca Raton, FL (2010)
5. Wong, T.Y., Mohamed, Q., Klein, R., et al.: Do retinopathy signs in non-diabetic individuals predict the subsequent risk of diabetes? Br. J. Ophthalmol. **90**, 301–303 (2006)
6. Fong, D.S., Aiello, L., Gardner, T.W., et al.: Diabetic retinopathy. Diabetes Care **26**, 226–229 (2003)
7. Dobrin, P.B., Schwarz, T.H., Baker, W.H.: Mechanisms of arterial and aneurismal tortuosity. Surgery **104**, 568–571 (1988)
8. Wenn, C.M., Newman, D.L.: Arterial tortuosity. Aust. Phys. Eng. Sci. Med. **13**, 67–70 (1990)
9. Dougherty, G., Varro, J.: A quantitative index for the measurement of the tortuosity of blood vessels. Med. Eng. Phys. **222**, 567–574 (2000)
10. Bor, A.S.E., Velthuis, B.K., Majoie, C.B., et al.: Configuration of intracranial arteries and development of aneurysms: a follow-up study. Neurology **70**, 700–705 (2008)
11. Klein, R., Meuer, S.M., Moss, S.E., et al.: Retinal aneurysm counts and 10-year progression of diabetic retinopathy. Arch. Ophthalmol. **113**, 1386–1391 (1995)
12. Kohner, E.M., Stratton, I.M., Aldington, S.J., et al.: Microaneurysms in the development of diabetic retinopathy (UKPDS 42). Diabetologia **42**, 1107–1112 (1999)

13. Hellstedt, T., Immonen I.: Disappearance and formation rates of microaneurysms in early diabetic retinopathy. Br. J. Ophthalmol. **80**, 135–139 (1996)
14. Kohner, E.M., Dollery, C.T.: The rate of formation and disappearance of microaneurysms in diabetic retinopathy. Eur. J. Clin. Invest. **1**, 167–171 (1970)
15. Goatman, K.A., Cree, M.J., Olson, J.A., et al.: Automated measurement of microaneurysm turnover. Invest. Ophthalmol. Vis. Sci. **44**, 5335–5341 (2003)
16. Phillips, R.P., Spencer, T., Ross, P.G., et al.: Quantification of diabetic maculopathy by digital imaging of the fundus. Eye **5**, 130–137 (1991)
17. Phillips, R., Forrester, J., Sharp, P.: Automated detection and quantification of retinal exudates. Graefe's Arch. Clin. Exp. Ophthalmol. **231**, 90–94 (1993)
18. Osareh, A., Shadgar, B., Markham, R.: A computational-intelligence-based approach for detection of exudates in diabetic retinopathy images. IEEE Trans. Inf. Tech Biomed. **13**, 535–545 (2009)
19. Preece, S.J., Claridge E. Monte Carlo modeling of the spectral reflectance of the human eye. Phys. Med. Biol. **47**, 2863–2877 (2002)
20. Cree, M.J., Gamble, E., Cornforth, D.J.: Colour normalisation to reduce inter-patient and intra-patient variability in microaneurysm detection in colour retinal images. In: APRS Workshop in Digital Imaging (WDIC2005), Brisbane, Australia, pp. 163–168 (2005)
21. Niemeijer, M., van Ginneken, B., Cree, M.J., et al.: Retinopathy online challenge: automatic detection of microaneurysms in digital color fundus photographs. IEEE Trans. Med. Imaging **29**, 185–195 (2010)
22. Baudoin, C.E., Lay, B.J., Klein, J.C.: Automatic detection of microaneurysms in diabetic fluorescein angiographies. Revue D'Épidémiologie et de Sante Publique **32**, 254–261 (1984)
23. Spencer, T., Olson, J.A., McHardy, K.C., et al.: An image-processing strategy for the segmentation and quantification in fluorescein angiograms of the ocular fundus. Comput. Biomed. Res. **29**, 284–302 (1996)
24. Cree, M.J., Olson, J.A., McHardy, K.C., et al.: A fully automated comparative microaneurysm digital detection system. Eye **11**, 622–628 (1997)
25. Frame, A.J., Undrill, P.E., Cree, M.J., et al.: A comparison of computer based classification methods applied to the detection of microaneurysms in ophthalmic fluorescein angiograms. Comput. Biol. Med. **28**, 225–238 (1998)
26. Streeter, L., Cree, M.J.: Microaneurysm detection in colour fundus images. In: Proceedings of the Image and Vision Computing New Zealand Conference (IVCNZ'03), Palmerston North, New Zealand, pp. 280–285 (2003)
27. Cree, M.J., Gamble, E., Cornforth, D.: Colour normalisation to reduce inter-patient and intra-patient variability in microaneurysm detection in colour retinal images. In: Proceedings of APRS Workshop on Digital Image Computing (WDIC2005), Brisbane, Australia, pp. 163–168 (2005)
28. Vincent, L.: Morphological grayscale reconstruction in image analysis: applications and efficient algorithms. IEEE Trans. Image Process. **2**, 176–201 (1993)
29. Dupas, B., Walter, T., Erginay, A., et al.: Evaluation of automated fundus photograph analysis algorithms for detecting microaneurysms, haemorrhages and exudates, and of a computer-assisted diagnostic system for grading diabetic retinopathy. Diabetes Metab. **36**, 213–220 (2010)
30. Quellec, G., Lamard, M., Josselin, P.M., et al.: Optimal wavelet transform for the detection of microaneurysms in retina photographs. IEEE Trans. Med. Imaging **27**, 1230–1241 (2008)
31. Niemeijer, M., Staal, J.S., van Ginneken, B., et al.: Comparative study of retinal vessel segmentation on a new publicly available database. Proc. SPIE 5370–5379 (2004)
32. Staal, J., Abramoff, M., Neimeijer, Mc, et al.: Ridge-based vessel segmentation in color images of the retina. IEEE Trans. Med. Imaging **23**, 501–509 (2004)
33. Early Treatment Diabetic Retinopathy Study Research Group: Grading diabetic retinopathy from stereoscopic color fundus photographs – an extension of the modified Airlie House classification. ETDRS report #10. Ophthalmology **98**, 786–806 (1991)

34. Hart, W.E., Goldbaum, M., Coté, B., et al.: Measurement and classification of retinal vascular tortuosity. Int. J. Med. Inform. **53**, 239–252 (1999)
35. Aslam, T., Fleck, B., Patton, N., et al.: Digital image analysis of plus disease in retinopathy of prematurity. Acta ophthalmol. **87**, 368–377 (2009)
36. Capowski, J.J., Kylstra, J.A., Freedman, S.F.: A numeric index based on spatial frequency for the tortuosity of retinal vessels and its application to plus disease in retinopathy of prematurity. Retina **15**, 490–500 (1995)
37. Wallace, D.K.: Computer-assisted quantification of vascular tortuosity in retinopathy of prematurity. Trans. Am. Ophthalmol. Soc. **105**, 594–615 (2007)
38. Owen, C.G., Rudnicka, A.R., Mullen, R., et al.: Measuring retinal vessel tortuosity in 10-year-old children: validation of the computer-assisted image analysis of the retina (CAIAR) program. Invest. Ophthalmol. Vis. Sci. **50**, 2004–2010 (2009)
39. Lotmar, W., Freiburghaus, A., Bracker, D.: Measurement of vessel tortuosity on fundus photographs. Graefe's Arch. Clin. Exp. Ophthalmol. **211**, 49–57 (1979)
40. Smedby, Ö., Högman, N., Nilsson, U., et al.: Two-dimensional tortuosity of the superficial femoral artery in early atherosclerosis. J. Vasc. Res. **30**, 181–191 (1993)
41. Saidléar, C.A.: Implementation of a Quantitative Index for 3-D Arterial Tortuosity. M.Sc. thesis, University of Dublin, 2002
42. Bullitt, E., Gerig, G., Pizer, S.M., et al.: Measuring tortuosity of the intracerebral vasculature from MRA images. IEEE Trans. Med. Imaging **22**, 1163–1171 (2003)
43. Grisan, E., Foracchia, M., Ruggeri, A.: A novel method for the automatic evaluation of retinal vessel tortuosity. IEEE Trans. Med. Imaging **27**, 310–319 (2008)
44. Johnson, M.J., Dougherty, G.: Robust measures of three-dimensional vascular tortuosity based on the minimum curvature of approximating polynomial spline fits to the vessel mid-line. Med. Eng. Phys. **29**, 677–690 (2007)
45. Dougherty, G., Johnson, M.J.: Clinical validation of three-dimensional tortuosity metrics based on the minimum curvature of approximating polynomial splines. Med. Eng. Phys. **30**, 190–198 (2008)
46. Dougherty, G., Johnson, M.J., Wiers, M.D.: Measurement of retinal vascular tortuosity and its application to retinal pathologies. Med. Biol. Eng. Comput. **48**, 87–95 (2010)
47. Sonka, M., Hlavac, V., Boyle, R.: Image Processing, Analysis, and Machine Vision, 3rd edn. Cengage Learning, Florence, KY (2007)
48. Dougherty G.: Digital Image Processing for Medical Applications. Cambridge University Press, Cambridge (2009) (a) pp. 259–263; (b) pp. 157–159; (c) pp. 296–301; (d) pp. 140–144
49. Meijering, E., Jacob, M., Sarria, J.C.F., et al.: Design and validation of a tool for Neurite tracing and analysis in fluorescence microscopy images. Cytometry A **58**, 167–176 (2004)
50. Xiong, G., Zhou, X., Degterev, A., et al.: Automated neurite labeling and analysis in fluorescence microscopy images. Cytometry A **69**, 494–505 (2006)
51. Zhang, Y., Zhou, X., Witt, R.M., et al.: Dendritic spine detection using curvilinear structure detector and LDA classifier. Neuroimage **36**, 346–360 (2007)
52. Fan, J., Zhou, X., Dy, J.G., et al.: An automated pipeline for dendrite spine detection and tracking of 3D optical microscopy neuron images of in vivo mouse models. Neuroinformatics **7**, 113–130 (2009)
53. Yuan, X., Trachtenberg, J.T., Potter, S.M., et al.: MDL constrained 3-D grayscale skeletonization algorithm for automated extraction of dendrites and spines from fluorescence confocal images. Neuroinformatics **7**, 213–232 (2009)
54. Sofka, M., Stewart, C.V.: Retinal vessel centerline extraction using multiscale matched filters, confidence and edge measures. IEEE Trans. Med. Imaging **25**, 1531–1546 (2006)
55. Sun, C., Vallotton, P.: Fast linear feature detection using multiple directional non-maximum suppression. J. Microsc. **234**, 147–157 (2009)
56. Lindeberg, T.: Feature detection with automatic scale selection. Int. J. Comput. Vis. **30**(2), 77–116 (1998)
57. Barrett, W.A., Mortensen, E.N.: Interactive live-wire boundary extraction. Med. Image Anal. **1**, 331–341 (1996)

58. Falcão, A.X., Udupa, J.K., Samarasekera, S., et al.: User-steered image segmentation paradigms: live wire and live lane. Graph. Models Image Process. **60**, 233–260 (1998)
59. Falcão, A.X., Udupa, J.K., Miyazawa, F.K.: An ultra-fast user-steered image segmentation paradigm: LiveWire on the fly. IEEE Trans. Med. Imaging **19**, 55–62 (2000)
60. Vallotton, P., Lagerstrom, R., Sun, C., et al.: Automated analysis of neurite branching in cultured cortical neurons using HCA-vision. Cytometry A **71**, 889–895 (2007)
61. Conrad, C., Gerlich D.W.: Automated microscopy for high-content RNAi screening. J. Cell Biol. **188**, 453–461 (2010)
62. Vallotton, P., Sun, C., Wang, D., et al.: Segmentation and tracking of individual *Pseudomonas aeruginosa* bacteria in dense populations of motile cells. In: Image and Vision Computing New Zealand, Wellington, New Zealand, 2009
63. Dougherty, G., Henebry, G.M.: Fractal signature and lacunarity in the measurement of the texture of trabecular bone in clinical CT images. Med. Eng. Phys. **23**, 369–380 (2001)
64. Family, F., Masters, B.R., Platt, D.: Fractal pattern formation in human retinal vessels. Physica D **38**, 98–103 (1989)
65. Daxer, A.: The fractal geometry of proliferative diabetic retinopathy: implications for the diagnosis and the process of retinal vasculogenesis. Curr. Eye Res. **12**, 1103–1109 (1993)
66. Avakian, A., Kalina, R.E., Sage, E.H., et al.: Fractal analysis of region-based vascular change in the normal and non-proliferative diabetic retina. Curr. Eye Res. **24**, 274–280 (2002)
67. Schepers, H.E., Van Beek, J.H.G.M., Bassingthwaighte, J.B.: Four methods to estimate the fractal dimension from self-affine signals. IEEE Eng. Med. Biol. **11**, 57–64 (1992)
68. MacGillivray, T.J., Patton, N.: A reliability study of fractal analysis of the skeletonised vascular network using the "box-counting" technique. Conf. Proc. IEEE Eng. Med. Biol. Soc. **1**, 4445–4448 (2006)
69. Cheung, N., Donaghue, K.C., Liew, G., et al.: Quantitative assessment of early diabetic retinopathy using fractal analysis. Diabetes Care **32**, 106–110 (2009)

Chapter 13
Medical Image Volumetric Visualization: Algorithms, Pipelines, and Surgical Applications

Qi Zhang, Terry M. Peters, and Roy Eagleson

13.1 Introduction

With the increasing availability of high-resolution datasets of 3D medical images, the development of volumetric image rendering techniques have become an important complement to classical surface-based rendering. Since volumetric visualization does not require that surfaces be selected from within the 3D volumes, the full volume dataset is maintained during the rendering process. These methods are based on a foundation of projecting rays through volumes, which have a range of opacity attributes, onto a viewing window. Volume rendering is computationally demanding, and the ever increasing size of medical image datasets means that brute-force algorithms are not feasible for interactive use.

More recently, further efficiencies have been attained by implementing many of these algorithms on graphics processing hardwares (GPUs). In this chapter, we describe volumetric visualization pipelines, and provide a comprehensive overview of rendering algorithms that use effective approximate models to compute volumetric scenes of medical applications. We review and implement several mainstream medical image visualization strategies and rendering pipelines, including multiplanar reformation (MPR), direct and indirect surface rendering (DSR and ISR) with shading, direct volume rendering (DVR), software-based raycasting, 2D and 3D texture mapping (3DTM), GPU-based raycasting, maximum intensity projection (MIP), X-ray based rendering, gradient estimation and different interpolation approaches, voxel classification, and optical composition schemes. We present an overview of these techniques, and also evaluate their image quality and rendering performance.

R. Eagleson (✉)
The University of Western Ontario, London, ON, Canada
e-mail: eagleson@uwo.ca

G. Dougherty (ed.), *Medical Image Processing: Techniques and Applications*, Biological and Medical Physics, Biomedical Engineering, DOI 10.1007/978-1-4419-9779-1_13,

Where space permits, we have also added some of our recent research results and new rendering and classifications algorithms. In particular, these include anatomical feature enhancement techniques, dynamic multimodality rendering, and interactive manipulation. We have implemented a GPU-based medical image manipulation and visualization system with these volume rendering enhancements. We compare the performance of our strategies with those obtained by implementation algorithms from the published literature. We also address the advantages and drawbacks of each in terms of image quality and speed of interaction.

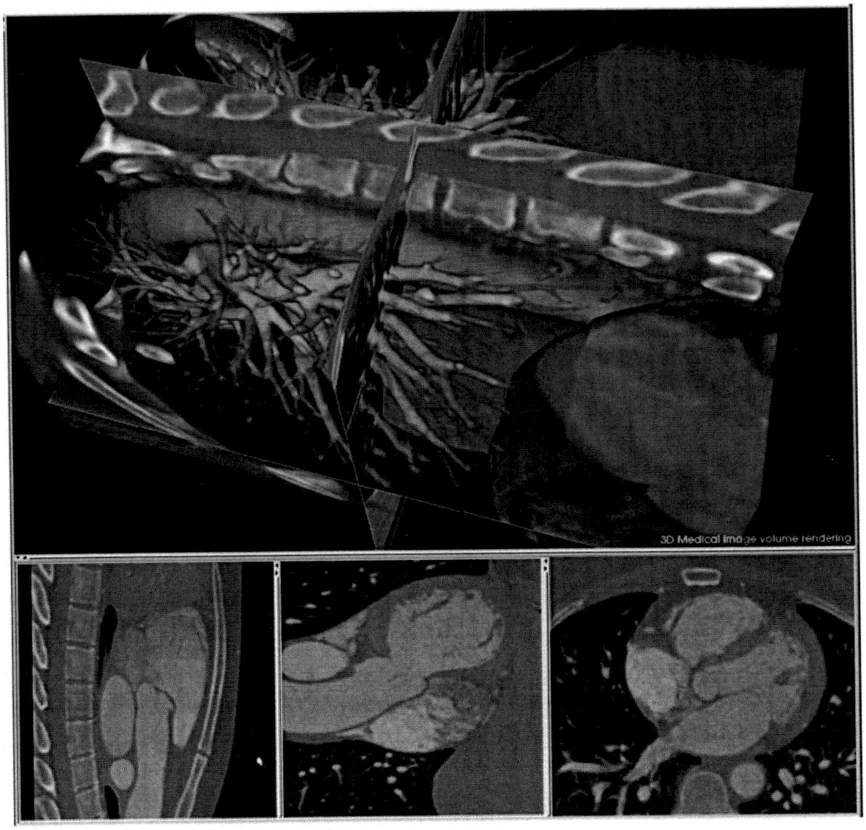

Fig. 13.1 MPR of 3D cardiac CT image. *Top*: three arbitrary cross-planes. *Bottom*: synchronized 2D displays of cross-planes

13.2 Volumetric Image Visualization Methods

Three principle rendering algorithms have been established for volumetric medical image visualization, that is multiplanar reformation (MPR; see Fig. 13.1), surface rendering (SR), and volume rendering (VR). As illustrated in the following sections, the rendering techniques can be categorized as direct and indirect rendering. Direct rendering includes DVR and DSR, while indirect rendering includes ISR. Here, we refer to both DSR and ISR as SR.

13.2.1 Multiplanar Reformation (2D slicing)

MPR is an image processing technique, which extracts 2D slices from a 3D volume using arbitrarily positioned orthogonal or oblique planes [1]. Although it is still a 2D method, it has the advantages of ease of use, high speed, and no information loss. The observer can display a structure of interest in any desired plane within the data set, and 4D MPR can be performed in real time using graphics hardware [2]. 2D multiplanar reformatting can readily complement 3D volume rendering where the 2D slices of MPR can be readily texture-mapped to cut-planes through a 3D volume.

13.2.2 Surface-Based Rendering

SR [3] is a common method of displaying 3D images. ISR can be considered as surface modeling, while DSR is a special case of DVR. ISR requires that the surfaces of relevant structure boundaries within the volume be identified a priori by segmentation and representation as a series of surface tiles using isosurface extracting such as marching cubes [4] or region growing [5], and can be accelerated by taking advantage of graphics processing unit (GPU) and geometry shaders [6, 7]. Such models reduce the amount of data to be displayed by several orders of magnitude, making it easy to manipulate the surfaces interactively with reasonable fidelity. For DSR, the surfaces are rendered directly from the volume without intermediate geometric representations, setting thresholds or using object labels to define a range of voxel intensities to be viewed. Only those voxels within this range, or which have labels, are selected and rendered with DVR. Surface reconstruction can also be improved further by employing GPUs [8].

A fully parallel isosurface extraction algorithm was presented by Zhang et al. [9]. In addition, a high degree of realism can be achieved with lighting models that simulate realistic viewing conditions. Figure 13.2a illustrates an implementation of ISR, while Fig. 13.3b shows the results of DSR for comparison.

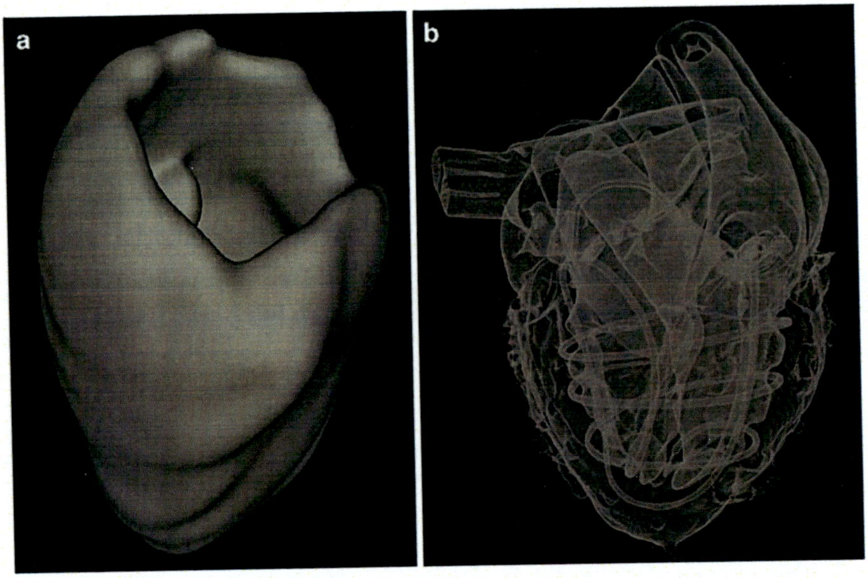

Fig. 13.2 (**a**) ISR of cardiac structures: myocardium (myo), and the left atrium and aorta (LAA), compared with (**b**) DSR of an MR cardiac volume and heart phantom

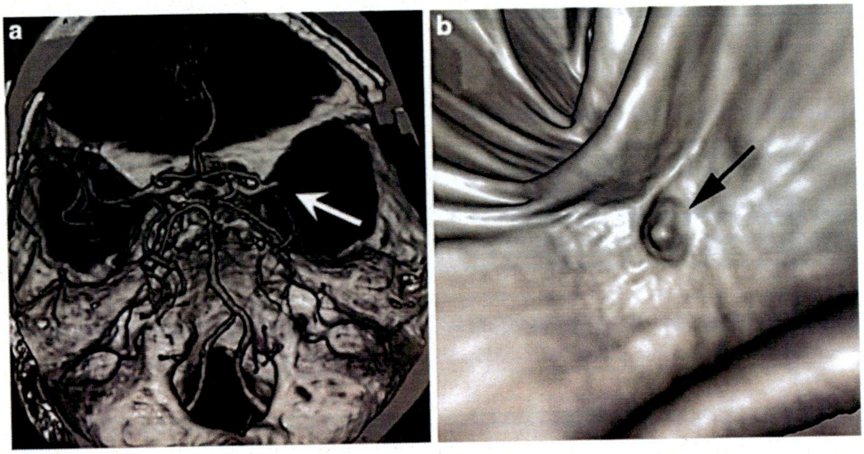

Fig. 13.3 SR applications: (**a**) image generated with MIP and shaded SR, demonstrating the proximal middle cerebral artery mainstem occlusion (*arrow*) [12]; (**b**). SR display of the endoluminal CT colonographic data, showing sessile morphology of the lesion (*arrow*) [13]

SR is often applied to contrast-enhanced CT data for displaying skeletal and vascular structures, and is also usually used in describing vascular disease and dislocations. In the process of detecting acute ischemic stroke, Schellinger and his colleagues [10] combined MIP with shaded SR to visualize proximal middle cere-

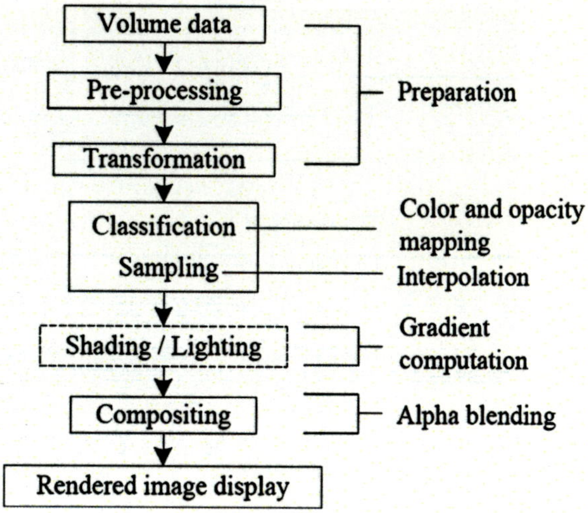

Fig. 13.4 Volume rendering pipeline and corresponding numerical operations

bral artery mainstem occlusion (Fig. 13.3a). SR has also been exploited to describe polyps within the 3D endoluminal CT colonographic images [11] (Fig. 13.3b).

We note that sometimes it is difficult to justify the accuracy and reliability of the images generated with shaded SR, that is the shiny surfaces might be misleading, causing the relation between image data and brightness in the resultant image becomes more complex, a property which could affect the diagnosis.

13.2.3 Volumetric Rendering

DVR displays the entire 3D dataset by tracing rays through the volume and projecting onto a 2D image, without computing any intermediate geometry representations [14–16]. This algorithm can be further divided into image-space DVR, such as software- [17, 18] and GPU-based raycasting [19], and object-space DVR, such as splatting [20,21], shell rendering [22], TM [23], and cell projection [24]. Shear-warp [25] can be considered as a combination of these two categories. In addition, MIP [26], minimum intensity projection (MinIP), and X-ray projection [27] are also widely used methods for displaying 3D medical images. This chapter now focuses its attentions on DVR, and the datasets discussed here are assumed to be represented on cubic and uniform rectilinear grids, such as are provided by standard 3D medical imaging modalities. Figure 13.4 describes the DVR pipeline with corresponding computations, which are described in detail in the next section that is followed by a discussion of traditional DVR algorithms. Figure 13.5 shows an example of our DVR results applied to an MR cardiac volume.

When compared with SR, the main advantage of DVR is that interior information is retained, and so provides more information about the spatial relationships of different structures [14]. In clinical applications, sensitivity and specificity are the

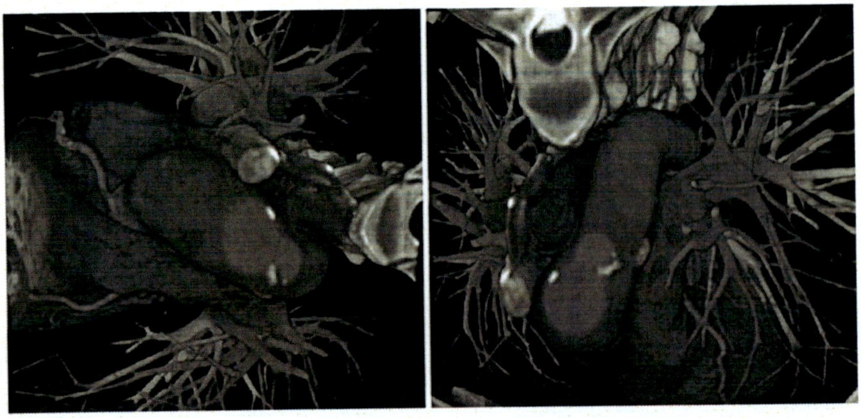

Fig. 13.5 Volume rendered images of volumetric human cardiac CT data set

major diagnostic criteria that must be considered. Even though DVR generally has high sensitivity and specificity for diagnosis [28], it is computationally intensive, so interactive performance is not always feasible. Another disadvantage is that it may be difficult to interpret the "cloudy" interiors that can result from the ray-tracing process. For detailed observation of specific lesions, slab imaging, where thick slices are rendered with DVR or MIP, has generally been used in clinical diagnosis [29]. In addition, the combination of cross-sectional MPR and DVR can significantly increase the interpretation rate of anatomical structures [30]. DVR has a wide range of clinical applications. For example, to perform renal donor evaluation, Fishman et al. [31] used DVR and MIP to display the renal vein of CT angiography (CTA) data, as shown in Fig. 13.6, for the DVR generated image, the left gonadal vein (large arrow) is well defined (a), while the locations of the renal vein and gonadal vein (arrow) are inaccurately depicted in the MIP generated image (b). Gill et al. [32] used DVR and MinIP to show the central airway and vascular structures of a 60-year-old man who underwent double lung transplantation for idiopathic pulmonary fibrosis, evaluating posttreatment in a noninvasive manner (Fig. 13.6c,d).

13.3 Volume Rendering Principles

The core component of DVR is to solve the volume rendering integral that describes the optical model. Different DVR techniques share similar components in the rendering pipeline, the main difference being the order in which they are applied, and the manner in which they traverse the volumetric data. Due to the rapid development of programmable GPUs, many CPU-based algorithms and techniques have been or can be implemented on GPUs. In this paper, for the texture-mapping based DVR, we refer to algorithms that use a fixed graphics pipeline, while for GPU-based raycasting, we refer to DVR implemented on GPUs.

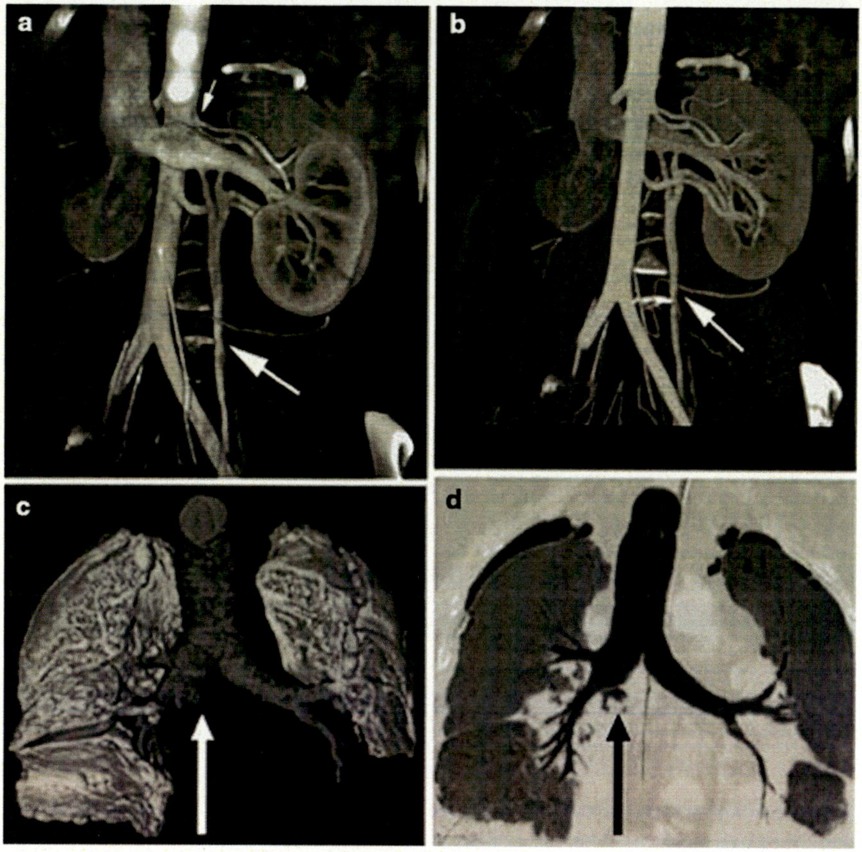

Fig. 13.6 DVR, MIP, and MinIP applications: (**a**) Coronal oblique DVR image of kidney and veins; (**b**) MIP display of the same data as (**a**) [31]. (**c**) DVR generated image of the central airway and vascular structures. (**d**) Coronal MinIP image of the same data as (**c**) [32]

13.3.1 Optical Models

The volume rendering integral is still often based on a model developed by Blinn [33] describing a statistical simulation of light passing through, and being reflected by, clouds of similar small particles. The optical models may be based on emission or absorption individually, or both, depending on the applications [34]. To reduce the computational cost, Blinn assumed that the volume is in a low albedo environment, in which multiple reflections and scattering of the particles are negligible. In this case, a light emission–absorption model is an optimal balance between realism and computational complexity, where every particle absorbs incoming light and emits light on its own without scattering between particles other than in the viewing ray direction [16]. Equation (13.1) demonstrates this procedure.

$$\frac{dI(\lambda)}{d\lambda} = c(\lambda)\tau(\lambda) - I(\lambda)\tau(\lambda) = \tilde{c}(\lambda) - \tilde{I}(\lambda). \tag{13.1}$$

The solution to this equation is given below, showing the intensity of each pixel.

$$I(D) = I_0 T(D) + \int_0^D \tilde{c}(\lambda) T(\lambda) d\lambda,$$

$$T(\lambda) = \exp\left(-\int_0^\lambda \tau(x) dx\right), \tag{13.2}$$

describes the volume transparency. The first term I_0 illustrates light coming from the background, and D is the extent of the ray over which light is emitted. The last term demonstrates the behavior of the volume emitting and absorbing incoming light. The source term $c()$ indicates the color change, and the extinction coefficient (tau)(D) defines the occlusion of light per unit length due to light scattering or extinction.

13.3.2 Color and Opacity Mapping

13.3.2.1 Voxel Classification

To display 3D medical images with DVR, the scalar values must first be mapped to optical properties such as color and opacity through transfer function (TF), a process referred to as voxel classification. Pre- and postclassification approaches differ with respect to the order in which the TF and sampling interpolation are applied [16, 35]. Pre-classification first maps every scalar value at the grid into color and opacity in a pre-processing step, where the color and opacity are assigned at the resampling points. However, for post-classification, we first sample the scalar value by interpolation, and then map the acquired values to colors and opacity through TFs. Both the pre-and post-classification operations introduce high frequency components via the nonlinear TFs [36–38]. Pre-classification suppresses this high-frequency information, so the rendered image appears blurry (Fig. 13.7a), while postclassification maintains all the high frequencies, but introduces "striping" artifacts in the final images (Fig. 13.7b).

To address the undersampling problem, Rottger et al. [39, 41] proposed a preintegrated classification algorithm for hardware-accelerated tetrahedra projection, which was first introduced by Max et al. [40]. Later, Engel et al. [36] applied this classification algorithm for 3D texture-mapping-based volume visualization of regular-grid volume data. Preintegrated classification separates the DVR integral into two parts, one for the continuous scalar value, and the other for the TF parameters c(colors) and tau(extinction). This algorithm renders the volume segment-by-segment, instead of point-by-point. In this manner, the Nyquist frequency for reconstructing the continuous signal is not increased by the TF nonlinear properties. Assuming there are $n+1$ sampling points along the viewing ray, then the segment

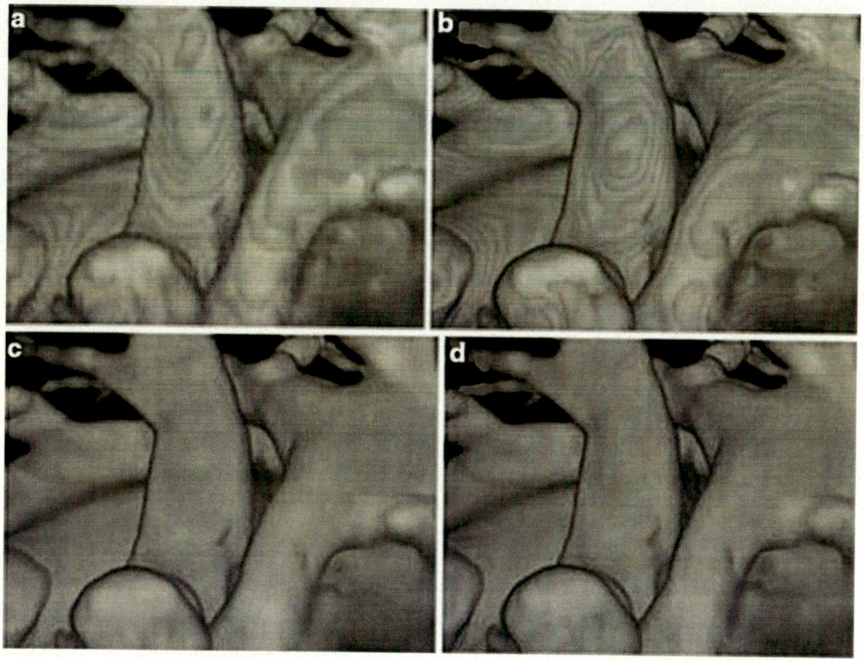

Fig. 13.7 DVR of cardiac vessels via different voxel classification algorithms: (**a**) preclassification; (**b**) postclassification; (**c**) preintegrated classification; (**d**) post color-attenuated classification

length d equals $D = n$, where D is the maximum ray length. For the ith segment, the front and back points are $\mathrm{sa} = \mathrm{s}(id)$ and $\mathrm{sb} = \mathrm{s}((i+1)d)$. The calculated opacity and color of this segment are then given by (13.3) and (13.4), respectively.

$$I(D)_{f\_b} = \sum_{i=0}^{n} \alpha_i C_i \prod_{j=0}^{i-1} T_j = \sum_{i=0}^{n} \alpha_i C_i \prod_{j=0}^{i-1} (1-\alpha_j) \tag{13.3}$$

$$I(D)_{b\_f} = \sum_{i=0}^{n} \alpha_i C_i \prod_{j=i+1}^{n} T_j = \sum_{i=0}^{n} \alpha_i C_i \prod_{j=i+1}^{n} (1-\alpha_j). \tag{13.4}$$

13.3.2.2 Transfer Function

The important step of voxel classification is implemented through a TF adjustment, which plays an important role in DVR. However, TF specification is a complex procedure and is a major obstacle for the widespread clinical use of DVR [42, 43]. In this section, we briefly review the TFs that are of crucial clinical importance. In DVR, the TF was typically used for tissue classification based on local intensities in the 3D dataset [44]. Multidimensional TF is efficient for multiple spatial feature detection, for example, Kniss et al. [45] designed such a TF and demonstrated its medical applications, while Higuera et al. [46] built a 2D TF to effectively visualize intracranial aneurysm structures. Abellan et al. [47] introduced a 2D fusion TF to facilitate the visualization of merged multimodal volumetric data, and a 3D TF was designed by Hadwiger et al. [48] to interactively explore feature classes within the industrial CT volumes, where individual features and feature size curves can be colored, classified, and quantitatively measured with the help of TF specifications. Furthermore, Honigmann et al. [49] designed an adaptive TF for 3D and 4D ultrasound display, and a default TF template was built to fit specific requirements. Similarly, Rezk-Salama et al. [50] proposed a template-based reproducible automatic TF design algorithm and applied it to medical diagnosis. Later, to facilitate the TF specification, these authors also introduced semantic models for TF parameter assignment, while a similar idea was used by Rautek et al. [51] to add a semantic layer in the TF design. As pointed by Freiman et al. [52], automation is important in TF design.

13.3.3 Composition

During the DVR process, a number of composition schemes are commonly employed, including simulating an X-ray projection (Fig. 13.8a) [53], MIP [54, 55], MinIP, and alpha blending. MIP and MinIP are widely used techniques in 3D CT and MR angiography. The salient features in the image are generally comprised by the voxels having the maximum (MIP) or minimum (MinIP) intensity along the viewing rays traversing through the object. MIP and MinIP images can be generated rapidly and can clearly display vessels, tumor, or bones [56], and the image generation has been accelerated by graphics hardware [57, 58]. Because no user input is necessary, MIP is a widely used 3D visualization option in radiology.

Local MIP, MinP, or closest vessel projection (CVP) are often used in slab imaging for vascular structure diagnosis [29, 59]. For vascular diagnosis, CVP is superior to MIP, however the user needs to set an appropriate threshold for the local maximum, which is determined by a specific dataset, making the application of CVP more difficult than MIP (which is shown in Fig. 13.8b). Alpha blending [17, 34] is a popular optical blending technique, often implemented by summing to discretize the continuous function (13.2), resulting front-to-back and back-to-front alpha blending, depending on the compositing order. The front-to-back and back-to-front alpha blending methods represent opposite rendering directions. Figure 13.8c describes the DVR result using alpha blending without shading, while Fig. 13.8d shows the result with shading.

13.3.4 Volume Illumination and Illustration

In volume illumination, the normal at every sampling point is calculated by interpolation using the intensity changes across that voxel. These approximated

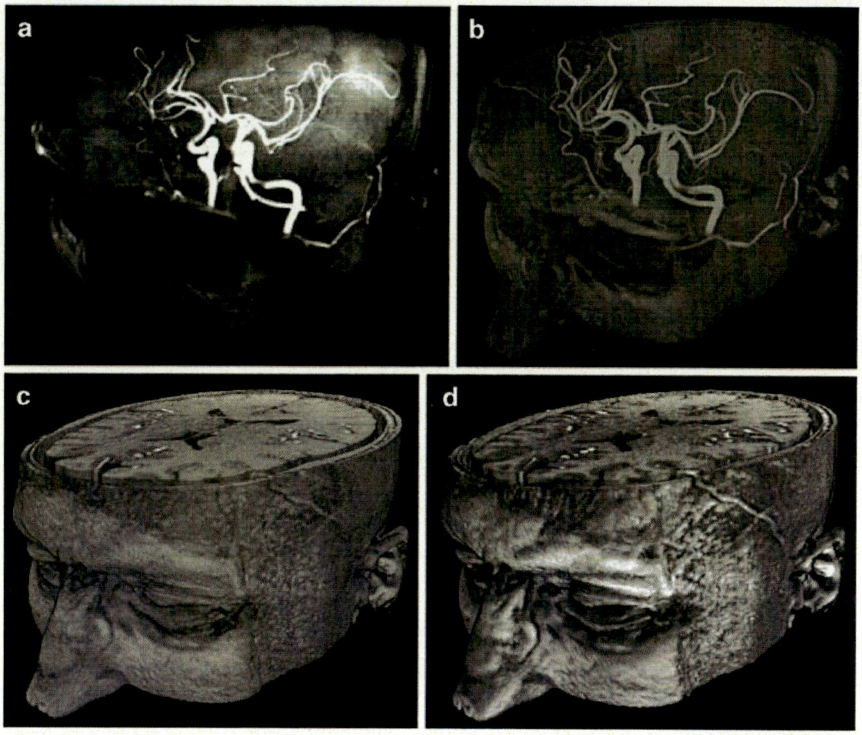

Fig. 13.8 DVR results of human head with angiographic contrast using different compositing techniques: (**a**) X-ray projection; (**b**) MIP; (**c**) alpha blending without shading; and (**d**) alpha blending with shading

voxel normals are then used in a Phong or Blinn-Phong model for shading computations, with the results being employed in the DVR composition. The shading computations may be accelerated using commodity graphics hardware [60]. In addition, the shading model can be used with volumetric shadows to capture chromatic attenuation for simulating translucent rendering [61], and can be combined with clipped volumes to increase the visual cues [62]. Figure 13.9 illustrates a DVR of MR and CT cardiac data sets with and without illumination, demonstrating that images with shading are visually more pleasing.

Lighting and illumination are important aspects of volumetric visualization. Examples of approaches employed include those by Rheingans and Ebert [63] who proposed an illumination method similar to volume shading, using nonphotorealistic rendering [64] to enhance physics-based DVR, and Lum and Ma [65] who accelerated this algorithm with multitexture-based hardware, and they also introduced a pre-integrated lighting with voxel classification-based DVR, resulting in decreased illumination artifacts [66]. To explore hidden structures and depict their spatial relations in volumes, Chan et al. [67] introduced a relation-aware visualization

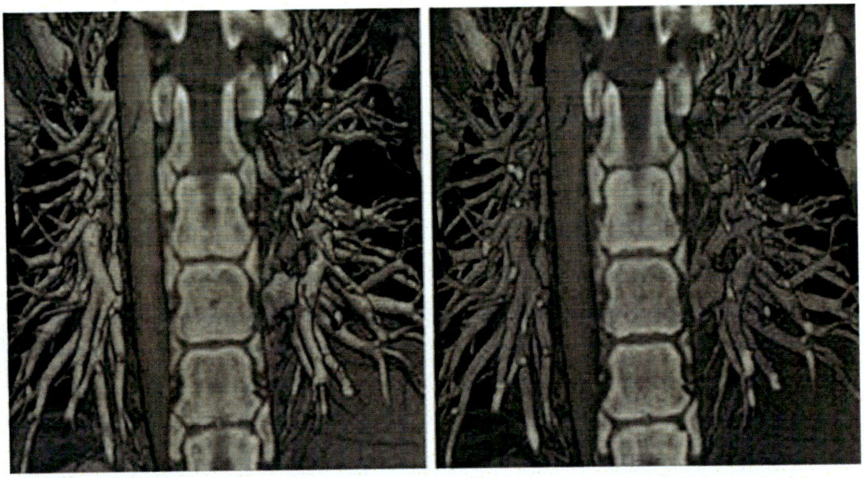

Fig. 13.9 DVR images of CT pulmonary data showing illumination (*right image*) compared with no illumination (*left image*)

pipeline, in which inner structural relationships were defined by a region connection calculus and were represented by a relation graph interface. In addition, Rautek et al. [68] gave a comprehensive review on illustrative visualization and envisioned their potential medical applications. Such illustrative techniques are sometimes integrated within commercial systems; however, there have been very few systematic studies done to verify the perceptual enhancement, nor to validate the clinical benefit [69].

13.4 Software-Based Raycasting

Raycasting [14] is a popular technique used to display a 3D dataset in two dimensions, in which the basic idea is to cast a ray from each pixel in the viewing plane into the volume, sampling the ray with a predetermined step in a front-to-back or back-to-front order using trilinear interpolation. In this process, a TF is used to map the scalar value to RGB color and opacity, which can be performed on every voxel in the volume before the sampling step (preclassification), or on the sampled scalar values along the casting ray after sampling, where post-, preintegrated, or postcolor attenuated classification can be used. Finally, the acquired optical values at these sampling points along the casting ray are composited using (13.5) or (13.6) to approximately compute the DVR integral (13.2), obtaining the corresponding final pixel color on the output image. Figure 13.10 illustrates the raycasting pipeline, and Fig. 13.11demonstrates this on four medical images.

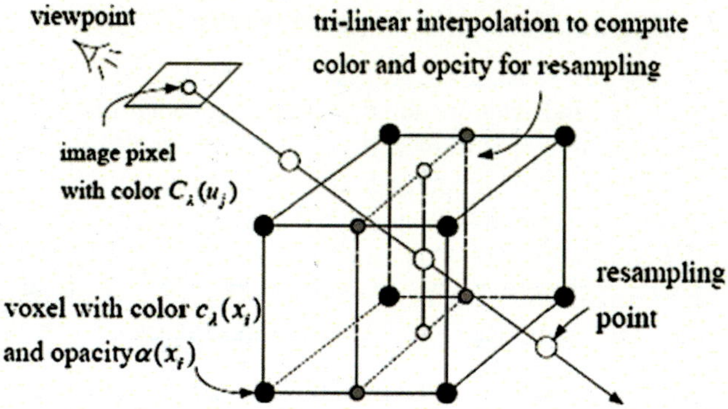

Fig. 13.10 Raycasting pipeline: sampling, optical mapping, and compositing

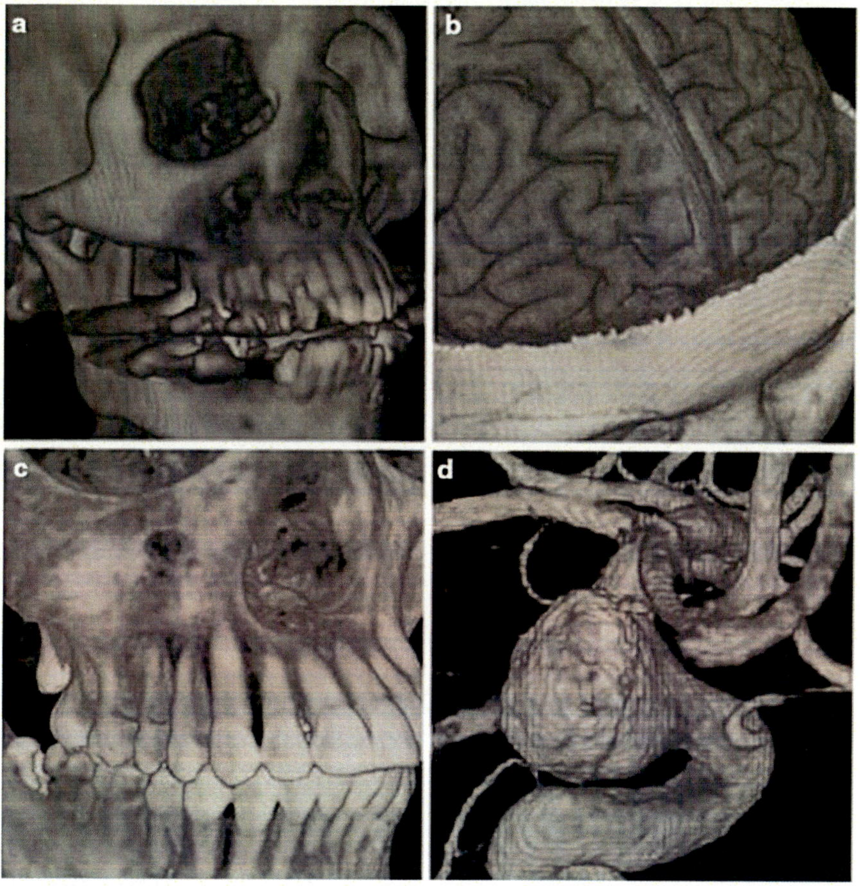

Fig. 13.11 Medical images rendered with software-based raycasting: (**a**) CT skull; (**b**) MR brain; (**c**) CT jaw, and (**d**) MR cerebral blood vessel

13.4.1 Applications and Improvements

Because of its rendering speed, raycasting has been not often used in clinical applications until now. However, it was a significant 3D display method in some applications ever when graphics hardware accelerated DVR was not commonly available. For example, Sakas et al. [70] used raycasting to display 3D ultrasound data of a fetus, and Hohne [71] and Tiede et al. [72] employed this algorithm for anatomical visualization. Many of the approaches to improve raycasting techniques have focused on schemes to eliminate unnecessary voxels from the computation.

13.5 Splatting Algorithms

Splatting is a popular DVR algorithm which was first proposed by Westhover [73–75] and was improved in terms of quality and speed by the research community over the years [76, 77]. This technique was developed to accelerate the speed of DVR at the expense of lower accuracy, and calculates the influence of each voxel in the volume on multiple pixels in the output image. This algorithm represents the volume as an array of overlapping basis functions called reconstruction kernels, which are commonly rotationally symmetric Gaussian functions with amplitudes scaled by the voxel values. This process is described in Fig. 13.12, and Fig. 13.13 presents examples of images rendered with the splatting algorithm.

13.5.1 Performance Analysis

Splatting is efficient because it reorders the DVR integral, making the preintegration of reconstruction kernels possible, so that each voxel's contribution to the integral

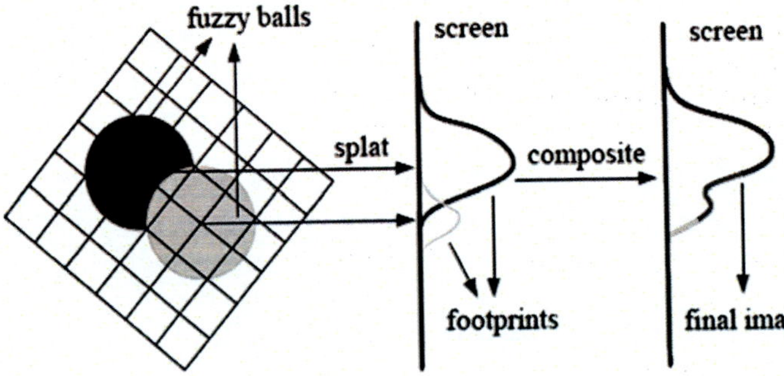

Fig. 13.12 Splatting pipeline: the optical model is evaluated for each voxel and projected onto the image plane, leaving a footprint (splat). Then these footprints are composited to create the final image

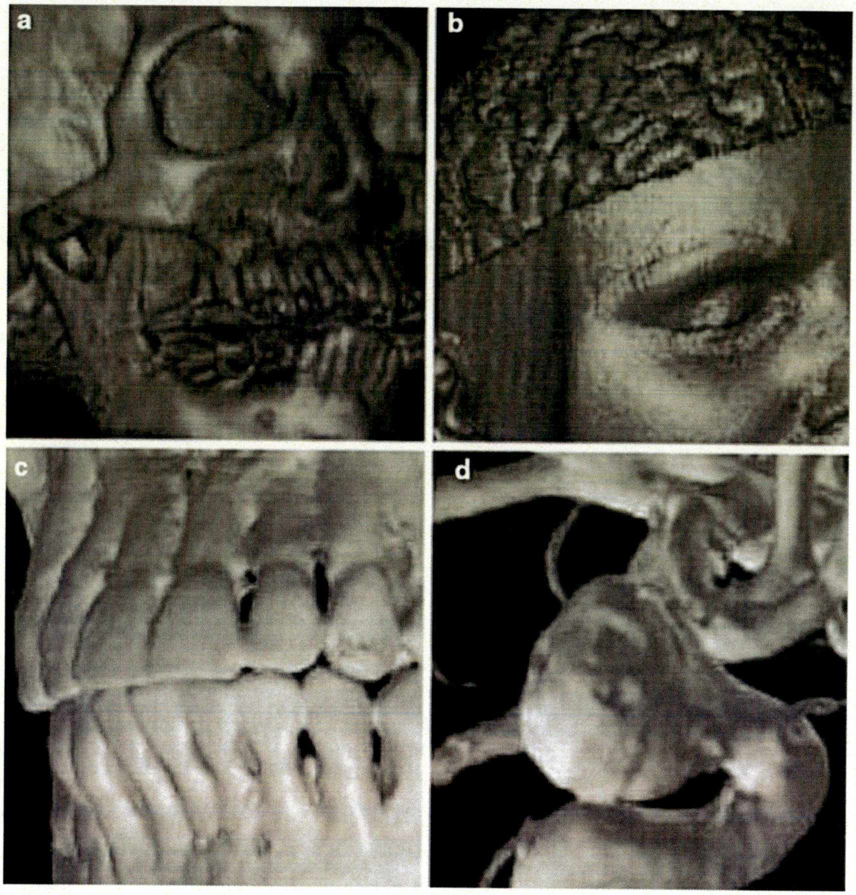

Fig. 13.13 Medical images rendered with splatting: (**a**) CT skull with shading; (**b**) MR brain with shading [76]; (**c**) CT jaw, and (**d**) MR cerebral blood vessel [77]

can be viewed separately. Another major advantage is that only voxels relevant to the image are projected and rasterized, so empty (transparent) regions can easily be skipped. However, because all of the splats are composited back-to-front directly without considering the kernel overlaps, the basic splatting algorithm is plagued by artifacts known as "color bleeding," where the colors of hidden objects or background appearing in the final image.

13.5.2 Applications and Improvements

Vega-Higuera et al. [78] exploited texture-accelerated splatting to visualize the neurovascular structures surrounded by osseous tissue in CTA data in real time.

Splatting was also employed by Birkfellner et al. [79] to generate digitally rendered radiographs (DRRs) rapidly in the iterative registration of medical images, and the authors later exploited graphics hardware to accelerate the splat-based creation of DRRs [80]. Audigier and his colleagues [81] used splatting with raycasting to guide the interactive 3D medical image segmentation, providing users with feedback at each iterative segmentation step. Since basic splatting algorithms suffer from "color bleeding" artifacts, Westover originally employed an axis-aligned sheet buffer to solve this problem. However, this technique needs to maintain three stacks of sheets and introduces "popping" artifacts. To address this issue, Mueller and his colleagues [82] aligned the sheet buffers parallel to the image plane instead of parallel to the axes, and they later accelerated this image aligned splatting algorithm with modern GPUs [83]. They also proposed a postshaded pipeline for splatting to improve the resultant image [77].

13.6 Shell Rendering

Shell rendering [84] is an efficient software-based hybrid of surface and volume rendering proposed by Udupa and Odhner. The shell rendering algorithm is based on a compact data structure referred to as a shell, which is a set of nontransparent voxels near the extracted object boundary with a number of attributes associated with each related voxel for visualization. The shell data structure can store the entire 3D scene or only the hard (binary) boundary. For a hard boundary, the shell is crisp and only contains the voxels on the object surface, and shell rendering degenerates to SR. For a fuzzy boundary, the shell includes voxels in the vicinity of the extracted surface, and shell rendering is identified as DVR. Figure 13.14 shows examples of shell SR and DVR.

13.6.1 Application and Improvements

Lei et al. [85] employed this algorithm to render the segmented structures of vessels and arteries of contrast-enhanced magnetic resonance angiography (CE-MRA) image. However, the explicit surface extraction creates errors. To address the problem, Bullitt et al. [86] selectively dilated the segmented object boundaries along all axes, and visualized the extracted fuzzy shell with raycasting. To accelerate the shell rendering speed, Falcao and his colleagues [87] added the shear-warp factorization to the shell data structure, and Botha and Post [88] used a splat-like elliptical Gaussian to compute the voxel contribution energy to the rendered image. Later, Grevera et al. [89] extended the point-based shell element to a new T-shell element comprised of triangular primitives for isosurface rendering, referred to as T-Shell rendering.

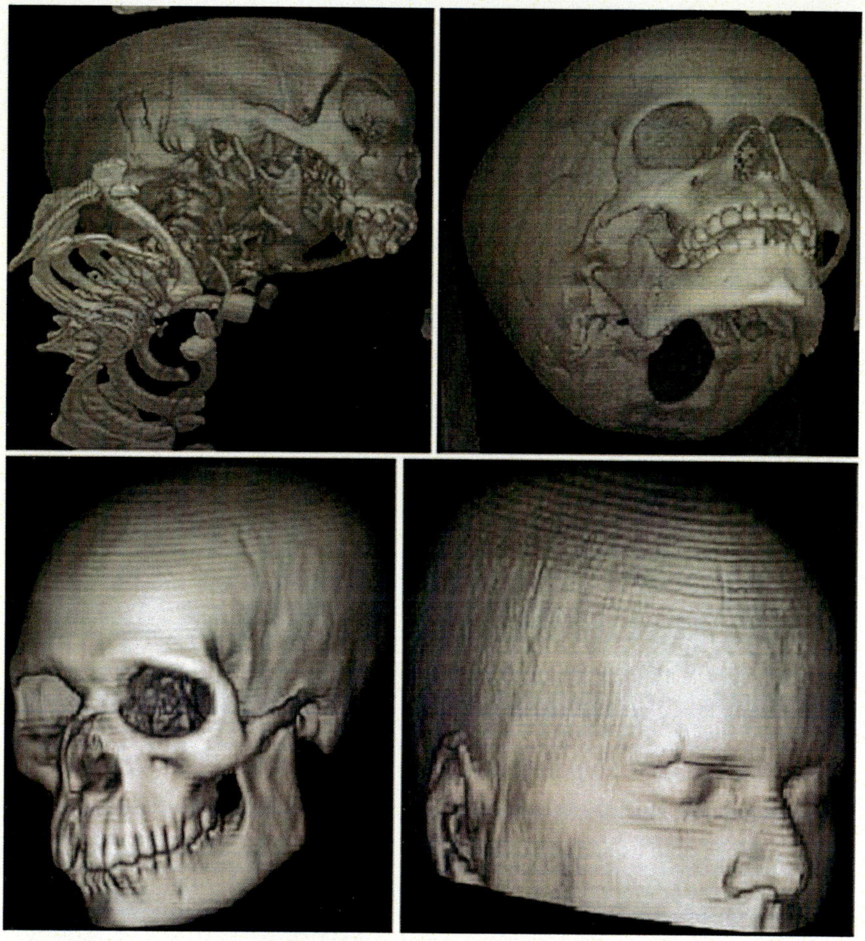

Fig. 13.14 Shell rendering examples. *Top row*, shell SR [174]: skull CT data (*left*) and CT data of a "dry" child's skull (*right*). *Bottom row*, shell DVR [175]: CT skull (*left*), and MR head (*right*)

13.7 Texture Mapping

The pioneering work of exploiting texture hardware for DVR was performed by Cullip and Neumann [90] and Cabral et al. [91]. When graphics hardware does not support trilinear interpolation, 2D texture mapping (2DTM) must be adopted. In this case, the volume is decomposed into three stacks of perpendicularly object-aligned polygons. For the current viewing direction, the stack whose slicing direction (normal) must be within 45 degrees of the current viewing direction is chosen for rendering. During rasterization, each of the polygon slices is textured with the image information obtained from the volume via bilinear interpolation. Finally, the textured slices are alpha-blended in a back-to-front order to produce the final image.

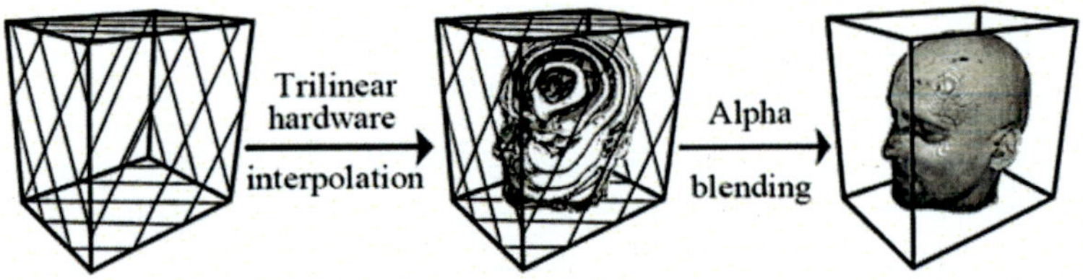

Fig. 13.15 The working pipeline of 3DTM for 3D head rendering

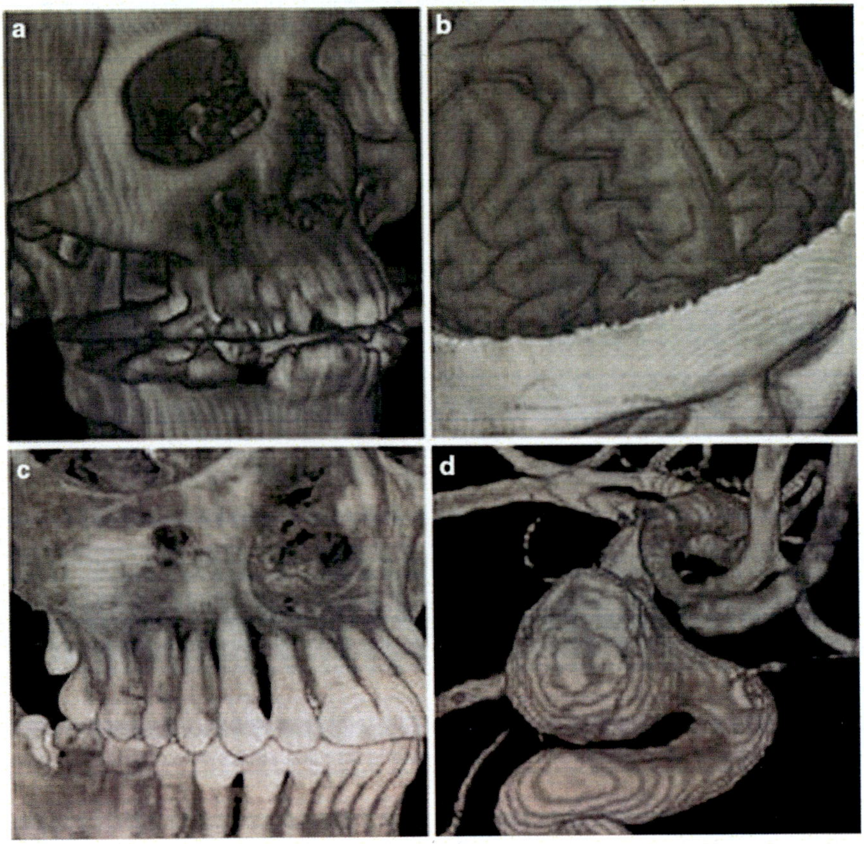

Fig. 13.16 Medical image rendered with 3DTM: (**a**) CT skull; (**b**) MR brain; (**c**) CT jaw, and (**d**) MR cerebral blood vessel

Figure 13.15 describes the working pipeline, and Fig. 13.16 illustrates the rendering results with this algorithm. 3DTM uploads the volume to the graphics memory as a single 3D texture, and a set of polygons perpendicular to the viewing direction is placed within the volume and textured with the image information by trilinear interpolation. Compared with 2DTM, there are no orientation limitations for the decomposed polygon slices, making the texture access more flexible. The compositing process is similar to 2DTM, in that the set of textured polygon planes are alpha-blended in a back-to-front order.

13.7.1 Performance Analysis

Compared with 3DTM, the main advantage of 2DTM is higher rendering speed and lower hardware requirements, since it uses efficient in-slice bilinear interpolations, instead of expensive trilinear interpolations. However, this algorithm is prone to aliasing artifacts at the edges of the slice polygons, and has to maintain three copies of the volume in graphics memory. The shift of viewing directions causes the algorithm to switch from one texture stack to another, resulting in the "popping" artifacts mentioned earlier. In 3DTM, the extracted slices can have arbitrary orientations and only a single copy of the data is required; therefore, there are no artifacts caused by the texture stack switching during the viewing direction changing process. In addition, the extracted slices are textured with trilinear instead of bilinear interpolation, so the mapped texture information has a higher accuracy. For orthographic projections, the viewport-aligned slices can be employed to achieve a consistent sampling step, but because the sampling distance cannot be uniform for projective views, "striping" artifacts are introduced. This limitation may be overcome using spherical rendering primitives [92], albeit with an increased computational cost.

13.7.2 Applications

In the published literature, there are three main types of medical applications for TM-based DVR. The first is multimodal medical image rendering and tissue separation. Sato et al. [93] designed a multidimensional TF to identify tissue structures in multimodal medical images generated with 3DTM. A two-level rendering technique was integrated with 3DTM by Hauser et al. [94], allowing different rendering techniques to be selected for different tissues based on segmentation information. Later, Hadwiger et al. [95] improved this 3DTM-based two-level DVR algorithm in the aspects of image quality and performance, minimizing the number of rendering passes and the computational cost of each pass, adding dynamic filtering, and using depth and stencil buffering to achieve correct compositing of objects created with different rendering techniques.

The second application area is in the display of specific tissues and organs for diagnosis and therapy. Holmes et al. [96] used 3DTM to achieve a real-time display of transurethral ultrasound (TUUS) for prostate treatment. This technique was also used by Etlik et al. [97] to find bone fractures, and also by Wenger et al. [98] to visualize diffusion tensor MRI (DTI) data in the brain. Wang et al. [99] exploited

3DTM to display 3D ultrasound for cardiac ablation guidance, and Sharp et al. [100] employed this technique to visualize the light diffusion in 3D inhomogeneous tissue to provide visual information relating to structures located beneath the skin surface. Lopera et al. [101] used this DVR approach to view a stenosis in arbitrary image planes, and employed the rendering results to demonstrate the subsequent arterial bypass.

The third area of application is in dynamic imaging, deformation, and 4D data display. Levin et al. [102] developed a software platform based on TM, which was used to render multimodality 4D cardiac datasets interactively. A similar technique was used by Lehmann et al. [103] to visualize the beating heart in real time, in which a hierarchical memory structure was employed to improve bandwidth efficiency. In addition, Yuan et al. [104] designed a TF for nonlinear mapping density values in the dynamic range medical volumes, while Correa and his colleagues [105] presented a 3DTM algorithm that sampled the deformed space instead of the regular grid points, simulating volume deformations caused by clinical manipulations, such as cuts and dissections.

13.7.3 Improvements

13.7.3.1 Shading Inclusion

In 1996, shading was first included into TM approaches by Van Gelder et al. [23]. Both diffuse and specular shading models were added to texture-based DVR by Rezk-Salama et al. [92] with the use of register combiners and paletted texture. In addition, Kniss et al. [61] proposed a simple shading model that captured volumetric light attenuation to produce volumetric shadows and translucency. The shading model was also used with depth-based volume clipping by Weiskopf and his colleagues [62] in a 3DTM pipeline. To decrease shading artifacts, Lum et al. [66] introduced preintegrated lighting into the TM-based DVR, resulting in decreased lighting artifacts. Recently, Abellan and Tost [106] defined three types of shadings for the TM-accelerated volume rendering of dual-modality dataset, that is emission plus absorption, surface shading, and the mixture of both shadings, with user-selected choice of shading model for a specific imaging modality.

13.7.3.2 Empty Space Skipping

As mentioned earlier, the DVR speed can often be improved if the empty spaces can be skipped during rendering procedure, for example, Li et al. [107–109] computed texture hulls of all connected nonempty regions, and they later improved the hull technique with "growing boxes" and an orthogonal binary space partitioning tree. Bethune and Stewart [110] proposed an adaptive slice DVR algorithm based on 3DTM. In this algorithm, the volume was partitioned into transparent and nontransparent regions, axis aligned bounding boxes were used to enclose the nontransparent ones, and an octree was used to encode these boxes for efficient empty space skipping. In addition, Keles et al. [111] exploited the slab silhouette maps (SSMs) and the early depth test to skip empty spaces along the direction normal to the texture slice.

13.8 Discussion and Outlook

DVR is an efficient technique to explore complex anatomical structures within volumetric medical data. Real-time DVR of clinical datasets needs efficient data structures, algorithms, parallelization, and hardware acceleration. The progress of programmable GPUs has dramatically accelerated the performance and medical applications of volume visualization, and opened a new door for future developments of real-time 3D and 4D rendering techniques. For DVR algorithms such as splatting, shell rendering, and shear-warp, and hardware accelerations have been proposed and implemented; however, most of these improvements are based on fixed graphics pipelines, and do not take full advantages of the programmable features of GPU. As mentioned previously, even if the TM-based DVR algorithms make use of graphics hardware in the volume-rendering pipeline, they differ considerably from the programmable GPU algorithms, such as raycasting, in that all of the volume rendering computations and corresponding acceleration techniques are implemented on the GPU fragment shaders. The TM-based DVR algorithms only use fixed graphics pipelines that include relative simple texture operations, such as texture extraction, blending, and interpolation.

In addition, to further enhance the visualization environment, DVR algorithms should be effectively combined with realistic haptic (i.e., force) feedback and crucial diagnostic and functional information extracted from medical data using new algorithms in the research fields such as machine learning and pattern recognition. Furthermore, innovative techniques of robotics, human–computer interaction, and computational vision must be developed to facilitate medical image exploration, interpretation, processing, and analysis.

Finally, we note that the current volume visualization algorithms should be integrated into standard graphical and imaging processing frameworks such as the Insight Segmentation and Registration Toolkit (ITK) [112] and the Visualization ToolKit (VTK) [113], enabling them to be used readily in clinical medical imaging applications. We also note that even though there have been many efficient and advanced techniques developed for all parts of the volume rendering pipeline, only a few simple ones have been developed for clinical visualization. The translation of these advances into clinical practice, nevertheless, remains a problem. However, we must ensure that such challenges are addressed so as not to introduce roadblocks with respect to related technologies as image-guided interventions.

Acknowledgments The authors thank the following people for allowing us to use their images as sample images: Marjo Götte, Novartis Institutes for BioMedical Research; Dr. Myles Fennell, Wyeth Research, Princeton, NJ, USA; Dr. Xiaokui Zhang, Helicon Therapeutics, Inc., USA; Prof. Pat Doherty, Kings College, London, UK; Dr. Jenny Gunnersen, Prof. Seong-Seng Tan, and Dr. Ross O'Shea Howard Florey Institute, Melbourne; Prof. Cynthia Whitchurch, UTS, Sydney.

References

1. Baek, S.Y., Sheafor, D.H., Keogan, M.T., DeLong, D.M., Nelson,R.C.: Two-dimensional multiplanar and three-dimensional volume-rendered vascular CT in pancreatic carcinoma: Interobserver agreement and comparison with standard helical techniques. Am. J. Roentgenol. **176**(6), 1467–1473 (2001)
2. Shekhar, R., Zagrodsky, V.: Cine MPR: interactive multiplanar reformatting of four-dimensional cardiac data using hardware-accelerated texture mapping. IEEE Trans. Inf. Technol. Biomed. **7**(4), 384–393 (2003)
3. Tatarchuk, N., Shopf, J., DeCoro, C.: Advanced interactive medical visualization on the GPU. J. Parallel Distrib. Comput. **68**(10), 1319–1328 (2008)
4. Buchart, C., Borro, D., Amundarain, A.: GPU local triangulation: an interpolating surface reconstruction algorithm. Comput. Graph. Forum **27**(3), 807–814 (2008)
5. Hadwiger, M., Sigg, C., Scharsach, H., Buhler, K., Gross, M., Realtime ray-casting and advanced shading of discrete isosurfaces. Comput. Graph. Forum **24**(3), 303–312 (2005)
6. Hirano, M., Itoh, T., Shirayama, S.: Numerical visualization by rapid isosurface extractions using 3D span spaces, J. Vis. **11**(3), 189–196 (2008)
7. Petrik, S., Skala, V.: Technical section: space and time efficient isosurface extraction, Comput. Graph. **32**(6), 704–710 (2008)
8. Kim, S., Choi, B., Kim, S., Lee, J.: Three-dimensional imaging for hepatobiliary and pancreatic diseases: emphasis on clinical utility. Indian J. Radiol. Imaging **19**, 7–15 (2009)
9. Hadwiger, M., Kniss, J.M., Rezk-salama, C., Weiskopf, D., Engel, K.: Real-Time Volume Graphics, 1st edn. A. K. Peters, Ltd., Natick (2006)
10. Schellinger, P.D., Richter, G., Kohrmann, M., Dorfler, A.: Noninvasive angiography magnetic resonance and computed tomography diagnosis of ischemic cerebrovascular disease. Cerebrovasc. Dis. 24(1), 16–23 (2007)
11. Park, S.H., Choi, E.K., et al.: Linear polyp measurement at CT colonography: 3D endoluminal measurement with optimized surface-rendering threshold value and automated measurement. Radiology **246**, 157–167 (2008)
12. Schellinger, P.D., Richter, G., Kohrmann, M., Dorfler, A.: Noninvasive angiography magnetic resonance and computed tomography in the diagnosis of ischemic cerebrovascular disease. Cerebrovasc. Dis. **24**(1), 16–23 (2007)
13. Park, S.H., Choi, E.K., et al.: Linear polyp measurement at CT colonography: 3D endoluminal measurement with optimized surface-rendering threshold value and automated measurement. Radiology **246**, 157–167 (2008)
14. Drebin, R., Carpenter, L., Hanrahan, P.: Volume rendering. In: Proceedings SIGGRAPH88, pp 65–74 (1988)
15. Johnson, C., Hansen, C.: Visualization Handbook. Academic, Orlando (2004)
16. Hadwiger, M., Kniss, J.M., Rezk-salama, C., Weiskopf, D., Engel, K.: Real-Time Volume Graphics, 1st edn. A. K. Peters, Natick (2006)
17. Levoy, M.: Display of surfaces from volume data. IEEE Comput. Graph. Appl. **8**(3), 29–37 (1988)
18. Höhne, K.H., Bomans, M., Pommert, A., Riemer, M., Schiers, C., Tiede, U.: Wiebecke G: 3D visualization of tomographic volume data using the generalized voxel model. Vis. Comput. **6**(1), 28–36 (1990)

19. Krüger, J., Westermann, R.: Acceleration techniques for GPU-based volume rendering. In: VIS'03: Proceedings of the 14th IEEE Visualization 2003 (VIS'03), IEEE Comp Society, Washington, DC, pp. 287–292 (2003)
20. Westover, L.: Interactive volume rendering. In; Proc. of the 1989 Chapel Hill Workshop on Volume visualization, ACM, pp. 9–16
21. Westover, L.: Footprint evaluation for volume rendering. In: SIGGRAPH'90: Proceedings of the 17th Annual Conference on Computer Graphics and Interactive Techniques, ACM, pp. 367–376 (1990)
22. Udupa, J.K., Odhner, D.: Shell rendering. IEEE Comput. Graph. Appl. **13**(6), 58–67 (1993)
23. Gelder, A.V., Kim, K.: Direct volume rendering with shading via three-dimensional textures. In: Proc. of the 1996 Symposium on Volume Visualization. IEEE, pp. 23–30 (1996)
24. Marroquim, R., Maximo, A., Farias, R.C., Esperança, C.: Volume and isosurface rendering with GPU accelerated cell projection. Comput. Graph. Forum **27**(1), 24–35 (2008)
25. Lacroute, P., Levoy, M.: Fast volume rendering using a shear-warp factorization of the viewing transformation. Comput. Graph. **28**(Annual Conference Series), 451–458 (1994)
26. Mroz, L., Hauser, H., Gröller, E.: Interactive high-quality maximum intensity projection. Comput. Graph. Forum **19**(3), 341–350 (2000)
27. Robb, R.: X-ray computed tomography: from basic principles to applications. Annu. Rev. Biophys. Bioeng. **11**, 177–182 (1982)
28. Ahmetoglu, A., Kosucu, P., Kul, S., et al.: MDCT cholangiography with volume rendering for the assessment of patients with biliary obstruction. Am. J. Roentgenol. **183**, 1327–1332 (2004)
29. Fishman, E.K., Horton, K.M., Johnson, P.T.: Multidetector CT and three-dimensional CT angiography for suspected vascular trauma of the extremities. Radiographics **28**, 653–665 (2008)
30. Singh, A.K., Sahani, D.V., Blake, M.A., Joshi, M.C., Wargo, J.A., del Castillo, C.F.: Assessment of pancreatic tumor respectability with multidetector computed tomography: semiautomated console-generated versus dedicated workstation-generated images. Acad. Radiol. **15**(8), 1058–1068 (2008)
31. Fishman, E.K., Ney, D.R., et al.: Volume rendering versus maximum intensity projection in CT angiography: what works best, when, and why. Radiographics **26**, 905–922 (2006)
32. Gill, R.R., Poh, A.C., Camp, P.C., et al.: MDCT evaluation of central airway and vascular complications of lung transplantation. Am. J. Roentgenol. **191**, 1046–1056 (2008)
33. Blinn, J.F.: Light reflection functions for simulation of clouds and dusty surfaces. Comput. Graph. **16**(3), 21–29 (1982)
34. Max, N.: Optical models for direct volume rendering. IEEE Trans. Vis. Comput. Graph. **1**(2), 99–108 (1995)
35. Rezk-Salama, C.: Volume Rendering Techniques for General Purpose Graphics Hardware. Ph.D. thesis, University of Siegen, Germany, 2001
36. Engel, K., Kraus, M., Ertl, T.: High-quality pre-integrated volume rendering using hardware-accelerated pixel shading. In: Eurographics/SIGGRAPH Workshop on Graphics Hardware'01, Annual Conf Series, pp. 9–16. Addison-Wesley, Boston (2001)
37. Kraus, M.: Direct Volume Visualization of Geometrically Unpleasant Meshes. Ph.D. thesis, Universität Stuttgart, Germany, 2003
38. Kraus, M., Ertl, T.: Pre-integrated volume rendering. In: Hansen, C., Johnson, C. (eds.) The Visualization Handbook, pp. 211–228. Academic, Englewood Cliffs, NY (2004)
39. Röttger, S., Kraus, M., Ertl, T.: Hardware-accelerated volume and isosurface rendering based on cell-projection. In: VIS'00: Proc. Conf. on Visualization'00, pp. 109–116. IEEE Computer Society, Los Alamitos (2000)
40. Max, N., Hanrahan, P., Crawfis, R.: Area and volume coherence for efficient visualization of 3D scalar functions. SIGGRAPH Comput. Graph. **24**(5), 27–33 (1990)
41. Roettger, S., Guthe, S., Weiskopf, D., Ertl, T., Strasser, W.: Smart hardware-accelerated volume rendering. In: Proc. of the Symp. on Data Visualisation 2003, pp. 231–238. Eurographics Assoc (2003)

42. Zhang, Q., Eagleson, R., Peters, T.M.: Rapid voxel classification methodology for interactive 3D medical image visualization. In: MICCAI (2), pp. 86–93 (2007)
43. Sato, Y., Westin, C.-F., Bhalerao, A., Nakajima, S., Shiraga, N., Tamura, S., Kikinis, R.: Tissue classification based on 3D local intensity structures volume rendering. IEEE Trans. Vis. Comp. Graph. **6** (2000)
44. Pfister, H., Lorensen, B., Bajaj, C., Kindlmann, G., Schroeder, W., Avila, L.S., Martin, K., Machiraju, R., Lee, J.: The transfer function bake-off. IEEE Comput. Graph. Appl. **21**(3), 16–22 (2001)
45. Kniss, J., Kindlmann, G.L., Hansen, C.D.: Multidimensional transfer functions for interactive volume rendering. IEEE Trans. Vis. Comput. Graph. **8**(3), 270–285 (2002)
46. Higuera, F.V., Hastreiter, P., Fahlbusch, R., Greiner, G.: High performance volume splatting for visualization of neurovascular data. In: IEEE Visualization, p. 35 (2005)
47. Abellán, P., Tost, D.: Multimodal volume rendering with 3D textures. Comput. Graph. **32**(4), 412–419 (2008)
48. Hadwiger, M., Laura, F., Rezk-Salama, C., Höllt, T., Geier, G., Pabel, T.: Interactive volume exploration for feature detection and quantification in industrial CT data. IEEE Trans. Vis. Comput. Graph. **14**(6), 1507–1514 (2008)
49. Petersch, B., Hadwiger, M., Hauser, H., Hönigmann, D.: Real time computation and temporal coherence of opacity transfer functions for direct volume rendering. Comput. Med. Imaging Graph. **29**(1), 53–63 (2005)
50. Rezk-Salama, C., Keller, M., Kohlmann, P.: High-level user interfaces for transfer function design with semantics. IEEE Trans. Vis. Comput. Graph. **12**(5), 1021–1028 (2006)
51. Rautek, P., Bruckner, S., Gröller, E.: Semantic layers for illustrative volume rendering. IEEE Trans. Vis. Comput. Graph. **13**(6), 1336 (2007)
52. Freiman, M., Joskowicz, L., Lischinski, D., Sosna, J.: A feature-based transfer function for liver visualization. CARS **2**(1), 125–126 (2007)
53. Robb, R.: X-ray computed tomography: from basic principles to applications. Annu. Rev. Biophys. Bioeng. **11**, 177–182 (1982)
54. Mroz, L., Hauser, H., Gröller, E., Interactive high-quality maximum intensity projection. Comput. Graph. Forum **19**(3) (2000)
55. Mora, B., Ebert, D.S.: Low-complexity maximum intensity projection. ACM Trans. Graph. **24**(4), 1392–1416 (2005)
56. Hoang, J.K., Martinez, S., Hurwitz, L.M.: MDCT angiography of thoracic aorta endovascular stent-grafts: pearls and pitfalls, Am. J. Roentgenol. **192**, 515–524 (2009)
57. Kiefer, G., Lehmann, H., Weese, J.: Fast maximum intensity projections of large medical data sets by exploiting hierarchical memory architectures. IEEE Trans. Inf. Technol. Biomed. **10**(2), 385–394 (2006)
58. Kye, H., Jeong, D.: Accelerated MIP based on GPU using block clipping and occlusion query. Comput. Graph. **32**(3), 283–292 (2008)
59. Rubin, G.D., Rofsky, N.M.: CT and MR Angiography: Comprehensive Vascular Assessment. Lippincott Williams & Wilkins, Philadelphia (2008)
60. Kautz, J.: Hardware lighting and shading: a survey. Comput. Graph. Forum **23**(1), 85–112 (2004)
61. Kniss, J., Premoze, S., Hansen, C.D., Shirley, P., McPherson, A.: A model for volume lighting and modeling. IEEE Trans. Vis. Comput. Graph. **9**(2), 150–162 (2003)
62. Weiskopf, D., Engel, K., Ertl, T.: Interactive clipping techniques for texture-based volume visualization and volume shading. IEEE Trans. Vis. Comput. Graph. **9**(3), 298–312 (2003)
63. Rheingans, P., Ebert, D.: Volume illustration: nonphotorealistic rendering of volume models. IEEE Trans. Vis. Comput. Graph. **7**(3), 253–264 (2001)
64. Winkenbach, G., Salesin, D.H.: Computer-generated pen-and-ink illustration. In: SIGGRAPH'94: Proc. of the 21st Annual Conference on Computer Graphics and Interactive Techniques, pp. 91–100. ACM, New York (1994)

65. Lum, E.B., Ma, K.-L.: Hardware-accelerated parallel nonphotorealistic volume rendering. In: NPAR'02: Proc. of the 2nd International Symposium on Non-Photorealistic Animation and Rendering, pp. 67–ff. ACM, New York (2002)
66. Lum, E.B., Wilson, B., Ma, K.-L.: High-quality lighting for preintegrated volume rendering: In: Deussen, O., Hansen, C.D., Keim, D.A., Saupe, D. (eds.) VisSym, Eurographics, pp. 25–34 (2004)
67. Chan, M.-Y., Qu, H., Chung, K.-K., Mak, W.-H., Wu, Y.: Relationaware volume exploration pipeline. IEEE Trans. Vis. Comput. Graph. **14**(6), 1683–1690 (2008)
68. Rautek, P., Bruckner, S., Groller, E., Viola, I.: Illustrative visualization: new technology or useless tautology? SIGGRAPH **42**(3), 1–8 (2008)
69. Preim, B., Bartz, D.: Visualization in Medicine: Theory, Algorithms, and Applications (The Morgan Kaufmann Series in Computer Graphics), 1st edn. Morgan Kaufmann, San Francisco, CA (2007)
70. Sakas, G., Schreyer, L.-A., Grimm, M.: Preprocessing and volume rendering of 3D ultrasonic data. IEEE Comput. Graph. Appl. **15**(4), 47–54 (1995)
71. KH Hohne, M., Bomans, A., Pommert, M., Riemer, C., Schiers, U., Tiede, G.: Wiebecke, 3D visualization of tomographic volume data using generalized voxel model. Vis. Comput. **6**(1), 28–36 (1990)
72. Tiede, U., Schiemann, T., Hohne, K.H.: Visualizing the visible human. IEEE Comput. Graph. Appl. **16**(1), 7–9 (1996)
73. Westover, L.: Interactive volume rendering. In: Proc 1989 Chapel Hill Workshop on Volume Visualization, pp. 9–16. ACM, New York (1989)
74. Westover, L.: Footprint evaluation for volume rendering. In: SIGGRAPH'90: Proc. of the 17th Annual Conference on Computer Graphics and Interactive Techniques, pp. 367–376. ACM, New York (1990)
75. Westover, L.: Splatting: A Parallel, Feed-Forward Volume Rendering Algorithm. PhD Thesis, Univ North Carolina, 1991
76. Lee, R.K., Ihm, I.: On enhancing the speed of splatting using both object- and image-space coherence. Graph. Models **62**(4), 263–282 (2000)
77. Meißner, M., Huang, J., Bartz, D., Mueller, K., Crawfis, R.: A practical evaluation of popular volume rendering algorithms. In: VVS'00: Proc. of the 2000 IEEE Symposium on Volume Visualization, pp. 81–90. ACM, New York (2000)
78. Higuera, F.V., Hastreiter, P., Fahlbusch, R., Greiner, G.: High performance volume splatting for visualization of neurovascular data. In: IEEE Visualization, p. 35 (2005)
79. Birkfellner, W., et al.: Fast DRR splat rendering using common consumer graphics hardware. Phys. Med. Biol. 50(9), N73–N84 2005
80. Spoerk, J., Bergmann, H., Wanschitz, F., Dong, S., Birkfellner, W., Fast DRR splat rendering using common consumer graphics hardware. Med. Phys. **34**(11), 4302–4308 (2007)
81. Audigier, R., Lotufo, R., Falcao, A.: 3D visualization to assist iterative object definition from medical images. Comput. Med. Imaging Graph. **4**(20), 217–230 (2006)
82. Mueller, K., Shareef, N., Huang, J., Crawfis, R.: High-quality splatting on rectilinear grids with efficient culling of occluded voxels. IEEE Trans. Vis. Comput. Graph. **5**(2), 116–134 (1999)
83. Neophytou, N., Mueller, K.: Gpu accelerated image aligned splatting. In: Kaufman, A.E., Mueller, K., Groller, E., Fellner, D.W., Moller, T., Spencer, S.N. (eds.) Volume Graphics. Eurographics Assoc, pp. 197–205 (2005)
84. Udupa, J.K., Odhner, D.: Shell rendering: IEEE Comput. Graph. Appl. **13**(6), 58–67 (1993)
85. Lei, T., Udupa, J.K., Saha, P.K., Odhner, D.: Artery-vein separation via MRA – an image processing approach. IEEE Trans. Med. Imaging **20**(8), 689–703 (2001)
86. Bullitt, E., Aylward, S.: Volume rendering of segmented image objects. IEEE Trans. Med. Imaging **21**(8), 998–1002 (2002)
87. Falcao, A.X., Rocha, L.M., Udupa, J.K.: A comparative analysis of shell rendering and shear-warp rendering. In: Proc. of SPIE Med. Imaging, vol. 4681, pp. 472–482. San Diego, CA (2002)

88. Botha, C.P., Post, F.H.: Shellsplatting: interactive rendering of anisotropic volumes. In: Proc. of the Symposium on Data Visualisation 2003, pp. 105–112. Eurographics Association (2003)
89. Grevera, G.J., Udupa, J.K., Odhner, D.: T-shell rendering and manipulation. In: Proc. of SPIE Med. Imaging, vol. 5744, pp. 22–33. San Diego, CA (2005)
90. Cullip, T.J., Neumann, U.: Accelerating Volume Reconstruction with 3D Texture Hardware. Technical report, Univ North Carolina Chapel Hill, 1994
91. Cabral, B., Cam, N., Foran, J.: Accelerated volume rendering and tomographic reconstruction using texture mapping hardware. In: Proc. of the 1994 Symposium on Volume Visualization, pp. 91–98. ACM, New York (1994)
92. Lamar, E.C., Hamann, B., Joy, K.I.: Multiresolution techniques for interactive texture-based volume visualization. IEEE Vis., 355–362 (1999)
93. Sato, Y., Westin, C.-F., Bhalerao, A., Nakajima, S., Shiraga, N., Tamura, S., Kikinis, R.: Tissue classification based on 3D local intensity structures for volume rendering. IEEE Vis. Comput. Graph. **6**(2), 160–180 (2000)
94. Hauser, H., Mroz, L., Bischi, G.I., Groller, E.: Two-level volume rendering. IEEE Trans. Vis. Comput. Graph. **7**(3), 242–252 (2001)
95. Hadwiger, M., Berger, C., Hauser, H.: High-quality two-level volume rendering of segmented data sets on consumer graphics hardware. In: VIS'03: Proceedings of the 14th IEEE Visualization 2003 (VIS'03), p. 40. IEEE Computer Society, Washington, DC (2003)
96. Holmes, D., Davis, B., Bruce, C., Robb, R.: 3D visualization, analysis, and treatment of the prostate using trans-urethral ultrasound. Comput. Med. Imaging Graph. **27**(5), 339–349 (2003)
97. Etlik, O¨., Temizoz, O., Dogan, A., Kayan, M., Arslan, H., Unal, O.: Three-dimensional volume rendering imaging in detection of bone fractures. Eur. J. Gen. Med. **1**(4), 48–52 (2004)
98. Wenger, A., Keefe, D.F., Zhang, S., Laidlaw, D.H.: Interactive volume rendering of thin thread structures within multivalued scientific data sets. IEEE Trans. Vis. Comput. Graph. 10(6), 664–672 (2004)
99. Wang, A., Narayan, G., Kao, D., Liang, D.: An evaluation of using real-time volumetric display of 3D ultrasound data for intracardiac catheter manipulation tasks. Comput. Med. Imaging Graph. 41–45 (2005)
100. Sharp, R., Adams, J., Machiraju, R., Lee, R., Crane, R.: Physics-based subsurface visualization of human tissue. IEEE Trans. Vis. Comput. Graph. **13**(3), 620–629 (2007)
101. Lopera, J.E., et al.: Multidetector CT angiography of infrainguinal arterial bypass. RadioGraphics **28**, 529–548 (2008)
102. Levin, D., Aladl, U., Germano, G., Slomka, P.: Techniques for efficient, real-time, 3D visualization of multi-modality cardiac data using consumer graphics hardware. Comput. Med. Imaging Graph. **29**(6), 463–475 (2005)
103. Lehmann, H., Ecabert, O., Geller, D., Kiefer, G., Weese, J.: Visualizing the beating heart: interactive direct volume rendering of high-resolution CT time series using standard pc hardware. In: Proc. of SPIE Med. Imaging, vol. 6141, pp. 614109–1–12. San Diego, CA (2006)
104. Yuan, X., Nguyen, M.X., Chen, B., Porter, D.H., Volvis, H.D.R.: High dynamic range volume visualization. IEEE Trans. Vis. Comput. Graph. **12**(4), 433–445 (2006)
105. Correa, C.D., Silver, D., Chen, M.: Feature aligned volume manipulation for illustration and visualization. IEEE Trans. Vis. Comput. Graph. 12(5), 1069–1076 (2006)
106. Abellan, P., Tost, D.: Multimodal volume rendering with 3D textures. Comput. Graph. **32**(4), 412–419 (2008)
107. Li, W., Kaufman, A.: Accelerating volume rendering with texture hulls. In: IEEE/SIGGRAPH Symposium on Volume Visualization and Graphics, pp. 115–122 (2002)
108. Li, W., Kaufman, A.E.: Texture partitioning and packing for accelerating texture-based volume rendering. In: Graphics Interface, p. 81–88 (2003)
109. Li, W., Mueller, K., Kaufman, A.: Empty space skipping and occlusion clipping for texture-based volume rendering. IEEE Vis. 317–324 (2003)
110. Bethune, C., Stewart, A.J.: Adaptive slice geometry for hardware-assisted volume rendering. J. Graph. Tools **10**(1), 55–70 (2005)

111. Keles, H.Y., Isler, V.: Acceleration of direct volume rendering with programmable graphics hardware. Vis. Comput. **23**(1), 15–24 (2007)
112. Ibanez, L., Schroeder, W., Ng, L., Cates, J., the Insight Software Consortium: The ITK Software Guide, 2nd edn. updated for itk v2.4 Nov 2005
113. Schroeder, W., Martin, K., Lorensen, B.: The Visualization Toolkit, 3rd edn. Kitware Inc., Clifton Park, (2004)

Chapter 14
Sparse Sampling in MRI

Philip J. Bones and Bing Wu

14.1 Introduction

The significant time necessary to record each resonance echo from the volume being imaged in magnetic resonance imaging (MRI) has led to much effort to develop methods which take fewer measurements. Faster methods mean less time for the patient in the scanner, increased efficiency in the use of expensive scanning facilities, improved temporal resolution in studies involving moving organs or flows, and they lessen the probability that patient motion adversely affects the quality of the images. Images like those of the human body possess the property of sparsity, that is the property that in some transform space they can be represented much more compactly than in image space. The technique of compressed sensing, which aims to exploit sparsity, has therefore been adapted for use in MRI. This, coupled with the use of multiple receiving coils (parallel MRI) and the use of various forms of prior knowledge (e.g., support constraints in space and time), has resulted in significantly faster image acquisitions with only a modest penalty in the computational effort required for reconstruction. We describe the background motivation for adopting sparse sampling and show evidence of the sparse nature of biological image data sets. We briefly present the theory behind parallel MRI reconstruction, compressed sensing and the application of various forms of prior knowledge to image reconstruction. We summarize the work of other groups in applying these concepts to MRI and our own contributions. We finish with a brief conjecture on the possibilities for future development in the area.

P.J. Bones (✉)
University of Canterbury, Christchurch, New Zealand
e-mail: phil.bones@canterbury.ac.nz

G. Dougherty (ed.), *Medical Image Processing: Techniques and Applications*, Biological and Medical Physics, Biomedical Engineering, DOI 10.1007/978-1-4419-9779-1_14,

14.1.1 Magnetic Resonance Imaging

MRI is the term used to represent the entire set of methods which apply the principles first developed for chemistry as nuclear magnetic resonance in such a way that the spatial variation of a property within an object is observed. A strong and extremely uniform magnetic field is applied to the object. Under the influence of the field those atoms in the object which have a magnetic spin property align in the direction of the magnetic field in one of two orientations such that a small net magnetization occurs. The atoms exhibit a resonance at a frequency, which is linearly dependent on the magnetic field strength. The resonance can be excited by means of a radiofrequency (RF) pulse at the appropriate frequency and the atoms which have been excited precess about an axis aligned with the direction of the magnetic field. After excitation, the precessing atoms relax back to equilibrium and in the process generate a small, but measurable, RF field – an "echo."

By imposing a gradient in magnetic field strength as a linear function of one of the Cartesian space coordinates, z say, it is possible to encode that spatial coordinate in the resonance frequencies of the spins. By considerable extension of this basic idea, signal processing of the signals recovered from echoes after a specific sequence of gradient impositions with respect to the x, y, and z directions, coupled with RF excitations, and echo signal acquisitions, allows the formation of an image of the interior of the object. Many excitation and acquisition sequences have been devised. Because of the relationship between resonance frequency and magnetic field strength, virtually all of them make measurements in *spatial frequency space*, or "k-space" as the MRI community generally refers to it. Moreover, tissues in the body can be characterized in terms of the time constants associated with the atomic resonances, known as "T1" and "T2". The differences between tissue responses help to make MRI effective in distinguishing between them. For a good overview of the basis of MRI, see [1] and for a comprehensive review of MRI sequences and algorithms, see [2].

While MRI also has applications in biological science and in the study of materials, it is its role in medicine that has led to a whole industry. The size of the annual meeting of the International Society for Magnetic Resonance in Medicine (ISMRM, http://ismrm.org) is testament to the extraordinary interest in the imaging modality. The reason that MRI has had such a profound effect on the practice of modern medicine is because of the exquisite detail that it has achieved in images of soft tissues within the human body. In this, it is quite complementary to X-ray computed tomography, which is particularly good at imaging harder tissues, notably bone. The two imaging modalities thus coexist and many patients are imaged using both for some diagnoses. Note that there are no known detrimental effects on the human body caused by MRI scanners of the sort in regular use, while the use of X-ray computed tomography is strictly limited by the dose of ionizing radiation the patient receives in such a scanner.

The use of MRI is restricted in one specific way: the physical processes involved in the excitation and reception of MR signals are inherently quite slow. Thus, the time taken to invoke a specific sequence and to measure the signals that are generated is measured in a matter of milliseconds per pulse. Note that this has nothing to do with the electronics associated with the scanner – faster hardware does not solve the problem. To take a complete set of measurements for, say, a 3D imaging study of the brain may take many minutes; some acquisition sequences require periods approaching 1 h [2]. As well as reducing the throughput of an expensive facility, the slowness of acquisition limits how useful MRI is in imaging organs where motion is intrinsic, most notably the heart and circulatory system; even when these organs are not specifically the target for a particular imaging study, their activity may affect the success of imaging nearby or associated organs. The efforts of the authors and of many others involved in MRI research are directed toward developing smarter algorithms to attempt to reconstruct useful images from fewer measurements and therefore in less time.

14.1.2 Compressed Sensing

The conventional wisdom in signal processing is that the sampling rate for any signal must be at twice the maximum frequency present in the signal. The Sampling Theorem is variously attributed to Whittaker, Nyquist, Kotelnikov, and Shannon and its conception represents a very significant landmark in the history of signal processing. However in the work performed in recent years related to signal compression, it has become obvious that the total amount of information which is needed to represent a signal or image to high accuracy is in many cases much less than that implied by the "Nyquist limit." This is nowhere more apparent than in the modern digital camera where quite acceptable images can be stored and recreated from a small fraction of the data volume that was associated with the original image sampling. The total amount of information acquired, 4 megapixels at 24-bit per pixel for example, may often be compressed to several hundred thousand bytes by the JPEG compression method without appreciable loss of image quality. The image property that lies behind this compressibility is "sparsity": the fact that under some transformation many of the data values in the space associated with the transform can be set to zero and the image reconstructed from the rest of the data values without appreciable effect. An image which is very sparse (under that transformation) is one for which the number of nonzero values is relatively low.

The technique of compressed sensing (also known as "compressive sensing") was introduced to exploit image sparsity [3, 4]. Consider a 2D image with N pixels represented by the vector $\mathbf{x}$ and suppose that it can be accurately represented by $K \ll N$ data values under the linear transformation $\mathbf{y} = \Phi\mathbf{x}$. Rather than measuring the N pixel values and then performing the transformation, we seek to make just M measurements $\mathbf{m}$, where $K \leq M \ll N$. Thus, $\mathbf{m} = \Psi\mathbf{y}$, where Ψ is a measurement matrix of dimension $M \times K$. While this might be of little direct benefit in the case cited above of a modern digital camera, for which the design of the sensor is most straightforwardly implemented as a regular 2D array of individual pixel detectors, there are many other applications, notably including MRI, for which making fewer

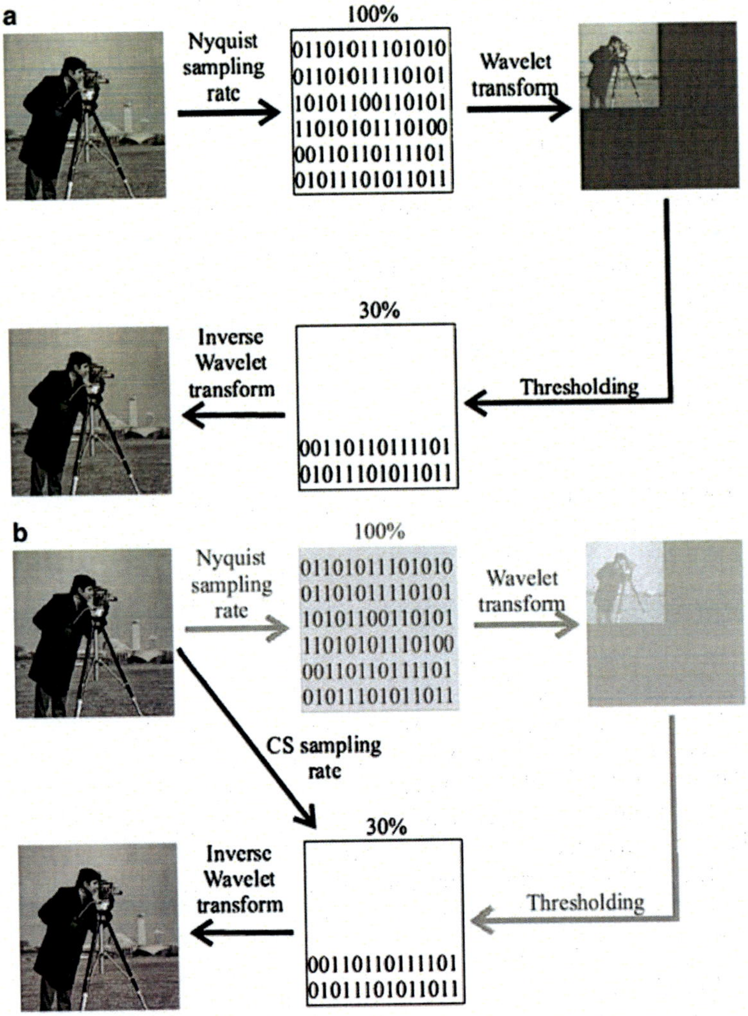

Fig. 14.1 Comparing (**a**) conventional image sensing and compression to (**b**) compressed sensing. In (**a**), the image is sampled at the Nyquist sampling rate and stored. Then all of the smallest coefficients in the image's wavelet transform are discarded to reduce the storage volume. In (**b**), the significant coefficients of the wavelet transform are directly estimated from a lesser number of samples of the image

measurements does offer an advantage. In the remainder of this chapter, we show how the exploitation of sparsity by means of compressed sensing methods has considerable potential in speeding up MRI image acquisition.

An illustration of how compressed sensing might work for the general optical imaging case is shown in Fig. 14.1. Suppose that the well-known cameraman image is being acquired with a conventional digital camera in Fig. 14.1a and a full set of pixels are being recorded and stored. Applying a linear transform, such as the discrete wavelet transform (DWT), to the image allows many of the DWT coefficients to be set to zero, resulting in a compressed form of storage. The compressed data set can be used to reconstruct a good likeness of the original image. The compressed sensing (CS) approach is illustrated in Fig. 14.1b: by some process many fewer samples are made of the original scene and the nonzero coefficients of the compressed image are directly estimated. Thus, the waste associated with full data measurement followed by compression is avoided in CS.

At the time of writing, the group at Rice University has established a very comprehensive bibliography of literature related to the theory and application of compressed sensing (see http://dsp.rice.edu).

14.1.3 The Role of Prior Knowledge

The term "prior knowledge" is most frequently associated with Bayesian inference, in which a *posterior* estimate based on gathered measurements is influenced by knowledge of *prior* distributions for the measurements and for the result. See Geman and Geman [5] and Hu et al. [6] for a full treatment of Bayesian estimation methods. Here, we use the term in a wider context, however. By prior knowledge is here meant any constraint which can be put on the estimate. For example if the interior of an object is being imaged that is known to exist within a given boundary, then that boundary, or possibly a conservative estimate of it, can be incorporated into the imaging process: this is an example of a "support constraint." In the situation cited, those pixels (or voxels, if 3D images are being considered), which lie outside the support region do not need to be estimated, suggesting that incorporating the prior knowledge may make the estimation process easier.

The spatial support constraint represents only one from a rich set of forms of prior knowledge, which may be applied to imaging problems in general and MRI in particular. Other examples include:

1. Knowledge that the image is sparse in some sense is itself a form of prior knowledge
2. Knowledge that the object being imaged is approximately piecewise homogeneous
3. Knowledge that the object has a smooth boundary or boundaries
4. Knowledge that changes in time occur smoothly
5. Knowledge that only a relatively small portion of the object undergoes changes with time

Some authors may argue that all of the examples of prior knowledge listed above may be able to be labeled as "sparsity," but here we use a more restricted definition: sparsity is the property that when the image is expressed in terms of some basis, many fewer coefficients are required to accurately represent the image than is implied by Nyquist sampling.

14.2 Sparsity in MRI Images

In this section, we explore the properties of MR images, which make them amenable to compressed sensing, show examples of how some common transforms can be used to exploit the sparsity, and introduce a novel nonlinear approach to promoting sparsity.

14.2.1 *Characteristics of MR Images (Prior Knowledge)*

As mentioned in the introduction, MRI measurements are made in the *k*-space domain. In some cases, the measurements may be at positions constrained by a regular Cartesian grid. Since an inverse Fourier transform is required to generate an image from the sampled *k*-space data, the regular Cartesian sample positions allow the straightforward use of the efficient FFT algorithm. However to achieve faster scanning or some signal processing advantages non-Cartesian sampling is frequently employed. Radial and spiral sampling [2] are quite common, for example. Some sampling strategies involve a higher density of samples near the center of *k*-space (i.e., concentrated in the area of lower spatial frequencies). In any case, there is a direct relationship between the extent of the *k*-space domain within which measurements are distributed and the resolution of the image obtained. Likewise, there is a direct relationship between the field-of-view (FOV) in the spatial domain and the effective spacing of samples in *k*-space [1].

The main source of noise in MR imaging is due to thermal fluctuations of electrolytes in the region being imaged which are detected by the receiver coil or coils. Electronic noise is inevitably present as well but may usually be of lesser order. Generally, the SNR increases as the square root of the acquisition time and linearly with the voxel size. Thus any moves to increase imaging speed and/or imaging resolution inevitably lead to a loss of SNR. Importantly, the noise statistics of each *k*-space sample is essentially equal [1]. Since the amplitude of samples near the origin of *k*-space is much greater than near the periphery, the SNR of these center samples is much better. This consideration often influences the design of a sampling scheme.

In many significant imaging situations, the object is known not to extend throughout the FOV. For example, in making a 2D axial plane image of the brain, the FOV is usually chosen to be a square or rectangular region that entirely contains the outline of the head. There is therefore part of the FOV which lies outside the head and which is therefore known not to contribute a significant signal to the measurements made. This support constraint is explicit. A support constraint may also be implicit: for example if it is known that a transformed version of the image is known to be nonzero only within an unspecified region which spans some known proportion of the transformed space.

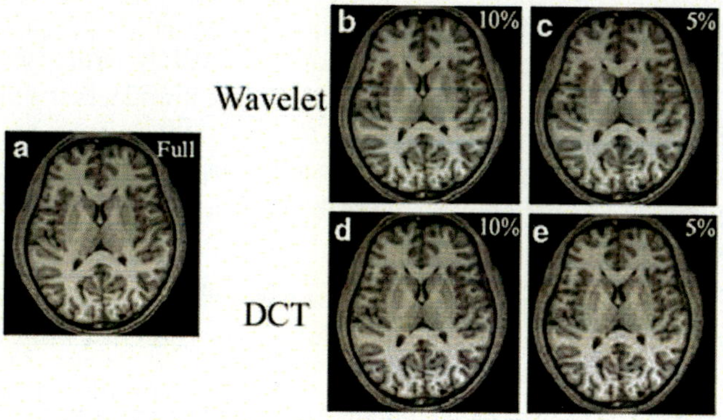

Fig. 14.2 Compressibility of an MR brain slice. The wavelet and DCT approximations of the original MR brain slice shown in (**a**), using only the 10% highest amplitude and the 5% highest amplitude coefficients of the transforms, are shown in (**b**) to (**e**). There appears to be little loss of information under such high levels of image compression

Constraints may also be temporal. While the relatively slow acquisition of MR data restricts its use in dynamic imaging tasks, the facts that measurements are made at precise timing instants and that objects under observation move relatively smoothly allows temporal constraints to be formulated and exploited.

Biological tissues comprise a large number of types. While at a microscopic level these tissues are generally inhomogeneous, at the resolution observed by MR techniques, each tissue type is relatively homogeneous. It is the difference in MR signal between tissue types which allows such useful anatomical information to be gleaned. The image as a whole therefore exhibits an approximately piecewise homogeneity, which can be exploited.

The forms of prior knowledge about the MR images discussed above can all be seen as evidence for expecting the images to be sparse.

14.2.2 Choice of Transform

The term implicit support was introduced in Sect. 14.2.1. This represents the property that under some transformation an image can be shown to be nonzero only within some unspecified part of the domain. The success of lossy image compression schemes based on the DCT and wavelet transforms indicate that these are useful sparsifying transforms. In Fig. 14.2, we illustrate the degree to which a typical MR image can be compressed by the two transforms. The image in Fig. 14.2a is formed for an axial slice of the brain with the full resolution afforded by the MRI sequence employed (256×256). The other parts of the figure show reconstructions with only a fraction of the transform coefficients retained. It is clear that under either of the two transforms a substantial reduction of data volume is possible before serious degradation of the image occurs. Note that the compression here is "lossy," in that there is always some degradation generated by setting small coefficients to zero, but not so much degradation that the image usefulness is seriously impaired.

The discrete cosine transform (DCT) was the transform of choice for many years for image compression [7]. It was the basis of the original JPEG standard. The properties of the DCT which make it a useful choice for image compression include:

1. It lends itself to representing small rectangular blocks of pixels
2. It can be shown to have a faster fall off in coefficient amplitude as frequency increases in comparison with the DFT
3. It is relatively efficient to compute via the FFT

The DWT has taken over from the DCT in certain regards [8]. The properties of the DWT which make it a useful choice for image compression include:

1. It naturally distributes information over a number of scales and localizes that information in space
2. It offers a wide range of basis function (wavelet families) from which to choose
3. It is inherently efficient to compute

The decision between the transforms is unlikely to be critical. The nature of the DWT, however, does render it better at representing boundaries in the image between two tissue types where the image function exhibits a step change. With the DCT, such a boundary necessarily injects some energy in high frequency components and adversely affects the sparsity in the transformed representation. The DWT with an appropriate choice of wavelet may perform better in this regard.

14.2.3 Use of Data Ordering

A quite distinctly different approach for increasing sparsity has recently been proposed. In 2008, Adluru and DiBella [9] and Wu et al. [10] independently proposed performing a sorting operation on the signal or image as part of the reconstruction process. The principle is presented in Fig. 14.3 for a 2D axial brain image. In Fig. 14.3a, the situation is shown whereby the image is transformed by the 2D DCT and then a compression occurs by setting all coefficients less than a given threshold to zero. The resulting reconstruction is similar to the original, but noticeably smoother due to the loss of some small amplitude high frequency components. In Fig. 14.3b, the image pixels are sorted from largest amplitude in the lower right to highest amplitude in the upper left to make the resulting function monotonic and the mapping required to do this is retained (denoted "R"). The same transformation and recovery operation after thresholding as in (a) is performed and a re-sorting (denoted "R^{-1}") is performed. Because the compression retains much of the shape of the image after sorting, the result has much higher fidelity than in

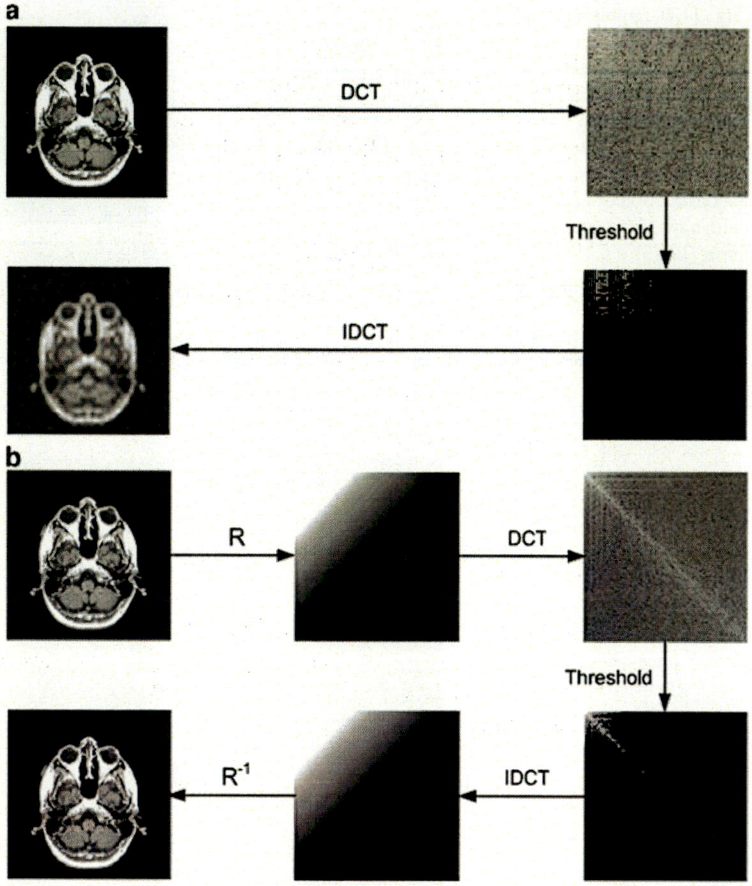

Fig. 14.3 Illustration of how a data ordering can achieve a higher sparsity for a 2D image. In (**a**), the signal is compressed by retaining only those DCT coefficients with amplitudes higher than a threshold. In (**b**), the image pixels are sorted to generate a monotonic function and then the same recovery operation is performed before a final resorting. Because the sorted data in (**b**) is more sparse, the recovery is of higher quality

Fig. 14.3a. We argue that many fewer coefficients need to be retained in the DCT of the sorted image than in the original, hence the more successful reconstruction.

Clearly, the process depicted in Fig. 14.3 requires knowledge of the original signal to derive R. The practical utility of what has been demonstrated is likely therefore to be questioned. However, we show in Sect. 14.4.2 that several methods to derive an approximate R are possible and that they lead to useful and practical algorithms for MR image recovery.

The advantage claimed for data ordering depends on the sorted data function being more sparse than the original. It is difficult if not impossible to prove this,

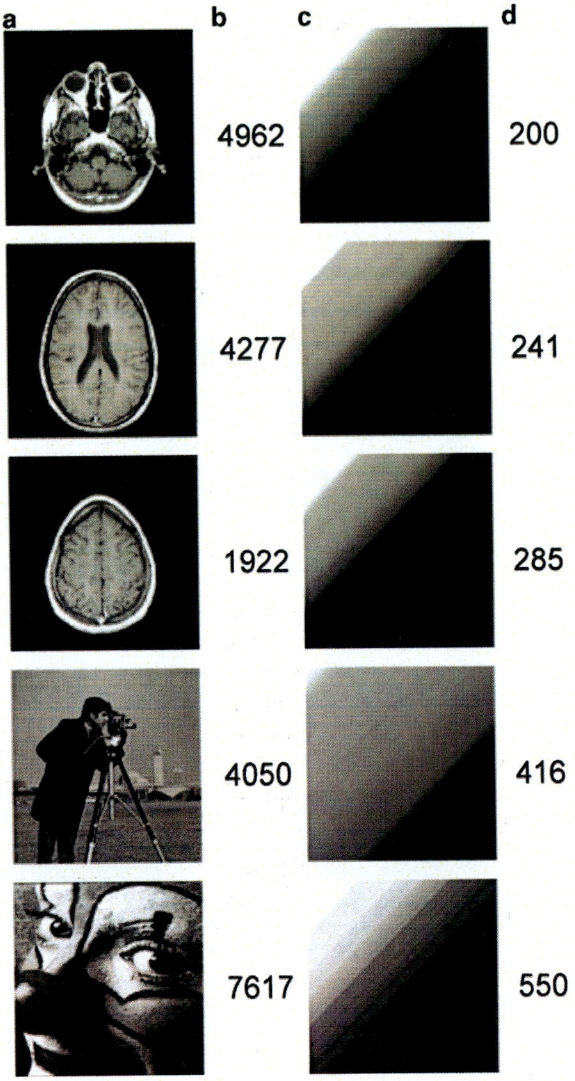

Fig. 14.4 Examples which show the higher sparsity that data ordering can achieve for 2D images: (**a**) a set of 5 original images; (**b**) the number of DCT coefficients needed to achieve a reconstruction of each image with a relative mean square error (RMSE) $\leq 1\%$; (**c**) the image after a data ordering operation; and (**d**) the number of DCT coefficients needed to achieve a reconstruction of the ordered image with RMSE$\leq 1\%$

but experiments indicate that in virtually all cases it is true. In Fig. 14.4, we show a set of 2D example images; three are typical brain axial slices (128×128 pixels) while the others are popular test images (200×200 pixels). A data sorting operation was performed on each image such that the highest intensity pixel was positioned in the top left corner and the lowest intensity pixel in the bottom right corner. A simple algorithm arranged the others in a type of raster which achieved the "wedge" images shown in column (c). Clearly, other algorithms for arranging the sorted pixels are possible. To the right of each image in columns (b) and (d) is the number of DCT coefficients that need to be retained to represent the image to within 1% relative mean square error (RMSE). It is clear that many fewer coefficients are required to represent the sorted images than the originals, with the ratio being around 10 to 1. Under the definition of sparsity we employ here, therefore, the sorting operation achieves a considerably sparser image.

14.3 Theory of Compressed Sensing

Recall that an image $\mathbf{x}$ can be called sparse if under the linear transformation $\mathbf{y} = \Phi\mathbf{x}$ just $K \ll N$ of the data values in $\mathbf{y}$ are enough to accurately represent the image, where N is the size of $\mathbf{x}$ (Sect. 14.1.2). Assume that $\mathbf{y}$ is of size N (under the sparsity assumption, $N - K$ of the elements y_i are close to zero). Therefore, Φ is $N \times N$. If we knew *a priori* which of the elements y_i may be neglected, we could reduce Φ to $K \times N$ and attempt to estimate $\mathbf{y}$ and recover an estimate of $\mathbf{x}$. However, in general it is not known which may be neglected and the data measured may be in a different space.

In MRI, data are measured in k-space and can be represented by a vector $\mathbf{d} = \mathrm{W}\mathbf{x}$, where W is the Fourier transform matrix. It is desirable to minimize the number of measurements and so we seek to reduce the size of $\mathbf{d}$ as much as possible, to M elements say $(\mathbf{d}_M)$, while recovering an acceptable quality of estimate of $\mathbf{x}$. A direct method would be to form a transformation Ψ of dimension $K \times M$ such that $\mathbf{y}' = \Psi\mathbf{d}$, with $\mathbf{y}'$ comprising only the important values of $\mathbf{y}$ being estimated. Again, however, such an approach requires prior knowledge of which of the elements y_i may be neglected.

The alternative approach used by many proponents of compressed sensing is to pose the problem in terms of an optimization. Before looking at this in detail, however, let us consider the nature of the measurement process.

14.3.1 Data Acquisition

We have established that speeding up the MRI can be achieved primarily by making fewer measurements. However, there is inevitably a cost incurred from making fewer measurements. First, fewer measurements with other properties of the scanning apparatus unchanged means a lower SNR for the data [11]. Second, undersampling in k-space causes aliasing in the image domain, that is simply inverting the relationship $\mathbf{d}_M = \mathrm{W}\mathbf{x}$ to estimate an image $\mathbf{x}' = \mathrm{W}^{-1}\mathbf{x}$ produces a heavily aliased image. Even if a more sophisticated image recovery process is adopted, it is clear that the choice of the sampling locations plays an important role.

The effect of noise can be ameliorated to some extent in post-processing, particularly if that noise is random in nature and its distribution throughout the image. The effect of aliasing can be reduced to acceptable levels by the use of prior

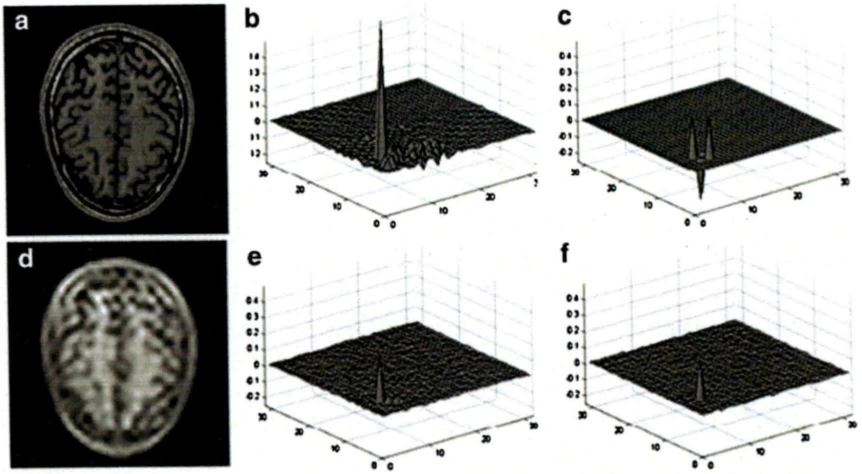

Fig. 14.5 The transform point spread functions (TPSFs) corresponding to one 2D DWT coefficient with random and regular sampling patterns. (**a**) Original 128 × 128 axial image; (**b**) TPSF for random *k*-space sampling; (**c**) TPSF for regular *k*-space sampling (showing clear aliasing); (**d**) low-resolution image formed from 1/16 *k*-space data; (**e**) TPSF for random k-space sampling with data ordering; and (**f**) TPSF for regular *k*-space sampling with data ordering

information. However, the pattern of undersampling plays a particularly important role. In CS, random sampling patterns are often employed [3, 12]. In Fig. 14.5, we illustrate the important roles that both random sampling patterns and data ordering can play. Figure 14.5a is the original 128 × 128 axial brain image formed from a fully sampled *k*-space dataset. Two subsets of the *k*-space samples were taken by a random pattern and a regular pattern. A DWT was formed (Debauchies-4 wavelets) in each case and one coefficient was chosen to be estimated from the undersampled *k*-space data. Figure 14.5b, c shows the estimates in the DWT domain for random and regular undersampling patterns, respectively. Lustig, Donoho, and Pauly [13] refer to this type of plot as the "transform point spread function" (TPSF). In Fig. 14.5b, the coefficient is estimated with relatively little aliasing, whereas in Fig. 14.5c the process generates several aliases for the coefficient. Many authors who have written on CS describe this as an "incoherence" property [3, 12, 13].

The remainder of Fig. 14.5 illustrates what happens when data ordering is introduced into the process, with the ordering based on the low-resolution reconstruction shown in Fig. 14.5d (formed from the center 1/16 of the *k*-space data). Figure 14.5e, f shows the TPSF for random and regular undersampling patterns, respectively, with data ordering and reordering included. In this case, little difference is seen between random and regular *k*-space undersampling patterns, with relatively minor aliasing occurring in both cases. We believe this indicates that the data ordering itself introduces the incoherence property. In Sect. 14.4.2 below, we relate our experience with a number of different sampling strategies.

14.3.2 Signal Recovery

Assuming that we have chosen a sampling strategy for the k-space data and a transform Φ under which the true image is expected to be sparse, we seek a solution $\mathbf{x}'$ as close as possible to $\mathbf{x}$, which is constrained in two ways:

1. The solution is consistent with the data $\mathbf{d}_M$
2. The solution is sparse under the transformation Φ

Condition 1 can be achieved in principle at least by minimizing the power of the error between the measurements and the values at those measurement points, which are predicted by the imaging model for the current image estimate, that is by minimizing the squared norm $||\mathbf{d}_M - \mathrm{W}_M\mathbf{x}'||_2$. Such squared norm minimizations have formed the backbone of image recovery for many years [14].

Condition 2 above implies a minimization of the quantity $||\Phi\mathbf{x}'||_0$, that is the number of nonzero elements in $\Phi\mathbf{x}'$. However, this minimization is computationally intractable [4, 15]. It turns out that a minimization of the quantity $||\Phi\mathbf{x}'||_1$, that is the first norm of the transformed image estimate, can achieve Condition 2 remarkably well [4,16]. The l_1 norm applied here has the effect of pushing negligible coefficients toward zero while retaining larger components accurately. This is in contrast with a squared norm which tends to penalize large coefficients.

As explained in the previous section, the random sampling patterns which offer advantages in CS do generate noise-like artifacts. Therefore in our experience, it is also useful to apply a further constraint:

3. The solution is piecewise smooth

Minimizing the total variation (TV), that is the sum of the magnitudes of differences between each pixel and its immediate neighbors, has been shown to be effective at meeting Condition 3. We denote the total variation for image vector $\mathbf{y}$, $\mathrm{TV}(\mathbf{y})$.

The minimization problem can now be posed: Find an estimate for the required image $\mathbf{x}'$ by minimizing

$$||\mathbf{d}_M - \mathrm{W}_M\mathbf{x}'||_2 + \lambda||\Phi\mathbf{x}'||_1 + \beta\,\mathrm{TV}(\mathbf{x}')$$

where λ and β are positive constants used to control the relative importance of the constraints being placed on the solution. A method such as conjugate gradient minimization is suitable to solve the problem. We have employed the *SparseMRI* package provided by [13] as part of the very comprehensive resource on compressed sensing provided by Rice University (see http://dsp.rice.edu).

14.4 Progress in Sparse Sampling for MRI

In this section, we briefly review the progress made to date in applying the principles of sparse sampling to MRI. We first review the important developments that have appeared in the literature. We believe that the biggest single contribution came from Lustig, Donoho, and Pauly [13]. This group has continued to make valuable contributions. Our own contributions, in the form of two new algorithms for applying sparse sampling in MRI, are then presented.

14.4.1 Review of Results from the Literature

Prior to the introduction of compressed sensing, exploiting signal sparseness by utilizing the l_1 norm constraint started in mid-1980s when Santosa, Symes, and Raggio [17] utilized an l_1 norm to recover a sparse series of spikes from the aliased representation that resulted from sub-Nyquist sampling in the Fourier domain. A similar experiment was implemented by Donoho [18] using soft thresholding, where the individuals in a sparse series of spikes were recovered sequentially in the order of the descending magnitude: the strongest component was first recovered and its aliasing effects were then removed to reduce the overall aliasing artifacts to allow the next strongest component to be recovered, and so on. These simple numerical experiments in fact have the same nature as the application of the modern compressed sensing technique in contrast-enhanced magnetic resonance angiography (CE-MRA). In CE-MRA, the contrast-enhanced regions to be recovered can be usefully approximated as isolated regions residing within a 2D plane, and hence the use of simple l_1 norm suffices in recovering the contrast-enhanced angiogram.

Another application of the l_1 norm before compressed sensing is in the use of TV filter [19], which imposes a l_1 norm in gradient magnitude images (GMI), or the gradient of the image. As discussed previously, l_1 norm promotes the strong components while penalizing weak components. In the operations on the GMI, the TV operator suppresses small gradient coefficients, whereas it preserves large gradient coefficients. The former are considered as noise to be removed, whereas the latter are considered to be part of the image features (edges) that need to be retained; hence, TV can serve as an edge-preserving denoising tool. TV itself can be employed as a powerful constraint for recovering images from undersampled data sets. In [20], TV is employed to recover MR images from undersampled radial trajectory measurements; Sidky and Pan [21] investigated the use of TV in recovering computed tomography images from limited number of projection angles.

The formal introduction of compressed sensing into MRI methods was made by Lustig, Donoho, and Pauly in 2007 [13]. Their key contribution is the explicit use of a different transform domain for appropriate application of the l_1 norm. Both the sparse set of spikes and the TV filter mentioned previously are special instances of the general transform-based compressed sensing setup. The authors identified the use of DWT and DCT as suitable transform bases for application in MR images, as evidenced by their sparse representation under DWT and DCT. A reconstruction framework was given, which converts the CS formulation into a convex optimization problem and hence allows for computational efficiency. The authors also spelt out that a key requirement in data measurement for successful compressed sensing recovery is to achieve incoherent aliasing. In MRI, such a requirement can be satisfied by employing a pseudorandom data acquisition pattern on Cartesian *k*-space grid. The idea of compressed sensing has received much attention thereafter and has been extended to the use of non-Cartesian trajectories such as radial [20] and spiral [22].

Another place where a high level of sparseness exists is the temporal domain of dynamic MRI. Not only does each individual temporal frame have inherent sparseness under the appropriate transform, and hence allows undersampling using compressed sensing, there also exists a high level of redundancy in the time domain due to the generally slow object variation over time. Such redundancy can be exploited in terms of its sparseness under appropriate transform to allow undersampling in the temporal domain. In fact, this idea has been well exploited prior to the use of compressed sensing: Fourier-encoding the overlaps using the temporal dimension (UNFOLD) [23], k-t broad-use linear acquisition speed-up technique (k-t BLAST) [24] and k-t focal underdetermined system solver (k-t FOCUSS, [25]) are all good examples of this. They all include an additional temporal frequency dimension in addition to the spatial frequency dimensions of the data set, and utilize the sparseness in the resulting spatial–temporal frequency domain. In fact, k-t FOCUSS was later shown to be a general implementation of compressed sensing by inherently utilizing the l_1 norm constraint. In [26], compressed sensing is explicitly applied in the additional temporal dimension to in MR cardiac imaging, and it was shown to outperform the k-t BLAST method.

In general, the application of compressed sensing to MRI is a recently emerged and fast developing area. The use of the l_1 norm in the appropriate domain, which is the core of compressed sensing, can be used as a general regularization constraint in many cases. One good example is in the conjoint use of parallel imaging, which normally faces the penalty of reduced SNR with reduced sampling density. The l_1 norm regularization intrinsically suppresses noise and hence offers complementary characteristics to those of parallel imaging [27].

14.4.2 Results from Our Work

Our own work in applying sparse sampling in MRI has led to the development of two new algorithms: Prior estimate-based compressed sensing (PECS) and sensitivity encoding with compressed sensing (SENSECS). PECS has been demonstrated in both brain imaging, that is imaging of a static structure, and in contrast-enhanced angiography, that is dynamic imaging as part of a pilot study on normal volunteers [28, 29]. SENSECS has been demonstrated in brain imaging [27]. We briefly summarize the work in the remainder of this section.

14.4.2.1 PECS

The success of compressed sensing is determined by the sparsity of the underlying signal. In our experience to date, the sparsest representation of a typical MRI

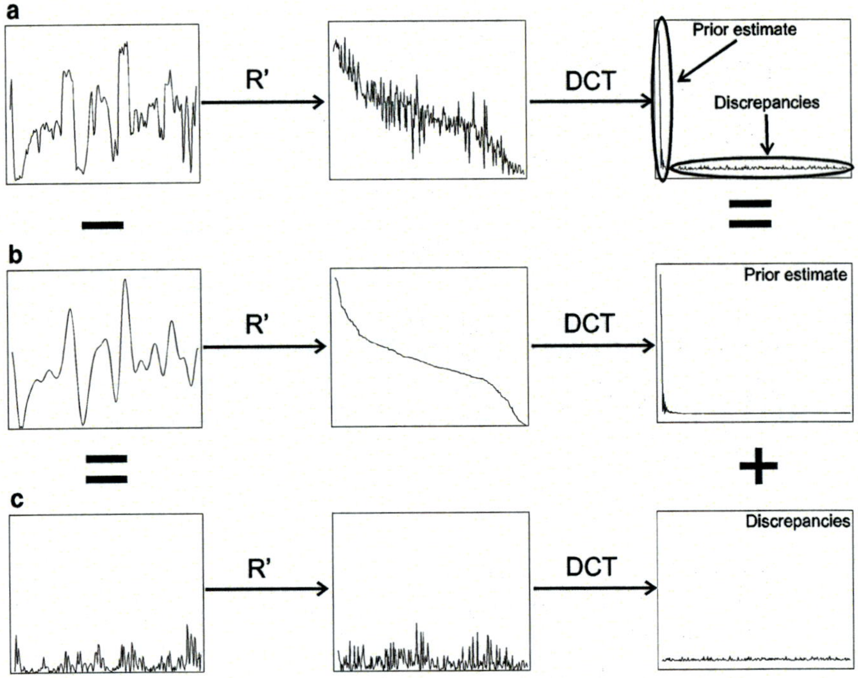

Fig. 14.6 Presentation of an argument for the approximate ordering process incorporating prior knowledge into PECS recovery. Figure (**b**) depicts the computation of an approximate data order from a low-resolution version of the signal. Figure (**a**) depicts the use of that sampling order on the original signal. Figure (**c**) depicts the effect of sorting on the discrepancies between the original signal and the prior knowledge. After transformation, the large amplitude coefficients retain information about the sorting prior, while the smaller coefficients are mainly associated with the discrepancies between the original and low-resolution signals

anatomical image is obtained by ordering the set of pixel (or voxel) amplitudes as described in Sect. 14.2.3. Assume that a set of undersampled k-space data have been collected with a particular MRI sequence with the purpose of forming a high-resolution image. In addition, a prior estimate is available of the image, for example a low-resolution image. In PECS, the prior estimate of the image is first used to derive a data ordering, R'; compressed sensing is used to recover an image from the measured k-space data, incorporating the approximate ordering (to promote sparsity) and a TV minimization (to promote piecewise smoothness in the resulting image). The process is illustrated in Fig. 14.6 for the case of a 1D signal. Recall from Sect. 14.2.3 that ordering the amplitudes of a signal into a monotonic function allows that signal to be represented by fewer coefficients under an appropriate transform (DCT or DWT, say). In this case the approximate ordering, R', is derived from a low resolution (low pass filtered) version of the signal. When R' is applied to the high resolution data, the result is a highly noisy signal which only approximates in form to a monotonic function. Under the transform the largest coefficients

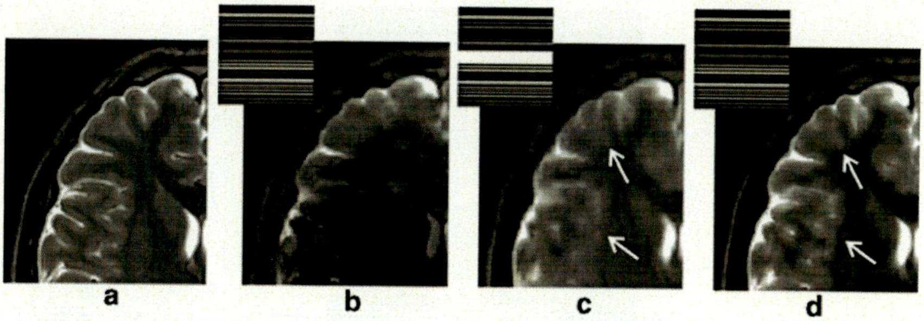

Fig. 14.7 Reconstructions comparing PECS with other methods: (**a**) *top left* quadrant of a reconstruction of an axial brain slice with full *k*-space sampling; (**b–d**) reconstructions at an acceleration factor of 4 (the sampling patterns used are shown in the *insets*). (**b**) CS with uniformly distributed randomly selected *k*-space samples; (**c**) CS with randomly selected *k*-space samples, but including the center 32 lines; and (**d**) PECS with ordering derived from a low-resolution approximation using just the center 32 lines of *k*-space. The *arrows* indicate areas where (**d**) shows better recovery than (**c**)

retain the form, while the errors tend to generate a noise-like spectrum with low amplitudes spread across many coefficients. After thresholding only the significant coefficients are nonzero and these retain the form of the true ordering. We argue that prior knowledge about the signal is thereby introduced by the application of the approximate data ordering [27].

A result for PECS is shown in Fig. 14.7. A 1.5T GE scanner equipped with an eight-channel head coil was used to obtain a 2D T2-weighted axial brain slice of a healthy adult volunteer. A fully sampled *k*-space data set (256×256) was obtained and then various forms of sampling patterns were applied in post processing to simulate the under-sampling required [29]. Note that in this case the under-sampling is applied in the single phase encoding direction (anterior–posterior). In Fig. 14.7a is shown the reconstruction for the slice utilizing the fully sampled *k*-space data; the top left quadrant is selected for more detailed study. To the right are three different reconstructions obtained from only one quarter of the *k*-space data, simulating an acceleration factor of 4 for the imaging process. In Fig. 14.7b, the reconstruction is by CS with a uniform sampling density, while in Fig. 14.7c an otherwise similar sampling pattern is altered to make the center 32 lines of *k*-space fully sampled. The improvement in (c) compared to (b) is obvious. A PECS reconstruction is shown in Fig. 14.7d, with the prior estimate used to generate R' being a low-resolution image formed from the center 32 *k*-space lines. The arrows indicate particular areas, where PECS in (d) performs better that CS in (c).

14.4.2.2 SENSECS

As the name implies, SENSECS combines SENSE with CS. A regular sampling pattern is employed which promotes the performance of SENSE, except that several

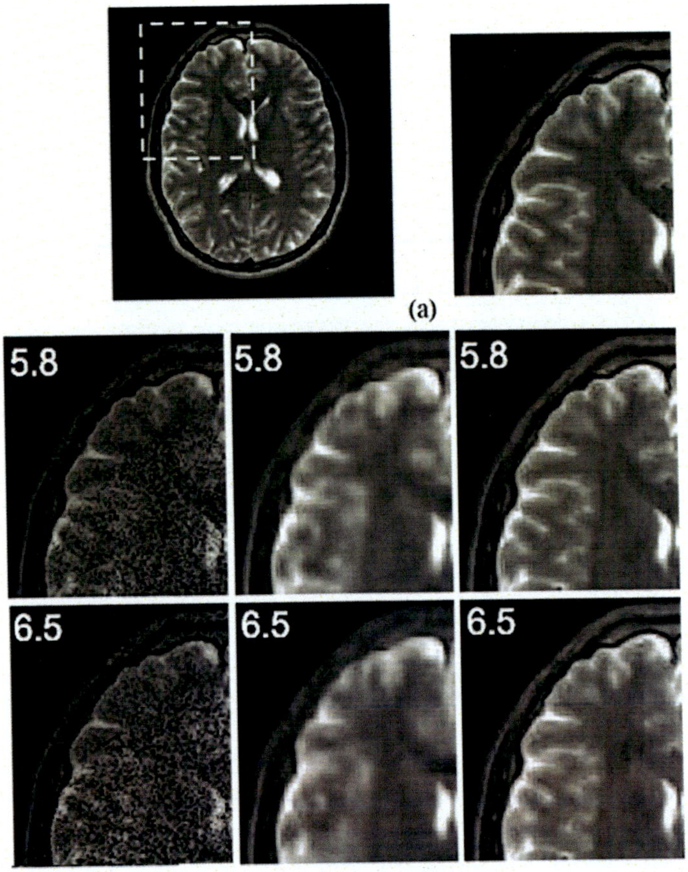

Fig. 14.8 Reconstructions comparing SENSECS with other methods at high acceleration factors (shown in the *top left* of the reconstructions): (**a**) reconstruction of an axial brain slice with full k-space sampling; in the lower section of the figure, the *left column* is SENSE, the center column is CS, and the right column is SENSECS. Note that the SENSECS reconstructions use the SENSE reconstructions to derive the data sorting order

additional lines are sampled at the center of k-space. SENSE is applied to achieve an intermediate reconstruction [11]. Because of the high acceleration factor being used, the image is likely to be quite noisy. The noise is in part at least due to the imperfect knowledge available of the sensitivities of the individual receiver coils. An approximate sorting order R' is derived from the intermediate SENSE-derived image. Then PECS is performed with the same set of k-space data and using R'.

Again, a single set of results is shown for the new algorithm in Fig. 14.8. The same set of data as described above in this section was used and the fully sampled k-space data was again undersampled in postprocessing [29]. High acceleration factors of 5.8 and 6.5 were simulated. In Fig. 14.8a is shown the reconstruction for the slice utilizing the fully sampled k-space data; the top left quadrant is selected for

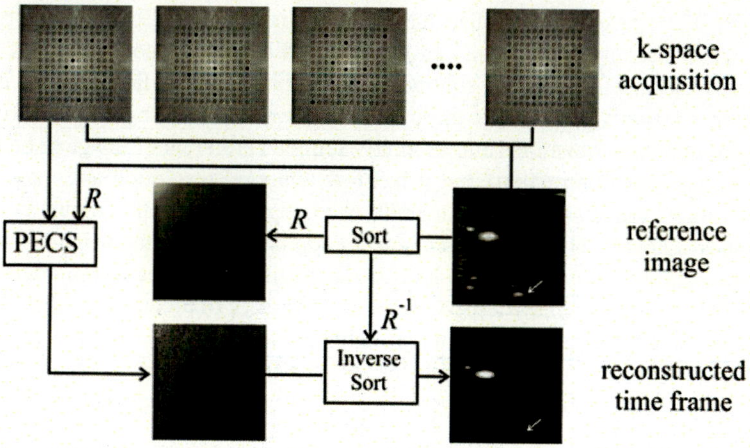

Fig. 14.9 The application of PECS to CE-MRA is illustrated. Samples in *k*-space samples are acquired progressively by means of different randomly selected subsets (*top row*). A combination of a set of the acquisitions is used to achieve a "reference" image which is sorted to derive R. PECS is applied to each of the (high acceleration factor) subsets using R. The result (*bottom right*) is a sequence of relatively high time resolution images. The *arrow* indicates a region in one of the output frames, which does not show an artifact which appears in the reference image

more detailed study. In each of the two lower rows are three different reconstructions obtained from a fraction of the *k*-space data, simulating the acceleration factors indicated. In the left column are shown the SENSE reconstructions, which are clearly noisy and unlikely to be diagnostically useful. CS reconstructions are shown in the center column; in this case, the sampling pattern employed was designed specifically for CS. The results are superior to the SENSE reconstructions, but somewhat blurred in appearance. SENSECS reconstructions are shown in the right column; they show better fidelity than the other reconstructions and diagnostically useful results up to at least an acceleration factor of 6.5. Note that the images obtained by SENSE were used to derive the sorting order here.

14.4.2.3 PECS Applied to CE-MRA

The PECS method has been extended to enable it to be applied to CE-MRA. There is a strong desire in CE-MRA to increase the temporal resolution, that is to increase the acceleration factor in image acquisition and reconstruction. The algorithm is depicted in Fig. 14.9. Note that the depiction is for 3D imaging, so the sampling patterns depicted correspond to the two phase encoding directions. In the first acquisition, a small number of *k*-space samples (corresponding to a high acceleration factor) are acquired at pseudo-random locations; a second acquisition takes the same number of samples, again randomly distributed, but at a different subset of *k*-space locations; the acquisitions proceed in this manner until all of k-space is filled, then the sequence is repeated. To improve sparsity, data for a "background" image can be collected before the contrast agent is injected, and those sample values subtracted from the corresponding samples collected as the contrast agent moves through the vessels. Combining the data from several contiguous acquisitions allows a fully sampled "reference" image to be made (after background subtraction), but it has low temporal resolution and may suffer from artifacts caused by the dynamically changing nature of the volume being imaged. It is adequate, however, to generate an approximate sorting order R. That ordering is used in applying PECS to individual acquisition frames (again after background subtraction) to achieve a high frame rate sequence of reconstructions. The arrows in the sample images lower right highlight how features, which appear in the relatively artifact filled reference image. Results with the method have been encouraging up to acceleration factors of 12 (for an eight-channel receiver coil system) [29].

14.5 Prospects for Future Developments

We have presented some preliminary and very encouraging results for incorporating a data ordering step in the recovery of MR images by compressed sensing. There remains considerable scope for putting this nonlinear processing on a firm theoretical footing. Candès and others have provided such rigor to the basic compressed sensing recovery of certain classes of image [3, 4], but no such attention has to our knowledge been directed at the data ordering and its use in incorporating prior knowledge.

We have demonstrated the exploitation of several forms of sparsity above. Briefly, this includes the sparsity achieved by ordering the image into a monotonic function, the use of a compressive transform such as DCT or DWT, and the subtraction of the contribution to signal from static structures in dynamic CE-MRA. Other authors have likewise exploited piecewise homogeneity. Given that CS is relatively new as a practical method in signal processing, it seems likely that other transforms may be available, or as yet undiscovered, which may allow more gains to be made. Our work and the work of many others in the area of applying sparse sampling in MRI suggests that it has a bright future and we should see the manufacturers of MRI scanning systems incorporating some of the algorithms based on sparse sampling soon.

References

1. McRobbie, D.W., Moore, E.A., Graves, M.J: MRI from Picture to Proton. Cambridge University Press, Cambridge (2003)
2. Bernstein, M.A., King, K.F., Zhou, X.J.: Handbook of MRI Pulse Sequences. Elsevier Academic, Amsterdam (2004)

3. Candes, E., Romberg, J., Tao, T.: Robust uncertainty principles: exact signal reconstruction from highly incomplete frequency information. IEEE Trans. Inf. Theory **52**, 489–509 (2006)
4. Baraniuk, B.: Compressive sensing. IEEE Signal Process. Mag. **24**, 118–121 (2007)
5. Geman, S., Geman, D.: Stochastic relaxation, Gibbs distributions, and the Bayesian restoration of images. IEEE Trans. Pat. Anal. Mach. Intel. **6**, 721–741 (1983)
6. Hu, X., Johnson, V., Wong, W.H., Chen, C-T.: Bayesian image processing in magnetic resonance imaging. Magn. Res. Imaging **9**, 611–620 (1991)
7. Strang, G.: The discrete cosine transform. SIAM Rev. **41**, 135–147 (1999)
8. Velho, L., Frery, A., Gomes, J.: Image Processing for Computer Graphics and Vision, Ch. 9: Multiscale Analysis and Wavelets. Springer, London (2008)
9. Adluru, G., DiBella, E.V.R.: Reordering for improved constrained reconstruction from undersampled k-space data. Int. J. Biomed. Imaging **2008**, 341684 (2008)
10. Wu, B., Millane, R.P., Watts, R., Bones P.J.: Applying compressed sensing in parallel MRI. In: Proc. 16th Ann. Meet. ISMRM, Toronto, p. 1480 (2008)
11. Pruessmann, K.P., Weiger, M., Scheidegger, M.B., Boesiger, P.: SENSE: sensitivity encoding for fast MRI. Magn. Reson. Med. **42**, 952–962 (1999)
12. Donoho, D.L.: Compressed sensing. IEEE Trans. Inf. Theory **52**, 1289–1306 (2006)
13. Lustig, M., Donoho, D., Pauly, J.M.: Sparse MRI: the application of compressed sensing for rapid MR imaging. Magn. Reson. Med. **58**, 1182–1195 (2007)
14. Jain, A.K.: Fundamentals of Digital Image Processing. Prentice-Hall, Englewood Cliffs, NJ (1989)
15. Candes, E., Romberg, J.: Sparsity and incoherence in compressive sampling. Inverse Probl. **23**, 969–985 (2007)
16. Candes, E., Wakin, M.B.: An introduction to comprehensive sampling. IEEE Signal Process. Mag. **25**, 21–30 (2008)
17. Santosa, F., Symes, W.W., Raggio, G.: Inversion of band-limited reflection seismograms using stacking velocities as constraints. Inverse Probl. **3**, 477–499 (1987)
18. Donoho, D.L.: De-noising by soft-thresholding. IEEE Trans. Inf. Theory **41**, 614–627 (1995)
19. Rudin, L.I., Osher, S., Fatemi, E.: Nonlinear total variation based noise removal algorithms. Physica D **60**, 259–268 (1992)
20. Block, K.T., Uecker, M., Frahm, J.: Undersampled radial MRI with multiple coils. Iterative image reconstruction using a total variation constraint. Magn. Reson. Med. **57**, 1086–1098 (2007)
21. Sidky, E.Y., Pan, X.: Image reconstruction in circular cone-beam computed tomography by constrained, total-variation minimization. Phys. Med. Biol. **53**, 47–77 (2008)
22. Seeger, M., Nickisch, H., Pohmann, R., Schölkopf, B.: Optimization of k-space trajectories for compressed sensing by Bayesian experimental design. Magn. Reson. Med. **63**, 116–126 (2010)
23. Madore, B., Glover, G.H., Pelc, N.J.: Unaliasing by fourier-encoding the overlaps using the temporal dimension (UNFOLD), applied to cardiac imaging and fMRI. Magn. Reson. Med. **42**, 813–828 (1999)
24. Taso, J., Boesiger, P., Pruessmann, K.P.: k-t BLAST and k-t SENSE: dynamic MRI with high frame rate exploiting spatiotemporal correlations. Magn. Reson. Med. 50, 1031–1042 (2003)
25. Jung, H., Sung, K., Nayak, K.S., Kim, E.Y., Ye, J.C.: k-t FOCUSS: a general compressed sensing framework for high resolution dynamic MRI. Magn. Reson. Med. **61**, 103–116 (2009)
26. Gamper, U., Boesiger, P., Kozerke, S.: Compressed sensing in dynamic MRI. Magn. Reson. Med. **50**, 1031–1042 (2003)
27. Wu, B., Millane, R.P., Watts, R., Bones, P.J.: Prior estimate-based compressed sensing in parallel MRI. Magn. Reson. Med. **65**, 83–95 (2011)
28. Wu, B., Bones, P.J., Millane, R.P., Watts, R.: Prior estimated based compressed sensing in contrast enhanced MRA. In: Proc. 18th Ann. Meet., ISMRM, Stockholm (2010)
29. Wu, B.: Exploiting Data Sparsity in Parallel Magnetic Resonance Imaging. PhD thesis, University of Canterbury, Christchurch, New Zealand, 2009

Chapter 15
Digital Processing of Diffusion-Tensor Images of Avascular Tissues

Konstantin I. Momot, James M. Pope, and R. Mark Wellard

15.1 Introduction

Diffusion is the process that leads to the mixing of substances as a result of spontaneous and random thermal motion of individual atoms and molecules. It was first detected by the English botanist Robert Brown in 1827, and the phenomenon became known as 'Brownian motion'. More specifically, the motion observed by Brown was *translational diffusion* – thermal motion resulting in random variations of the position of a molecule. This type of motion was given a correct theoretical interpretation in 1905 by Albert Einstein, who derived the relationship between temperature, the viscosity of the medium, the size of the diffusing molecule, and its *diffusion coefficient* [1]. It is translational diffusion that is indirectly observed in MR diffusion-tensor imaging (DTI). The relationship obtained by Einstein provides the physical basis for using translational diffusion to probe the microscopic environment surrounding the molecule.

In living systems, translational diffusion is vital for the transport of water and metabolites both into and around cells. In the presence of a *concentration gradient*, diffusion results in the mixing of substances: The molecules of a compound on average tend to move from areas of high concentration into areas of low concentration, resulting in a net transport of the compound in the direction of the gradient. A classic example of this is the spontaneous mixing of a dyestuff into a stationary solvent.

Diffusive mass transport can serve as the basis for the measurement of molecular diffusion: a concentration gradient is artificially created, and its equilibration with time observed (Fig. 15.1). This method of measuring diffusion is not always physically relevant because a concentration gradient is neither required for diffusion

K.I. Momot (✉)
Queensland University of Technology, Brisbane, Australia
e-mail: k.momot@qut.edu.au

G. Dougherty (ed.), *Medical Image Processing: Techniques and Applications*, Biological and Medical Physics, Biomedical Engineering, DOI 10.1007/978-1-4419-9779-1_15,

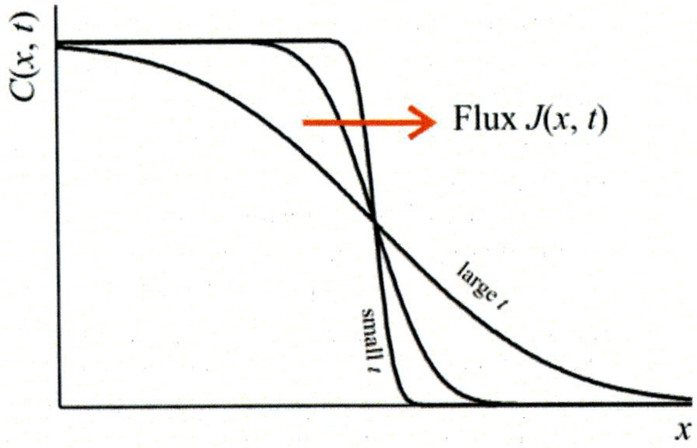

Fig. 15.1 Diffusion in the presence of a concentration gradient $C(x,t)$ gives rise to a net flux or flow of particles $J(x,t)$ from high to low concentration

nor always present. The majority of DTI applications are based on the diffusion of water, whose concentration is essentially uniform in extracellular and intracellular microenvironments of living organisms. Diffusion of molecules of the same substance in the absence of a concentration gradient is known as '*self-diffusion*'. It is self-diffusion that is observed in DTI. Self-diffusion can be measured by the technique of Pulsed Field Gradient Nuclear Magnetic Resonance (PFG-NMR), which is exquisitely sensitive to the microstructural environment of nuclear spins. (Other examples of applications of magnetic resonance to tissues can be seen in Chapters 5, 9, and 10.) In recent years, PFG-NMR has been increasingly combined with magnetic resonance imaging (MRI) to study diffusion of water protons in biological tissues for diagnosis of stroke and multiple sclerosis, for white matter fiber tracking in the brain, muscle fiber tracking, and other applications.

While no concentration gradient is necessary for DTI, the notion of a concentration gradient is instructive for understanding how DTI works. In an isotropic medium such as bulk water, the process of diffusion is itself isotropic and can be described by a scalar diffusion coefficient D. If we were to "label" a subset of molecules, the flux of the labeled molecules would be governed by Fick's first law of diffusion:

$$\mathbf{J}(\mathbf{r},t) = -D\,\nabla C(\mathbf{r},t) \equiv -D\left(\mathbf{i}\frac{\partial C}{\partial x} + \mathbf{j}\frac{\partial C}{\partial y} + \mathbf{k}\frac{\partial C}{\partial z}\right). \tag{15.1}$$

Here, $C(\mathbf{r},t)$ is the spatial concentration profile of the labeled molecules; D is the diffusion coefficient; and $\mathbf{J}$ is the flux of particles, defined as the amount of substance that flows through a unit area per unit time. The meaning of (15.1) is that in isotropic media the flux occurs strictly in the direction of the concentration gradient. Combining (15.1) with the conservation of mass and the assumption that D is independent of concentration yields Fick's second law of diffusion or the diffusion equation:

$$\frac{\partial\, C(\mathbf{r},t)}{\partial t} = D\,\nabla^2 C(\mathbf{r},t) \equiv D\left(\frac{\partial^2 C}{\partial x^2} + \frac{\partial^2 C}{\partial y^2} + \frac{\partial^2 C}{\partial z^2}\right). \tag{15.2}$$

Diffusion in biological tissues is substantially different from isotropic diffusion. Tissues are intrinsically heterogeneous: there are barriers to free diffusion of water molecules arising from the presence of macromolecules, organelles, cell membranes, and larger scale structures. As a result, diffusion of water molecules in many tissues is both *restricted* and *anisotropic*.

Restricted diffusion results in measurements of the diffusion coefficient giving results that are dependent on the *timescale* of the diffusion interval Δ over which the measurement is performed. This is known as an 'apparent diffusion coefficient' (ADC). Besides Δ, the ADC is dependent on the nature and the length scale of the obstructions and is generally smaller than the self-diffusion coefficient of bulk water ($D_0 = 2.3 \cdot 10^{-9}\,\mathrm{m^2\,s^{-1}}$ at 25°C). For example, the ADC of water confined between parallel, semipermeable barriers approximately equals D_0 at $\Delta << d^2/D_0$, where d is the separation between the barriers, but decreases to $D_0/(1+1/P)$ at $\Delta >> d^2/D_0$ (where P is the permeability of the barriers) [2].

Anisotropic diffusion means that the diffusing molecules encounter *less restriction in some directions than others*. Diffusion can be anisotropic when the tissue possesses some form of global alignment. Two well-known examples of anisotropic tissues are the white matter of the brain and the heart muscle. In muscles, the global alignment arises from the elongated form of the muscle cells forming muscle fibers. In white matter, the anisotropy arises from the fact that nerve fiber tracts follow specific pathways. In both these cases, the cellular structures preferentially restrict the diffusion of water in the direction perpendicular to the fibers. Diffusion is also anisotropic in the two tissues that are the focus of this chapter: articular cartilage (AC) and the eye lens. In AC, the anisotropic restrictions to diffusion are imposed by the aligned collagen fibers that form the biomacromolecular "scaffold" of the tissue. In the crystalline eye lens, the restrictions are imposed by the fiber cells.

To take account of anisotropic diffusion, a common approach is to re-write the diffusion equation in terms of a *diffusion tensor*:

$$\mathbf{J}(\mathbf{r},t) = -\mathbf{D} \cdot \nabla C(\mathbf{r},t), \tag{15.3}$$

where the diffusion tensor $\mathbf{D}$ is represented by a symmetric and real 3×3 matrix:

$$\mathbf{D} = \begin{pmatrix} D_{xx} & D_{xy} & D_{xz} \\ D_{xy} & D_{yy} & D_{yz} \\ D_{xz} & D_{yz} & D_{zz} \end{pmatrix}. \tag{15.4}$$

In the anisotropic case, Fick's second law becomes:

$$\frac{\partial C}{\partial t} = \nabla \cdot \mathbf{D} \cdot \nabla C \equiv \begin{pmatrix} \frac{\partial}{\partial x} & \frac{\partial}{\partial y} & \frac{\partial}{\partial z} \end{pmatrix} \begin{pmatrix} D_{xx} & D_{xy} & D_{xz} \\ D_{xy} & D_{yy} & D_{yz} \\ D_{xz} & D_{yz} & D_{zz} \end{pmatrix} \begin{pmatrix} \partial/\partial x \\ \partial/\partial y \\ \partial/\partial z \end{pmatrix} C. \tag{15.5}$$

Note that while the diagonal elements of the diffusion tensor (DT) scale concentration gradients and fluxes that are in the same direction, the off-diagonal

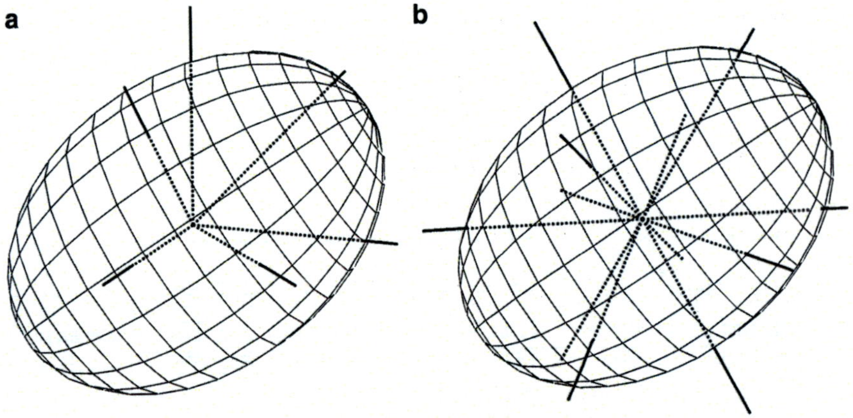

Fig. 15.2 Diffusion ellipsoid as a visual representation of the diffusion tensor. The straight lines radiating from the center of the ellipsoid illustrate two possible choices of the diffusion sampling directions, as discussed in Sects. 15.2.2 and 15.2.5

elements couple fluxes and concentration gradients in orthogonal directions. This is because in the anisotropic case the distribution of diffusional displacements of molecules tends to follow the geometry of the restricting barriers. This is the physical basis for using DTI to measure the microscopic morphology of the tissue. In Sects. 15.2.4 and 15.4, we discuss applications of DTI to the eye lens and AC, respectively, as examples.

A convenient way of representing the DT is the diffusion ellipsoid, which is illustrated in Fig. 15.2. The shape of the ellipsoid represents the directional asymmetry of the average displacements of the diffusing molecules. The directions of the principal axes of the ellipsoid characterize the orientation of the DT, which in turn represents the spatial anisotropy of the restricting barriers imposed by the tissue.

In the isotropic case, the DT is a diagonal matrix:

$$\mathbf{D} = \begin{pmatrix} D & 0 & 0 \\ 0 & D & 0 \\ 0 & 0 & D \end{pmatrix}, \tag{15.6}$$

where D is the isotropic diffusion coefficient. In this case, (15.5) reverts to (15.2), and the ellipsoid in Fig. 15.2 becomes a sphere.

15.2 Acquisition of DT Images

15.2.1 Fundamentals of DTI

DT images can be obtained using Nuclear Magnetic Resonance (NMR). NMR measures the frequency of precession of nuclear spins such as that of the proton (^{1}H), which in a magnetic field $\mathbf{B}_0$, is given by the Larmor equation:

$$\omega_0 = \gamma B_0. \tag{15.7}$$

The key to achieving spatial resolution in MRI is the application of time-dependent magnetic field gradients that are superimposed on the (ideally uniform) static magnetic field $\mathbf{B}_0$. In practice, the gradients are applied via a set of dedicated 3-axis gradient coils, each of which is capable of applying a gradient in one of the orthogonal directions (x, y, and z). Thus, in the presence of a magnetic field gradient $\mathbf{g}$,

$$\mathbf{g} = \left(\frac{\partial B_z}{\partial x}, \frac{\partial B_z}{\partial y}, \frac{\partial B_z}{\partial z}\right) \tag{15.8}$$

the magnetic field strength, and hence the precession frequency become position dependent. The strength of the magnetic field experienced by a spin at position $\mathbf{r}$ is given by:

$$B = B_0 + \mathbf{g} \cdot \mathbf{r} \tag{15.9}$$

The corresponding Larmor precession frequency is changed by the contribution from the gradient:

$$\omega(\mathbf{r}) = \frac{\partial \phi(\mathbf{r})}{\partial t} = \gamma(B_0 + \mathbf{g} \cdot \mathbf{r}). \tag{15.10}$$

The precession frequency ω is the rate of change of the *phase*, ϕ, of a spin – that is, its precession angle in the transverse plane (Fig. 15.3). Therefore, the time-dependent phase ϕ is the integral of the precession frequency over time. In MRI, we switch gradients on and off in different directions to provide spatial resolution, so the gradients are *time dependent* and the phase of a spin is given by:

$$\phi(\mathbf{r},t) = \int_0^t \omega(\mathbf{r},t')\mathrm{d}t' = \gamma B_0 t + \gamma \int_0^t \mathbf{g}(t') \cdot \mathbf{r}\mathrm{d}t'. \tag{15.11}$$

We observe the phase relative to the reference frequency given by (15.7). For example if the gradient is applied in the x direction in the form of a rectangular pulse of amplitude g_x and duration δ the additional phase produced by the gradients is

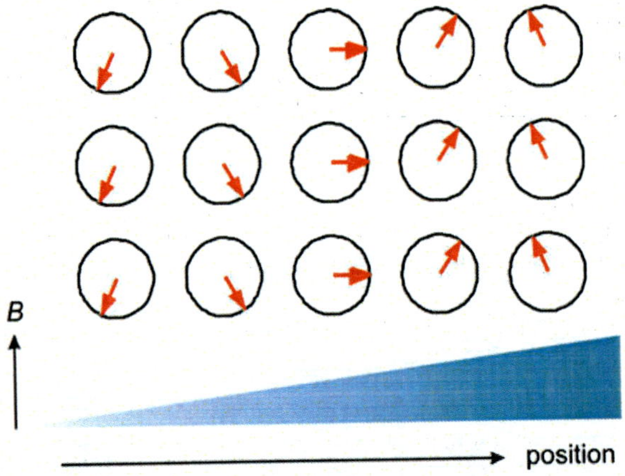

Fig. 15.3 The effect of a magnetic field gradient on precession of spins. A constant magnetic field gradient **g** (illustrated by the blue ramp) applied in some arbitrary direction changes the magnetic field at position **r** from $\mathbf{B}_0$ to a new value $\mathbf{B} = \mathbf{B}_0 + \mathbf{g} \cdot \mathbf{r}$. The gradient perturbs the precession of the spins, giving rise to an additional position-dependent phase ϕ', which may be positive or negative depending on whether the magnetic field produced by the gradient coils strengthens or weakens the static magnetic field B_0

$$\phi'(\mathbf{r},t) = \gamma \int_0^{\delta} g_x(t')x\mathrm{d}t' = \gamma\delta g_x x = 2\pi k_x x, \tag{15.12}$$

where the "spatial frequency" $k_x = \gamma\delta g_x/2\pi$ is also known as the "k value". It plays an important role in the description of spatial encoding in MRI and can be thought of as the frequency of spatial harmonic functions used to encode the image.

In MRI to achieve spatial resolution in the plane of the selected slice (x, y), we apply gradients in both x and y directions sequentially. The NMR signal is then sampled for a range of values of the corresponding spatial frequencies k_x and k_y.

For one of these gradients (g_x, say), this is achieved by keeping the amplitude fixed and incrementing the time t at which the signal is recorded (the process called 'frequency encoding').

In the case of the orthogonal gradient (g_y), the amplitude of the gradient is stepped through an appropriate series of values. For this gradient, the appropriate spatial frequency can be written:

$$k_y = \gamma \int_0^{\delta} g_y(t')\mathrm{d}t' = \gamma\delta g_y/2\pi. \tag{15.13}$$

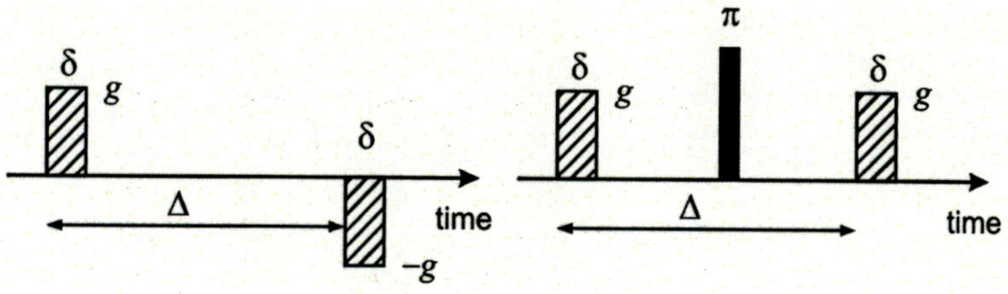

Fig. 15.4 Gradient pulse pairs used for diffusion attenuation. The first gradient sensitizes the magnetisation of the sample to diffusional displacement by winding a magnetization helix. The second gradient rewinds the helix and thus enables the measurement of the diffusion-attenuated signal. The two gradients must have the same amplitude if they are accompanied by the refocusing RF π pulse; otherwise, their amplitudes must be opposite

The MR image is then generated from the resulting two-dimensional data set $S(k_x, k_y)$ by Fourier transformation:

$$S(x,y) = \int\int S(k_x, k_y)\mathrm{e}^{-2\pi i(k_x x + k_y y)}\mathrm{d}k_x\mathrm{d}k_y. \tag{15.14}$$

The Fourier transform relationship between an MR image and the raw NMR data is analogous to that between an object and its diffraction pattern.

15.2.2 *The Pulsed Field Gradient Spin Echo (PFGSE) Method*

Consider the effect of a gradient pair consisting of two consecutive gradient pulses of opposite sign shown in Fig. 15.4 (or alternatively two pulses of the same sign separated by the 180° RF pulse in a 'spin echo' sequence).

It is easy to show that spins moving with velocity **v** acquire a net phase shift (relative to stationary spins) that is *independent of their starting location* and given by:

$$\phi(\mathbf{v}) = -\gamma \mathbf{g} \cdot \mathbf{v}\delta\Delta, \tag{15.15}$$

where δ is the duration of each gradient in the pair and Δ is the separation of the gradients. Random motion of the spins gives rise to a *phase dispersion* and attenuation of the spin echo NMR signal.

Stejskal and Tanner [3] showed in the 1960s that, for a spin echo sequence this additional attenuation (Fig. 15.5) takes the form:

$$S(\Delta, g) = S_0\mathrm{e}^{-TE/T_2}\mathrm{e}^{-D\gamma^2 g^2 \delta^2(\Delta-\delta/3)}. \tag{15.16}$$

The first term is the normal echo attenuation due to transverse (spin-spin) relaxation. By stepping out the echo time *TE*, we can measure T_2.

The second term is the diffusion term. By incrementing the amplitude of the magnetic field gradient pulses (g), we can measure the self-diffusion coefficient D.

For a fixed echo time TE, we write:

$$S = S_0'\mathrm{e}^{-bD} = S_0\mathrm{e}^{-TE/T_2}\mathrm{e}^{-bD}, \tag{15.17}$$

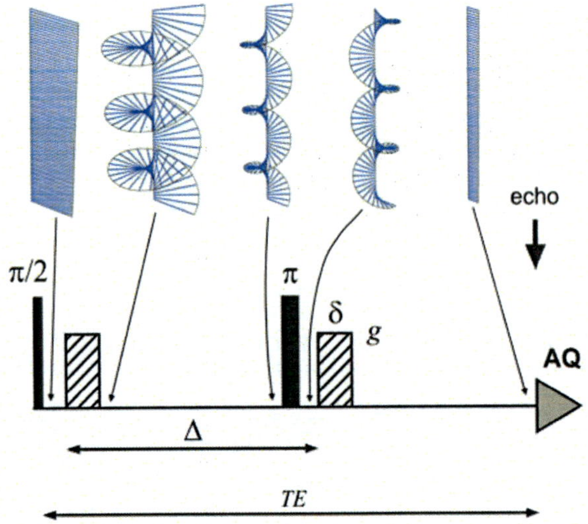

Fig. 15.5 A pulsed field gradient spin echo (PGSE) sequence showing the effects of diffusive attenuation on spin echo amplitude

where

$$b = \gamma^2 g^2 \delta^2 \left(\Delta - \frac{\delta}{3}\right). \tag{15.18}$$

The ADC is then given by:

$$ADC = -\ln\left(\frac{S}{S'_0}\right) \Big/ b \tag{15.19}$$

For the case of anisotropic diffusion described by a diffusion tensor **D**, the expression for the echo attenuation in a PFG spin echo experiment becomes:

$$S(\Delta, g) = S'_0 \mathrm{e}^{-\gamma^2 \mathbf{g}\cdot\mathbf{D}\cdot\mathbf{g}\delta^2(\Delta-\delta/3)}, \tag{15.20}$$

where $\mathbf{g} = (g_x, g_y, g_z)$ is the gradient vector, and the scalar product $\mathbf{g}\cdot\mathbf{D}\cdot\mathbf{g}$ is defined analogously to (15.5).

Overall, if diffusion is anisotropic, the echo attenuation will have an *orientation dependence* with respect to the measuring gradient **g**. Gradients along the *x*, *y*, and *z* directions sample, respectively, the *diagonal* elements D_{xx}, D_{yy}, and D_{zz} of the DT. In order to sample the off-diagonal elements, we must apply gradients in *oblique directions* – that is combinations of g_x and g_y or g_y and g_z, etc. Because the DT is symmetric, there are just 6 independent elements. To fully determine the DT therefore requires a minimum of 7 separate measurements – for example:

$$\begin{pmatrix} g_x \\ g_y \\ g_z \end{pmatrix} = \begin{pmatrix} 0 \\ 0 \\ 0 \end{pmatrix}, \begin{pmatrix} g \\ 0 \\ 0 \end{pmatrix}, \begin{pmatrix} 0 \\ g \\ 0 \end{pmatrix}, \begin{pmatrix} 0 \\ 0 \\ g \end{pmatrix}, \frac{1}{\sqrt{2}}\begin{pmatrix} g \\ g \\ 0 \end{pmatrix}, \frac{1}{\sqrt{2}}\begin{pmatrix} g \\ 0 \\ g \end{pmatrix}, \frac{1}{\sqrt{2}}\begin{pmatrix} 0 \\ g \\ g \end{pmatrix}. \tag{15.21}$$

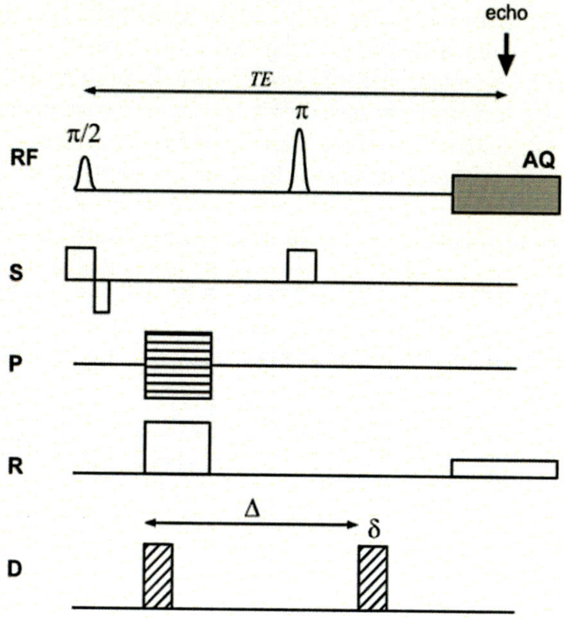

Fig. 15.6 Spin echo diffusion imaging pulse sequence. "RF" denotes the RF pulses and acquisition. Gradient pulses: *S*, slice selection; *P*, encoding in the Phase direction; *R*, encoding in the Read direction; *D*, diffusion gradients

This choice of diffusion gradient directions is illustrated in Fig. 15.2a. We shall refer to a data set measured with this set of gradients as the *minimal diffusion-tensor dataset*. As seen below, this is neither the only nor the best choice of DTI gradient directions. Other gradient combinations exist that achieve optimal signal-to-noise ratio (SNR) in the resulting DT images and/or optimal gradient amplifier efficiency (see Sect. 15.2.5). The first measurement with all gradients off is required to determine S'_0.

15.2.3 Diffusion Imaging Sequences

Diffusion gradients can readily be incorporated in a conventional spin echo MRI sequence as follows (Fig. 15.6).

The sequence is repeated for the appropriate different combinations of gradients g_x, g_y, and g_z to yield a set of 7 different diffusion weighted images. These are then used to calculate the elements of the DT, pixel by pixel, to yield 6 images representing the three diagonal elements and 3 off-diagonal elements of the DT. (Because of the symmetry of the DT, the off-diagonal elements are duplicated in the 3×3 DT image). Once obtained the DT must be diagonalized to obtain the eigenvalues and eigenvectors. For more details, see, for example Basser and Jones [4].

For a given DTI imaging sequence and available MRI hardware, the effects of T_2 relaxation can be minimized by making more efficient use of available gradient power to maximize b values and reduce the minimum echo time TE. For example by ensuring that gradients are applied simultaneously along two axes at the maximum amplitude for each individual axis, the resultant gradient amplitude is increased by a factor of $\sqrt{2}$, while by employing all three basic gradients in an icosahedral arrangement it is possible to increase the maximum amplitude by Fibonacci's golden ratio: $(1+\sqrt{5})/2$ (see e.g. [5] and references therein). This choice of diffusion gradient directions is illustrated in Fig. 15.2b.

For clinical applications of DTI, patient motion can be a serious problem because even relatively small bulk motions can obscure the effects of water diffusion on the NMR signal. In such applications, it is common to employ spin echo single shot echo planar imaging (SS-EPI) sequences that incorporate diffusion weighting in order to acquire an entire DWI data set in a fraction of a second (albeit at somewhat reduced spatial resolution when compared with more conventional spin echo imaging sequences). Such SS-EPI sequences also have the added advantage of a relatively high SNR per unit scanning time, allowing a complete DTI data set to be acquired in 1–2 min. Further improvements in acquisition time and/or SNR can be achieved by combining such sequences with parallel imaging techniques and/or partial Fourier encoding of k-space (see e.g. [6] and references therein).

15.2.4 Example: Anisotropic Diffusion of Water in the Eye Lens

We have used the PFGSE method to measure the components of the DT for water (H_2O) in human eye lenses [7]. In this case, we were measuring diffusion on a timescale of ~20 ms corresponding to diffusion lengths $\ell = \sqrt{2Dt} \cong 10\,\mu\text{m}$ with $D = 2.3 \cdot 10^{-9}\text{m}^2\text{s}^{-1}$ for bulk water at 20°C and $t = 20\,\text{ms}$. This is comparable to the cell dimensions. Since the cells are fiber-like in shape (i.e., long and thin) with diameter ~8 μm, we might expect to observe diffusion anisotropy on this timescale.

Note that four of the off-diagonal elements in the (undiagonalized) DT are almost zero. This implies that in this example diagonalization (see Figs. 15.7 and 15.8) involves a simple rotation of axes about the normal to the image plane.

If we assume cylindrical symmetry for the cell fibers within a voxel, then $\varepsilon = 0$ and in the principal axes frame we can describe the diffusion in terms of a 2×2 tensor:

$$\mathbf{D}' \equiv \begin{pmatrix} D_\perp & 0 \\ 0 & D_{//} \end{pmatrix}. \tag{15.22}$$

What is more if we choose the image plane to correspond to the center of symmetry of the lens, we only require one angle θ to describe the orientation of the principal axis of the DT with respect to the gradients g_x and g_z, say. Consequently, we only require four images to calculate $D_{//}$, $D_\perp$ and θ, corresponding to gradients of 0, g_x, g_z, and $\frac{1}{\sqrt{2}}(g_x + g_z)$.

The next problem is how to display the data, since even in this case of cylindrical symmetry and a 2×2 DT, we have 3 parameters to display for each pixel! The

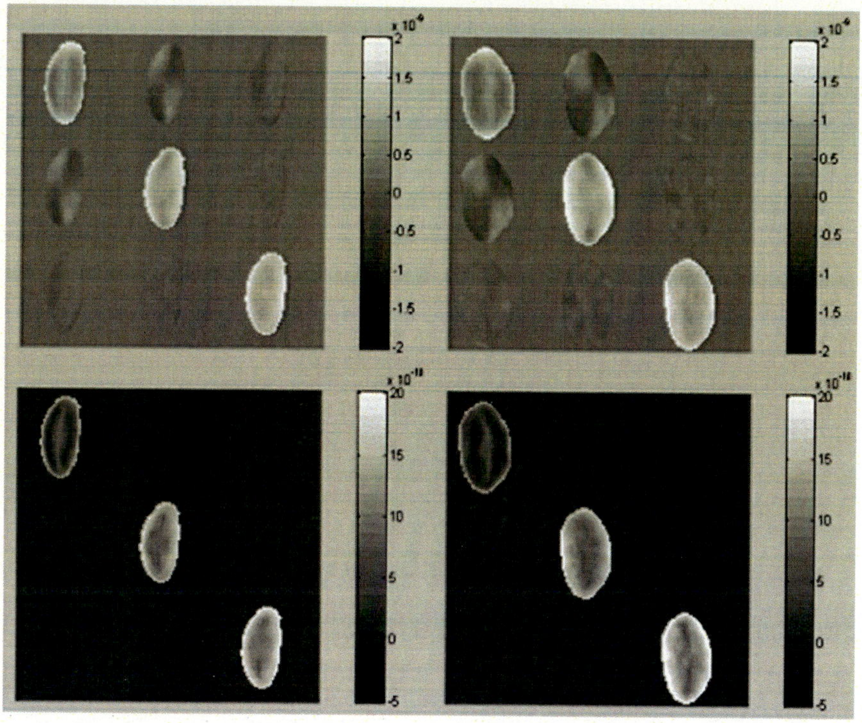

Fig. 15.7 Diffusion tensor images of human eye lenses in vitro from a 29-year-old donor (left column) and an 86-year-old donor (right column) [7]. Top row images are of the raw (undiagonalized) diffusion tensor; those in the bottom row are after diagonalization

method we have developed using MatLab is to display for each pixel a pair of orthogonal lines whose lengths are proportional to $D_{//}$ and $D_{\perp}$, respectively, with the direction of the larger component defining the angle θ, viz (Fig. 15.8).

More generally, if the DT does not display cylindrical symmetry, there are 6 parameters to define per pixel (three eigenvalues and three Euler angles defining the directions of the eigenvectors relative to the laboratory frame). In such cases, it may be necessary to map the principal eigenvalues, the orientations of the eigenvectors, the fractional anisotropy (FA), and the mean eigenvalues (see below) as separate diffusion maps or images in order to visualize the full DT.

15.2.5 Data Acquisition

In situations where time is limited by the need to minimize motion artifacts or to achieve adequate patient throughput, it may be practical only to acquire data for the minimum number of diffusion gradient combinations required to define the DT. In other cases, it may be necessary to employ signal averaging to reduce

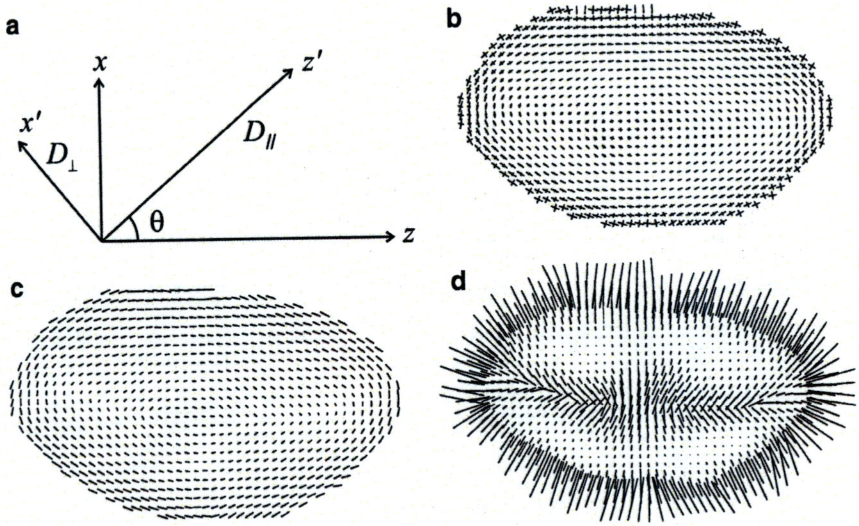

Fig. 15.8 2D diffusion tensor images of a human eye lens from a 29-year-old donor: (**a**) axes of the principal components $D_{//}$ and $D_{\perp}$ of the diagonalized diffusion tensor with respect to the directions of the diffusion gradients; (**b**) quiver plot showing both principal components on the same scale; (**c**) and (**d**) plots of $D_{//}$ and $D_{\perp}$, respectively

'sorting bias' (see below) and/or to acquire data for additional gradient directions to improve precision in measuring the eigenvalues and eigenvectors of the DT and derived parameters such as the FA. Even for the case where the number of gradient directions is restricted to the minimum value [6], significant improvements in precision of DTI-derived parameters can be achieved by appropriate choice of those directions [8].

Several authors have investigated optimum strategies for measuring diffusion parameters in anisotropic systems using MRI [4, 5, 8–13]. Jones et al. [9] derived expressions for the optimum diffusion weighting (b values) and the optimum ratio of the number of signal acquisitions acquired with high diffusion weighting (N_H) to the number (N_L) with low or minimum diffusion weighting, for which $b \sim 0$. (Note that for an imaging sequence $b = 0$ is generally not strictly achievable due to the influence of the imaging gradients, which produce some diffusive attenuation of the signal.) If the effects of transverse relaxation (T_2) are ignored, they found $b = 1.09 \times 3/\mathrm{Tr}(D)$ and $N_H = 11.3 \cdot N_L$, where $\mathrm{Tr}(D) = D_{xx} + D_{yy} + D_{zz}$ is the trace of the DT and b here refers to the difference in diffusion weighting between high and low values (assuming the latter is non-zero). This result applies provided that the diffusion is not too anisotropic (so that diffusive attenuation is similar in all directions). It compares with the situation of minimum overall imaging time in which each of the 7 combinations of gradient magnitude and direction is applied only once, for which clearly $N_H = 6N_L$ and according to Jones et al. [9] the optimum $b = 1.05 \cdot 3/\mathrm{Tr}(D)$. However, these results must be modified to take account of the effects of T_2 relaxation, which results in additional signal attenuation since it is necessary to operate with a finite echo time TE to allow sufficient time to apply the gradients. For example, in the case of white matter in the human brain, for which $T_2 \sim 80\,\text{ms}$, Jones et al. [9] find that it is necessary to reduce both the b value and the ratio N_H/N_L to $\sim 77\%$ of the asymptotic (long T_2) values quoted above.

Chang et al. [14] used a first order perturbation method to derive analytical expressions for estimating the variance of diffusion eigenvalues and eigenvectors as well as DTI derived quantities such as the trace and FA of the DT, for a given experimental design and over a useful range of SNRs. They also validated their results using Monte Carlo simulations.

A number of authors have compared the merits of applying diffusion gradients in more than the minimum six directions. Some reports [10, 12] have suggested there may be no advantage in using more than the minimum number of sampling directions provided that the selected orientations point to the vertices of an icosahedron [11]. However, a more recent Monte Carlo analysis [5] supports earlier suggestions [13, 15] that ∼20–30 unique and evenly distributed sampling directions are required for robust estimation of mean diffusivity, FA and DT orientation. Batchelor et al. [11] conclude that 'the recommended choice of (gradient) directions for a DT-MRI experiment is ... the icosahedral set of directions with the highest number of directions achievable in the available time.'

The use of multiple sets of magnetic field gradient directions is of particular importance for applications involving fiber tracking in the brain. Fiber tracking or 'Tractography' is used to infer axonal connectivity in the white matter of the brain [16–19]. It relies on the fact that the myelin sheaths surrounding neuronal fibers in the white matter restrict water diffusion perpendicular to the direction of the fiber bundles, while diffusion parallel to the nerve fibers is relatively unrestricted. Consequently, the eigenvectors corresponding to the largest eigenvalues reflect the (average) fiber direction within a voxel. By analyzing the directions of the principal eigenvectors in adjacent voxels, it is possible to trace the fiber tracts and infer connectivity between different regions of the brain. The situation becomes more complicated if two or more fiber bundles with significantly different directions intersect or cross within a voxel due to partial volume effects. (Typical voxel dimensions in DTI $\sim$ 1–3 mm are much larger than the individual white matter tracts $\sim$1–10 μm). Behrens et al. [20] estimate that one-third of white matter voxels in the human brain fall into this category. In such cases, the use of a single DT will yield a principal diffusion eigenvector that represents a weighted average of the individual fiber directions and as such will not correspond to the direction of any of the individual fiber bundles. This problem can be at least partially alleviated by acquiring data for multiple gradient directions using high angular resolution diffusion imaging (HARDI) and employing spherical tomographic inversion methods [21] or constrained spherical deconvolution (CSD) techniques [22] to model the resulting DWI data in terms of a set of spherical harmonics rather than a single DT. HARDI techniques employ stronger diffusion weighting gradients (b-values $\geq 3{,}000\,\text{s/mm}^2$) compared with those $\sim 1{,}000\,\text{s/mm}^2$ more routinely employed in clinical DTI. Recently, Tournier et al. [23] using such methods have shown in a DWI phantom that it is possible to resolve two fiber orientations with a crossing angle as small as 30°.

15.3 Digital Processing of DT Images

The raw data set obtained from a DTI measurement described in Sect. 15.2 contains one or more zero-gradient images and six or more diffusion-weighted images corresponding to distinct diffusion directions. To render this data in a form amenable to interpretation, the following processing steps are usually performed:

(I) For each voxel in the image, the six independent components of the DT (DT) are calculated. The tensor obtained in this step is the so-called *laboratory-frame* DT: it is linked to laboratory-based coordinate axes, which may be defined as the directions of the hardware X, Y, Z gradient coils or the Read, Phase, and Slice directions of the image.

(II) The laboratory-frame DT can then be diagonalized. The diagonalization procedure yields:

   (i) The principal diffusivities or*eigenvalues* D_1, D_2 and D_3 of the DT;

   (ii) The orientation of the principal axes or*eigenvectors* of the DT with respect to the laboratory frame.

This represents the DT in the 'sample' frame linked with the physical alignment order in the tissue. The relationship between the laboratory-frame and the diagonalized DT is illustrated in Fig. 15.9 and discussed in detail later in this section.

Steps (I) and (II) can be regarded as the primary DTI processing. These steps are common to all DTI processing and carried out irrespective of the tissue imaged.

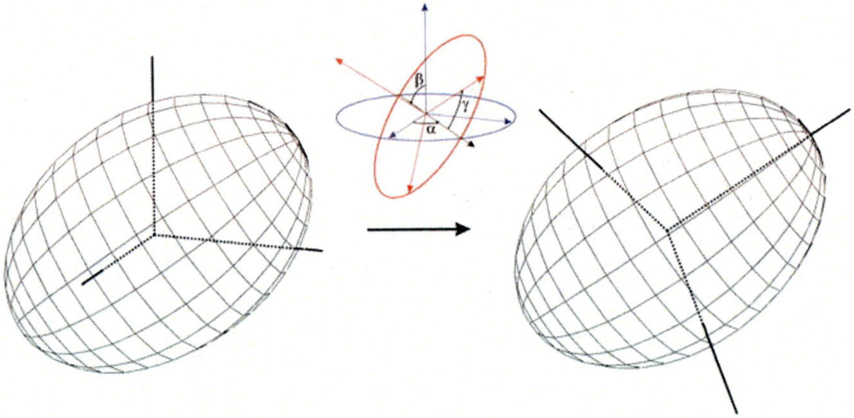

Fig. 15.9 Diagonalization of the diffusion tensor involves finding the rotation of the coordinate frame that aligns the coordinate axes with the principal axes of the ellipsoid

(III) In "secondary" processing, the DT image obtained in step (II) is represented as a voxel-by-voxel map of one or more of the following parameters:

Direction of the principal eigenvector.
Angle between the principal eigenvector and a specified axis.
Principal eigenvalue (maximum diffusivity).
Mean eigenvalue (mean diffusivity).
Fractional anisotropy.
The nonaxial anisotropy parameters of the DT.

The user must decide what DT parameters best enable visualization of the image acquired.

(IV) In "tertiary" processing, the information from individual voxels is translated into "global" characteristics describing the image as a whole. An example of such analysis is the nerve fiber tracking used in DTI of the brain or the spinal cord. The voxels of the image are grouped into tracts such that the principal eigenvectors of the voxels within a tract form continuous "flow lines" representing a large bundle of axons.

Unlike the primary DTI processing, the secondary and tertiary processing is organ- or tissue dependent. The choice of the processing approaches and the DT metrics is determined by the morphology of the tissue and the information sought about the tissue. In avascular tissues, the objective is to characterize the overall alignment order in the tissue rather than identify individual fibers. (The latter is not possible because of the huge number of fibers within a single voxel.) Examples of secondary processing of DT images of cartilage will be presented in Sect. 15.4.

In the following, we provide an overview of the basic principles and the mathematics underlying DT image processing. The processing techniques are described without reference to a specific platform and are generally applicable.

15.3.1 Primary DTI Processing: Calculation of the Laboratory-Frame DT

In Sect. 15.2, the signal intensity was represented as a function of the diffusion gradient as shown in (15.20). This representation provides an intuitive and visual explanation of the diffusive attenuation of the signal in DT images. In practice, it is more convenient to base DTI processing on the so-called *B matrix*. Equation(15.20) can be rewritten as follows (15.24):

$$\ln\left[\frac{S(\mathbf{g})}{S_0}\right] = -\sum_{i=1}^{3}\sum_{j=1}^{3} b_{ij}D_{ij} \equiv -\mathbf{b}:\mathbf{D}, \tag{15.23}$$

where the indices i, j take the values of x, y, or z. The B matrix, **b**, is a 3×3 real symmetric matrix. In the spin-echo experiment, its values are given by

$$b_{ij} = \gamma^2 g_i g_j \delta^2 (\Delta - \delta/3), \tag{15.24}$$

where g_i, g_j are the components of the diffusion gradient vector **g**. The B matrix is an extension of the quantity b introduced in (15.18) to multiple gradient directions.

There are two main advantages to using the B matrix rather than the gradient vectors for processing of DT images. First, the functional form of the signal attenuation is dependent on the DTI pulse sequence used. Equation(15.20) applies to the basic spin-echo pulse-sequence with rectangular diffusion gradients. The attenuation expression is different if a different pulse sequence or nonrectangular diffusion gradients are used [25]. Calculation of the attenuation factor can be difficult and time-consuming for the general pulse sequence [26]. Fortunately, the attenuation equation is amenable to algorithmic, software-based calculation. When the attenuation factor is kept in the simple and general form given by (15.23), any pulse sequence-specific factors can be incorporated into the B matrix as part of the algorithm. The software of most modern MRI spectrometers is capable of automatic calculation of the B matrix for any pulse sequence installed on the spectrometer, eliminating the need for the operator to perform this time-consuming calculation manually.

The second advantage of using the B matrix is that it facilitates accounting for contribution to the diffusive attenuation due to the imaging gradients. This source usually leads to much smaller attenuation than the diffusion gradients. However, it can be important when an accurate DT is sought or when imaging at high spatial resolution. As with diffusion-gradient attenuation factors, the spectrometer software can automatically build all the pulse sequence-specific corrections to the diffusion attenuation factor into the B matrix. Once the B matrix for each diffusion gradient is known, the calculation of the DT can be performed in a way that is independent of the measurement method. Automatic calculation of the B matrix means that DTI processing is greatly simplified from the operator's point of view.

Equation (15.23) yields the signal attenuation for a known B matrix and a known DT. In DTI measurements, where the DT is not known a priori, the inverse problem must be solved: the DT needs to be determined from a set of $N_G \geq 7$ measurements of the signal intensity. In this inverse problem, the inputs are N_G distinct 3×3 B matrices (one B matrix for each diffusion gradient vector) and the corresponding N_G measured signal values. The DT is the output. In DT *imaging*, this problem is solved for each voxel in the image, yielding a separate DT for each voxel (see Fig. 15.10).

In practice, two typical scenarios are encountered:

(1) The diffusion gradient directions correspond to the "pure" elements of the laboratory-frame DT: D_{xx}, $D_{xy}, \ldots$, as shown in (15.21) and Fig. 15.2a.

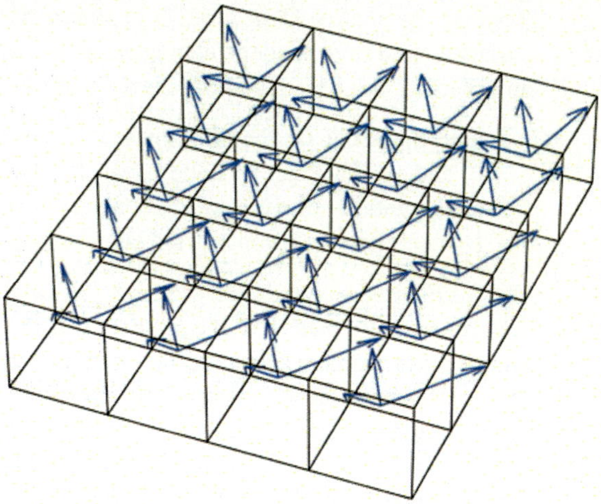

Fig. 15.10 Schematic illustration of a DTI dataset. Each voxel in the image is characterized by a unique diffusion tensor: three eigenvalues (the principal diffusivities) and three mutually perpendicular eigenvectors. In this illustration, the lengths of the eigenvectors are proportional to the corresponding eigenvalues

In this scenario, the diagonal elements of the laboratory-frame DT are simply the diffusivities along the respective gradient directions:

$$D_{\mathrm{ii}} = -\frac{1}{b}\ln\left(\frac{S_{\mathrm{ii}}}{S_0}\right) \quad i = x, y, z. \tag{15.25}$$

The off-diagonal elements are given by [27]:

$$D_{\mathrm{xy}} = -\frac{1}{2b}\left(\ln\frac{S_{\mathrm{xx}}}{S_0} + \ln\frac{S_{\mathrm{yy}}}{S_0}\right) + \frac{1}{b}\ln\frac{S_{\mathrm{xy}}}{S_0}, \text{etc.} \tag{15.26}$$

Equations (15.25) and (15.26) are applicable only in the special case when the gradient directions are given by (15.21). This special case is very instructive for beginners because it visually and simply illustrates the meaning of the diagonal and the off-diagonal elements of the DT.

(2) The second scenario is a data set containing more than the minimal number of diffusion gradient directions, as illustrated in Fig. 15.2b.

In this case, the signal corresponding to each direction depends on a combination of several (potentially all) elements of the DT. The DT is determined using least-squares fitting of (15.23) to all the measured signal values simultaneously:

(i) Create a vector of length N_G containing the signal values from the N_G measurements: $\mathbf{s} = (S_1 \ldots S_{NG})$.
(ii) For each $n = 1 \ldots N_G$, calculate $y_n = -\ln(S_n)$;
(iii) Set up the linearized least-squares fit equation:

$$y_n = A + \sum_{i=1}^{3}\sum_{j=1}^{3}(b_{ij})_n D_{ij}. \tag{15.27}$$

Because the matrix **D** in (15.27) is symmetric ($D_{ij} \equiv D_{ji}$), the LSF involves 7 parameters: 6 independent elements of the symmetric DT and the 7th is the amplitude of the nonattenuated signal.

(iv) Find the set of D_{ij} that minimizes the sum of the squared differences between s_n and y_n. This can be done using the standard linear LSF procedure [28] or mathematical software packages such as Mathematica or Matlab. The elements D_{ij} comprise the reconstructed laboratory-frame DT.

The LSF-based approach of scenario (2) is generally applicable: it can be used with an arbitrary pattern of the gradient directions (including the optimal-sampling patterns discussed in Sect. 15.2) as well as the minimal $6+1$ dataset. The zero-gradient measurement S_0 is crucially important in both scenarios. However, in the LSF-based approach, the zero-gradient measurements do not have a special status: the least-squares fitting procedure treats them on par with diffusion-attenuated points. Nevertheless, the importance of the zero-gradient measurements can be recognized by assigning a greater LSF weight to them than to diffusion-attenuated measurements.

As discussed earlier, one advantage of the LSF-based approach is that it allows the diffusive attenuation due to imaging gradients to be accounted for easily. Its other advantage is that, when redundant measurements are available (i.e., when more than the minimal set of $6+1$ measurements was made), it enables an estimation of the standard errors of the DT elements. This can be done as part of the LSF and does not require additional computation time. In the absence of redundant measurements, the seven parameters can always be adjusted to fit the 7 "minimal" measurements exactly; therefore, this advantage is realized only when redundant measurements are available.

15.3.2 Diagonalization of the DT

The laboratory-frame DT is difficult to interpret directly because its off-diagonal elements lack a straightforward physical meaning. The off-diagonal elements can be negative; therefore, they are *not* simply the diffusivities along the directions given by (15.21) (any diffusivity must be positive).

To enable a physical interpretation, the laboratory-frame DT is usually subjected to diagonalization. In the first approximation, diagonalization can be visualized as a 3D rigid-body rotation that aligns the laboratory-frame coordinate axes with the principal axes of the DT ellipsoid, as shown in Fig. 15.9. Such a rotation is described by the *Euler angles* α, β, γ, which relate the orientation of the principal axes of the DT to the laboratory axes. The lengths of the principal axes correspond to the principal diffusivities (also known as the DT eigenvalues). The directions of the principal axes relative in the laboratory frame are known as the DT eigenvectors. DT eigenvectors tend to represent the alignment order in the tissue and therefore provide a means of visualizing the tissue microstructure.

Diagonalization may also involve improper rotations - rotations combined with permutations of the coordinate axes or inversion of the signs of the axes. This is because there is no physical distinction between the positive and the negative direction of DT eigenvectors. In general, diagonalization is represented by a *unitary transformation*:

$$\mathbf{D}' = \mathbf{U}(\alpha,\beta,\gamma)\,\mathbf{D}\mathbf{U}^{+}(\alpha,\beta,\gamma), \tag{15.28}$$

where **U** is a unitary matrix, defined as a matrix whose Hermitian conjugate equals its inverse: $\mathbf{U}\mathbf{U}^{+} = \mathbf{1}$. Rotational transformations illustrated in Fig. 15.9 are a subset of unitary transformations.

In general, a given DT can be diagonalized by more than one matrix **U**. **U** can be found using the standard algorithms such as Jacobi diagonalization [28]. Packages such as Mathematica or Matlab contain built-in diagonalization functions that can be used for this purpose.

A general property of unitary transformations is that they conserve the sum of the diagonal elements (the *trace* of the matrix). Therefore, the trace of the DT remains unchanged under a transformation given by (15.28). This means that the mean diffusivity can be found from the laboratory-frame DT without diagonalization:

$$D_{\text{av}} = \frac{1}{3}(D_1 + D_2 + D_3) = \frac{1}{3}(D_{\text{xx}} + D_{\text{yy}} + D_{\text{zz}}). \tag{15.29}$$

In the experimental setting, the measured signal inevitably contains a contribution from random noise, which can distort the elements of the DT. In the limit of strong noise, the distortion can be sufficiently large to make some of the diagonal elements or the eigenvalues of the DT negative. In this case, the measurement should be considered unreliable and the DT in the given voxel discarded. Alternatively, the DT can be calculated using an algorithm that enforces its positive-definiteness [29].

15.3.3 Gradient Calibration Factors

Another important factor from the experimental standpoint is the need for gradient calibration factors. On many NMR spectrometers, diffusion gradient amplitudes are set as percentages of the maximum amplitude; however, the absolute amplitude corresponding to "100%" may differ between the x, y, and z gradient coils. In this case, it is useful to introduce unitless calibration factors relating the actual and the nominal amplitude of each gradient:

$$\mathbf{g}^{\text{real}} = \begin{pmatrix} g_x^{\text{real}} \\ g_y^{\text{real}} \\ g_z^{\text{real}} \end{pmatrix} = \mathbf{C}\cdot\mathbf{g}^{\text{nom}} = \begin{pmatrix} C_x & 0 & 0 \\ 0 & C_y & 0 \\ 0 & 0 & C_z \end{pmatrix} \cdot \begin{pmatrix} g_x^{\text{nom}} \\ g_y^{\text{nom}} \\ g_z^{\text{nom}} \end{pmatrix}. \tag{15.30}$$

The gradient calibration matrix, **C**, can be incorporated into the B matrix: in the coordinate system of the hardware gradients, the actual and the nominal matrices are related as $\mathbf{b}^{\text{real}} = \mathbf{C} \cdot \mathbf{b}^{\text{nom}} \cdot \mathbf{C}$, where $\mathbf{b}^{\text{nom}}$ is calculated from the uncalibrated gradient values. It is important to note that **C** is not a unitary matrix – rather, it is a rescaling matrix that scales different b_{ij}'s by the appropriate factors.

In a different coordinate system (say, the RPS coordinates), the B matrix can be recalibrated according to

$$\mathbf{b}'^{\text{real}} = (\mathbf{UCU}^{+}) \cdot (\mathbf{Ub}^{\text{nom}}\mathbf{U}^{+}) \cdot (\mathbf{UCU}^{+}) = \mathbf{C}' \cdot \mathbf{b}'^{\text{nom}} \cdot \mathbf{C}', \tag{15.31}$$

where the $'$ refers to the RPS coordinates.

An alternative approach is to make use of an isotropic region of the sample, for example the saline surrounding the anisotropic tissue. In an isotropic region, the diffusion attenuation should depend only on the b value (i.e., the trace of the B matrix) and not on the direction of the diffusion gradient. By comparing the attenuation factors of (15.17) corresponding to different gradient directions, one can empirically introduce scalar calibration factors for each gradient direction. This approach is often more robust than that given by (15.31).

15.3.4 Sorting Bias

Each eigenvalue of the DT is associated with a 3D vector that represents the characteristic direction corresponding to that diffusivity, as illustrated in Fig. 15.10. The greatest eigenvalue and the corresponding eigenvector are referred to as the principal eigenvalue and the principal eigenvector. The second largest diffusivity is referred to as the secondary eigenvalue (secondary eigenvector).

In the experimental context identifying the correct order of the eigenvalues is not completely straightforward because of the presence of noise in the images. Noise leads to the so-called *sorting bias*, which can be understood as follows. Suppose that two voxels, A and B, contain physically identical tissue and are therefore characterized by an identical underlying DT, D^{True}, with eigenvalues $D_1{}^{\text{True}} \geq D_2{}^{\text{True}} \geq D_3{}^{\text{True}}$. The apparent DT is a combination of the underlying DT and a contribution due to noise:

$$\begin{aligned} D_{1\text{A}} &= D_{1\text{A}}^{\text{true}} + \Delta D_{1\text{A}} \quad & D_{1\text{B}} &= D_{1\text{B}}^{\text{true}} + \Delta D_{1\text{B}} \\ D_{2\text{A}} &= D_{2\text{A}}^{\text{true}} + \Delta D_{2\text{A}} \quad & D_{2\text{B}} &= D_{2\text{B}}^{\text{true}} + \Delta D_{2\text{B}} \\ D_{3\text{A}} &= D_{3\text{A}}^{\text{true}} + \Delta D_{3\text{A}} \quad & D_{3\text{B}} &= D_{3\text{B}}^{\text{true}} + \Delta D_{3\text{B}} \end{aligned} \tag{15.32}$$

where $\Delta D_{1\text{A}} \ldots \Delta D_{3\text{B}}$ are contributions from noise. Therefore, although the underlying DT in the two voxels is the same, the experimentally measured tensors in voxels A and B usually differ due to the random nature of the noise contribution. Suppose that, in a particular instance, $\Delta D_{1\text{A}}$ and $\Delta D_{2\text{B}}$ are negative, while $\Delta D_{1\text{B}}$ and

ΔD_{2A} are positive. If the noise is sufficiently large, or the DT anisotropy small, the order of the eigenvalues in voxel A may be reversed: $D_{1A} < D_{2A}$ but $D_{1B} > D_{2B}$. If the sorting of the eigenvalues is based only on the magnitude of the diffusivity, then the eigenvalues and the eigenvectors in voxel A will be assigned incorrectly: D_{2A} will be taken as the principal eigenvalue and D_{1A} as the secondary eigenvalue. This sorting bias has two main consequences:

(1) It results in an overestimation of the principal eigenvalue and underestimation of the secondary eigenvalue. This happens because the diffusivity-based sorting fails to take into account the possibility of negative ΔD_{1A}, which introduces an inherent bias into the distribution of the eigenvalues;
(2) In the example above, the direction of the principal DT eigenvector in voxel A will be off by 90° because the eigenvalues are misidentified. Therefore, sorting bias also introduces disjoint voxels in an eigenvector map.

The basic principles of techniques that minimize sorting bias can be understood based on the following idea. If the morphology of the tissue varies slowly from one voxel to another, then it can be assumed that the corresponding eigenvectors in neighboring voxels should have similar directions. Conversely, in the biased example described above, the apparent principal eigenvectors in voxels A and B would be nearly perpendicular. Therefore, in order to minimize sorting bias, the eigenvalues and eigenvectors need to be treated as pairs, and the sorting of eigenvalues needs to take into account the directions of the corresponding eigenvectors. A number of approaches exist that alleviate (but do not completely eliminate) sorting bias [30].

15.3.5 Fractional Anisotropy

For a prolate DT ($D_1 > D_2 \approx D_3$), the FA is defined as

$$\begin{aligned} \mathrm{FA} &= \sqrt{\frac{3}{2}}\,\frac{\sqrt{(D_1-\bar{D})^2+(D_2-\bar{D})^2+(D_3-\bar{D})^2}}{\sqrt{D_1^2+D_2^2+D_3^2}} \\ &= \frac{1}{\sqrt{2}}\,\frac{\sqrt{(D_1-D_2)^2+(D_2-D_3)^2+(D_3-D_1)^2}}{\sqrt{D_1^2+D_2^2+D_3^2}}. \end{aligned} \tag{15.33}$$

This definition is appropriate for diffusion between long fibers (such as in AC) or withiners (e.g., within nerve fiber tracts). In the case of extreme anisotropy, the FA given by (15.33) equals 1, while in the perfectly isotropic case $\mathrm{FA} = 0$.

For an oblate DT ($D_1 \approx D_2 > D_3$), the appropriate definition of the FA is

$$\mathrm{FA} = \sqrt{\frac{2}{3}}\,\frac{\sqrt{(D_1-\bar{D})^2+(D_2-\bar{D})^2+(D_3-\bar{D})^2}}{\sqrt{D_1^2+D_2^2+D_3^2}}. \tag{15.34}$$

The FA given by (15.34) can be used for diffusion between confining planes (e.g., diffusion of water molecules in the aqueous domain of lamellar lipid bilayers) and also has the range between 1 (extreme anisotropy) and 0 (isotropic limit).

The value of FA represents the amount of restriction imposed on diffusional displacement of water molecules by the solid component of the tissue (e.g., collagen fibers or cell walls). The value of FA depends on both the relative volume fraction occupied by the solid domain and the degree of alignment of the fibers or cells. FA is therefore a useful morphological metric of the tissue. Specific examples of the relationship between FA and the morphology of the tissue are presented in Sect. 15.4.

The theoretical value of the FA defined according to (15.33) and (15.34) in the isotropic case is zero. In practice, the presence of noise in MR signal leads to a positive FA even when the underlying eigenvalues of the true DT are equal. The origin of this is fundamentally the same as the origin of sorting bias discussed above. If ${D_{1A}}^{true} = {D_{2A}}^{true} = {D_{3A}}^{true}$, the measured eigenvalues D_{1A}, D_{2A} and D_{3A} would almost always be different due to the presence of noise, as shown in (15.32). By combining (15.32) and (15.33), it is easily seen that the measured FA in this case given by

$$\mathrm{FA}_{\mathrm{noise}} = \sqrt{\frac{3}{2}}\frac{\Delta D}{D}. \tag{15.35}$$

Equation(15.35) represents a "noise" FA that is observed in isotropic parts of the sample such as water or saline surrounding the anisotropic tissue. Its magnitude depends on the conditions of the measurement but typically lies in the range 0.01–0.1 [31–33]. Nonzero FA due to noise is also observed in Monte Carlo simulations of the DT, where it is inversely proportional to the square root of the ensemble size [28,34]. Noise FA should be taken as a baseline when interpreting the values of FA in tissue. In the limit of low noise ($\Delta D/D << 1$), the experimentally measured FA is the sum of the "true" underlying FA ($\mathrm{FA}_{\mathrm{true}}$) and the noise contribution given by (15.35):

$$\mathrm{FA} = \mathrm{FA}_{\mathrm{true}} + \mathrm{FA}_{\mathrm{noise}} \tag{15.36}$$

15.3.6 Other Anisotropy Metrics

The FA definitions of (15.33) and (15.34) are usually used to characterize axially symmetric tensors (when two of the eigenvalues are equal or nearly equal to each other). In the asymmetric case, the following model-free parameters can be applied to characterize the DT anisotropy:

$$\eta = \frac{1}{3}\left[D_1 - \frac{(D_2 + D_3)}{2}\right] \tag{15.37}$$

$$\varepsilon = \frac{D_2 - D_3}{2}. \tag{15.38}$$

In the case of axial symmetry, $\varepsilon = 0$.

15.4 Applications of DTI to Articular Cartilage

In Sect. 15.2.4, we discussed two ways of presenting DT images of the eye lens: maps of individual DT elements and eigenvector maps. In this section, we focus on another avascular tissue, articular cartilage (AC) [31–33]. We discuss several types of DTI parameter maps used by us for visualizing the DT in this tissue. Different types of parameter maps emphasize different aspects of the DT, and the choice of the type of map to be used is determined by what characteristics of the tissue microstructure need to be gleaned from the images.

15.4.1 Bovine AC

Figure 15.11 shows a spin echo MR image from a sample of bovine patellar AC (with bone attached) recorded at a magnetic field strength B_0 of 16.4 T. The sample, immersed in Fomblin® oil (which gives no ^{1}HNMR signal), was oriented with the normal to the articular surface at 55° to the static magnetic field to: (1) optimize the SNR, and (2) suppress the characteristic banding seen in conventional MR images of AC to ensure relatively uniform signal intensity throughout the cartilage [31]. Diffusion-weighted images were acquired with the minimal set of diffusion gradients using a spin-echo pulse sequence with the following acquisition parameters: echo time, 18 ms; repetition time 700 ms; average b value 1,550 s mm^{-2}; 2 ms diffusion gradients; 12 ms diffusion interval; 10 × 12.8 mm field of view; 50 μm in-plane resolution and 400μm slice thickness. Two images were acquired without diffusion gradients, one of which is shown in Fig. 15.11. Total acquisition time was 14 h 38 m.

The magnitude of the FA is shown in Fig. 15.12a with black representing the smallest FA. The direction of the principal diffusion eigenvector within the voxels is incorporated into the map in Fig. 15.12b using color. Figure 15.13 shows the average FA as a function of distance from the articular surface.

In Fig. 15.14, the principal eigenvectors are scaled by their eigenvalue to enable visualization of how the collagen fibers 'direct' the diffusion of water perpendicular to the supporting bone in the radial zone. The fibers are less ordered in the transitional zone and align parallel to the articular surface in the superficial zone. This figure shows the eigenvectors from two contiguous slices.

Fig. 15.11 A raw SE image of an excised sample of bovine articular cartilage at 16.4 T

15.4.2 Human AC

The image in Fig. 15.15 was recorded at 7 T from a sample of human right lateral tibia, obtained from a 57-year-old male undergoing complete knee replacement. This region was the only remaining cartilage in the joint and was described by the surgeon as being in poor condition. Acquisition parameters: echo time, 13.3 ms; repetition time 2,000 ms; 2 ms diffusion gradient duration; 8 ms diffusion interval; average b value 1,075 mm^{-2}; 20×20 mm field of view, with a 156 μm in-plane isotropic voxel dimension and 2 mm slice thickness. Total acquisition time was 19 h.

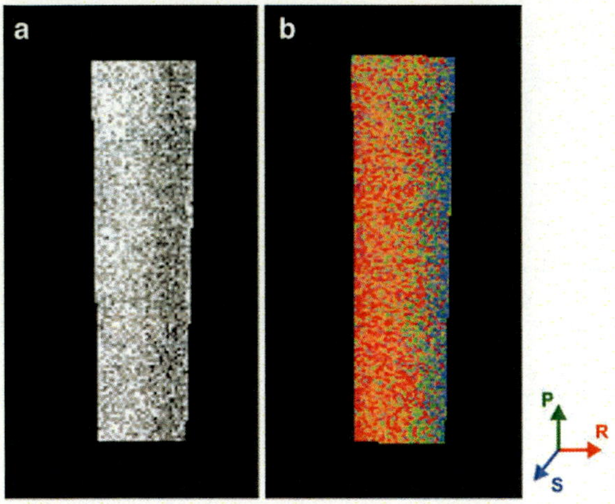

Fig. 15.12 (**a**) Fractional anisotropy map of the sample shown in Fig. 15.11. Black corresponds to FA = 0; white to FA = 0.15. (**b**) Directional FA map of the same sample. The colors denote the direction of the principal DT eigenvector: Read, Phase, and Slice gradient directions are shown in red, green, and blue, respectively. Color intensity reflects the magnitude of the FA

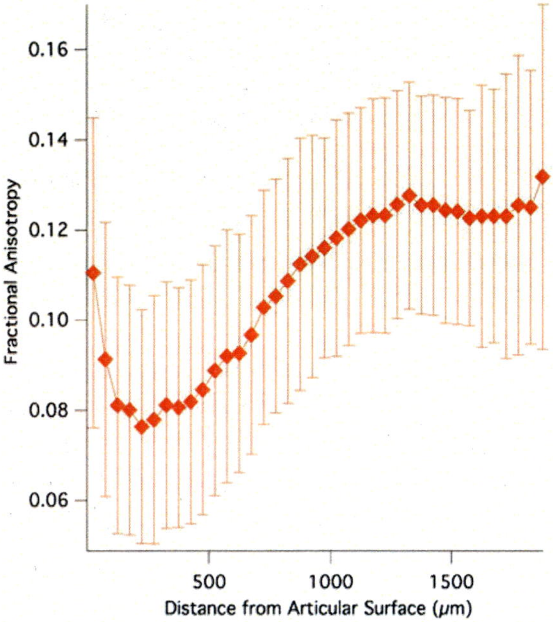

Fig. 15.13 The average fractional anisotropy in the same sample plotted as a function of distance from the articular surface

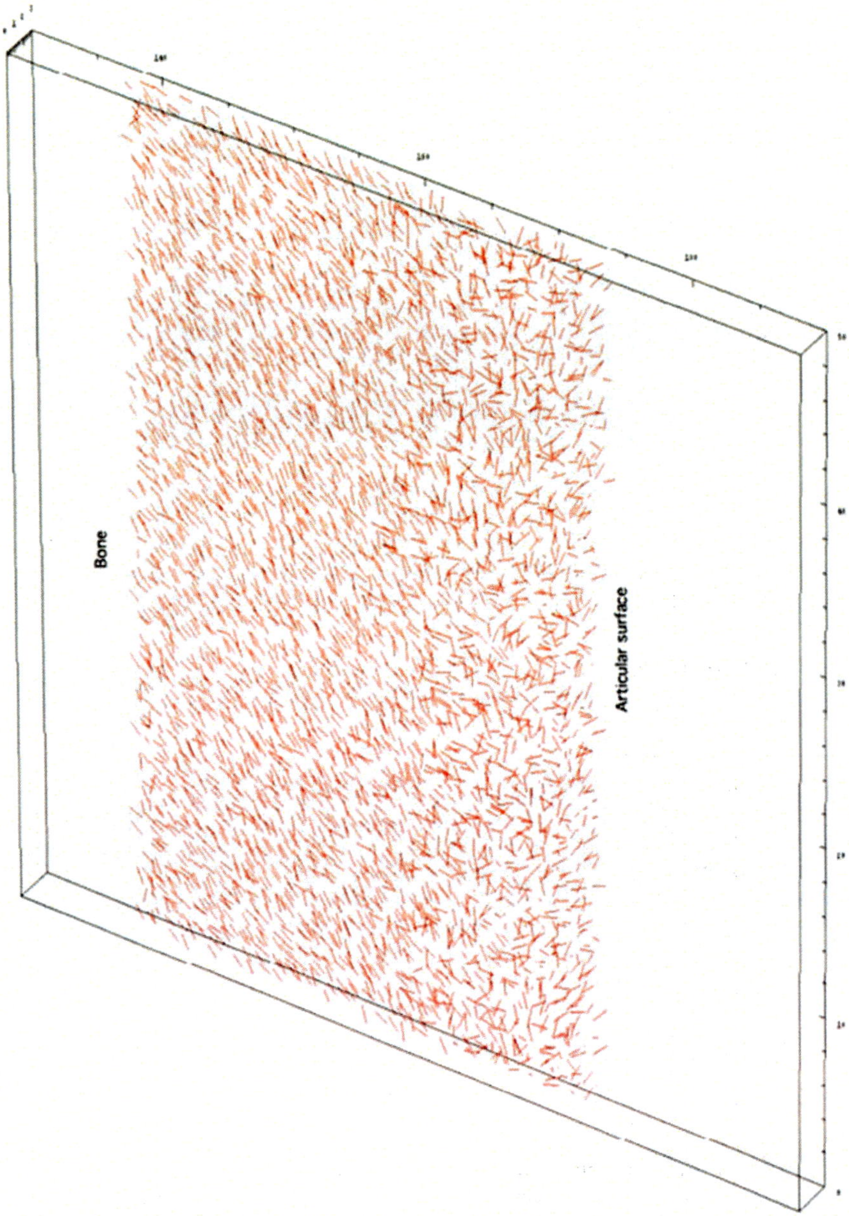

Fig. 15.14 A quiver plot showing the directions of the principal DT eigenvectors in the same cartilage sample

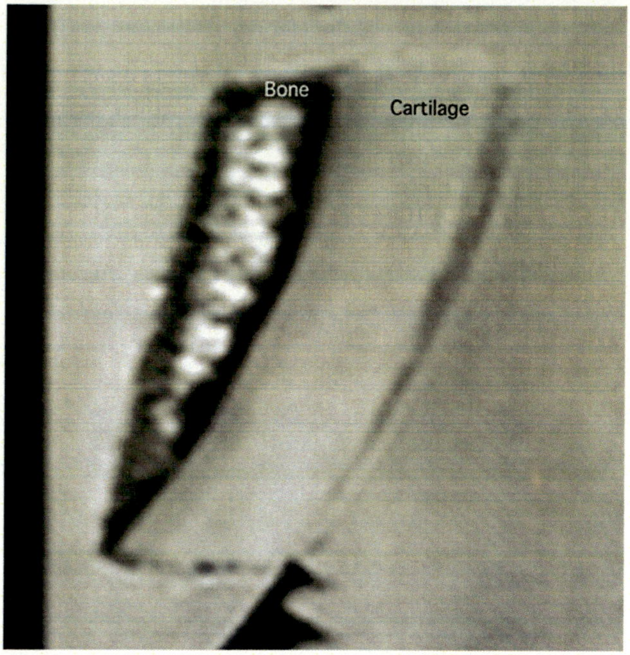

Fig. 15.15 MR image of human cartilage recorded at 7 T in vitro

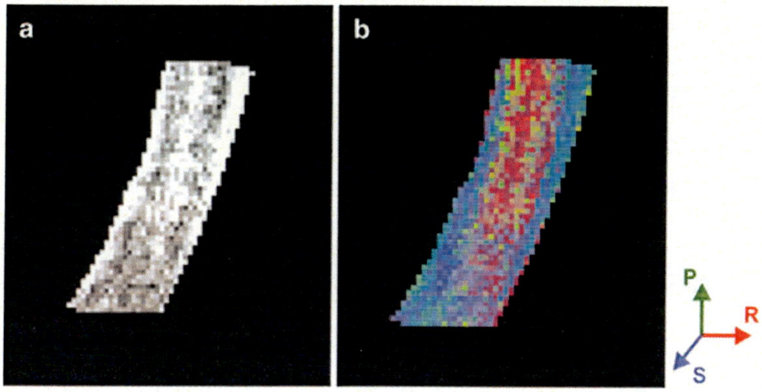

Fig. 15.16 The conventional (**a**) and the directional (**b**) FA maps of the same human cartilage sample. In (**b**), the principal eigenvector direction is represented by colors: *red*, left-right (Read); *blue*, up-down (Phase); and *Green*, in-out (Slice)

Figure 15.16 shows the conventional (a) and the directional (b) FA maps for the human cartilage sample shown in Fig. 15.15. The color coding in the directional map is identical to Fig. 15.12b.

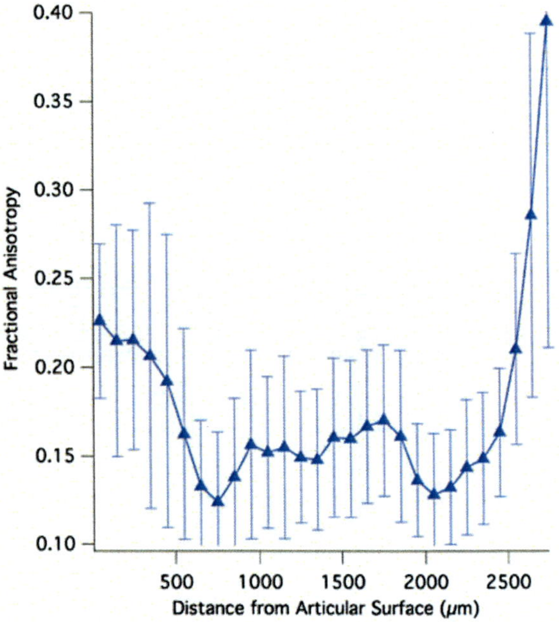

Fig. 15.17 The average FA plotted against depth from the articular surface

The profile of average FA (± std dev) as a function of distance from the articular surface for the human cartilage sample is shown in Fig. 15.17. The FA is within the expected range for cartilage of (0.04–0.28) [34], except for the region near the supporting bone, where calcification is likely to contribute to an increase in the observed FA.

Figure 15.18 shows a 'quiver' plot for a single slice of the same human cartilage sample in which the principal eigenvector is represented by a line, proportional in length to the principal eigenvalue.

In addition to DTI processing with the Matlab or Mathematica software packages utilized by us, DTI data can be processed using proprietary software from the scanner manufacturers if available, or transformed data to a common format, such as DICOM, Analyse or NIFTI and processed using one of the readily available shareware diffusion processing packages.

Acknowledgments This research was supported under Australian Research Council's Discovery Projects funding scheme (project number DP0880346). We thank the University of Queensland node of the National Instrumentation Facility for access to the 16.4 T microMRI spectrometer and Dr Gary Cowin for assistance with data acquisition. We thank Mr Garth Brooks and Mrs Stacey Manson (Teys Bros Pty. Ltd., Beenleigh, Australia) for providing samples of bovine patellar cartilage. We thank Prof Ross Crawford for providing human cartilage samples and Mrs Sally de Visser for acquiring the DTI data sets used to generate Figs. 15.15–15.18.

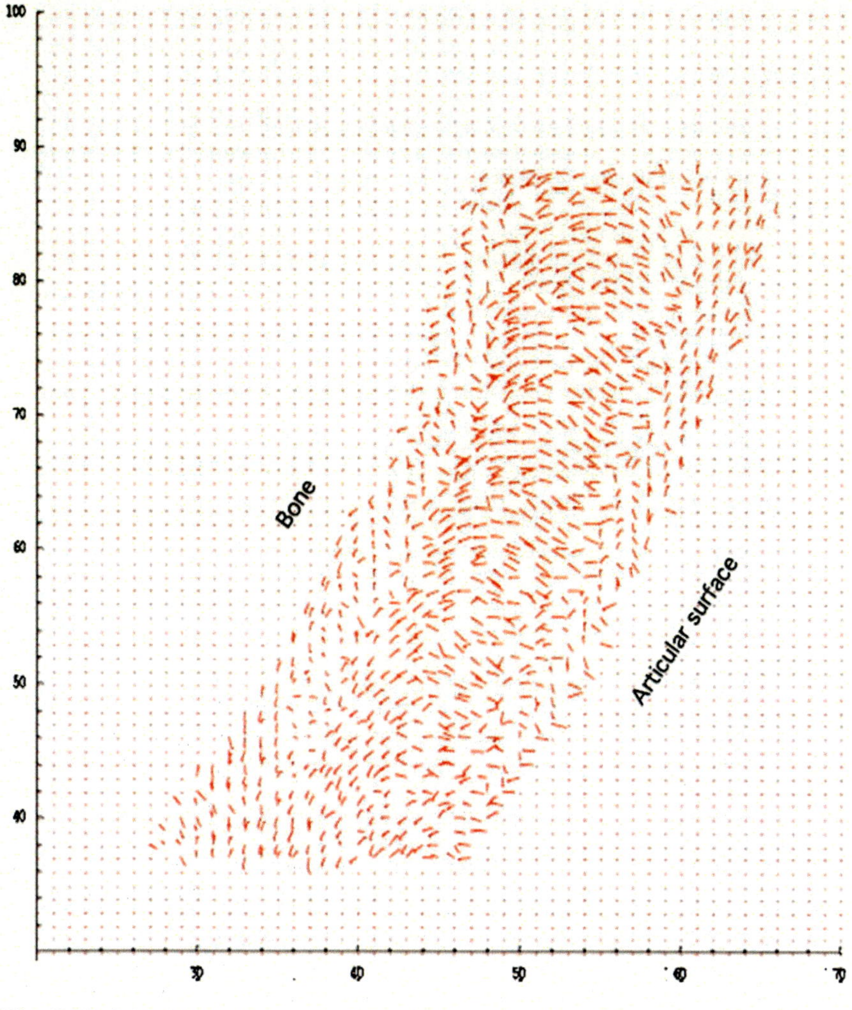

Fig. 15.18 Quiver plot showing the principal DT eigenvector for each voxel in the sample

References

1. Einstein, A.: Zur allgemeinen molekularen Theorie der Wärme. Annalen der Physik **14**(S1), 154–163 (2005)
2. Tanner, J.E.: Transient diffusion in a system partitioned by permeable barriers. Application to NMR measurements with a pulsed field gradient. J. Chem. Phys. **69**(4), 1748–1754 (1978)
3. Stejskal, E.O., Tanner, J.E.: Spin diffusion measurements: Spin echoes in the presence of a time-dependent field gradient. J. Chem. Phys. **42**, 288–292 (1965)
4. Basser, P.J., Jones, D.K.: Diffusion-tensor MRI: theory, experimental design and data analysis – a technical review. NMR Biomed. **15**(7–8), 456–467 (2002)
5. Jones, D.K.: The effect of gradient sampling schemes on measures derived from diffusion tensor MRI: A Monte Carlo study. Magn. Reson. Med. **51**(4), 807–815 (2004)

6. Mukherjee, P., Chung, S.W., Berman, J.I., Hess, C.P., Henry, R.G.: Diffusion tensor MR imaging and fiber tractography: Technical considerations. Am. J. Neuroradiol. **29**(5), 843–852 (2008)
7. Moffat, B.A., Pope, J.M.: Anisotropic water transport in the human eye lens studied by diffusion tensor NMR micro-imaging. Exp. Eye. Res. **74**(6), 677–687 (2002)
8. Papadakis, N.G., Xing, D., Huang, C.L.H., Hall, L.D., Carpenter, T.A.: A comparative study of acquisition schemes for diffusion tensor imaging using MRI. J. Magn. Reson. **137**(1), 67–82 (1999)
9. Jones, D.K., Horsfield, M.A., Simmons, A.: Optimal strategies for measuring diffusion in anisotropic systems by magnetic resonance imaging. Magn. Reson. Med. **42**(3), 515–525 (1999)
10. Batchelor, P.G.: Optimisation of Direction Schemes for Diffusion Tensor Imaging. St Malo, France (2002)
11. Batchelor, P.G., Atkinson, D., Hill, D.L.G., Calamante, F., Connelly, A.: Anisotropic noise propagation in diffusion tensor MRI sampling schemes. Magn. Reson. Med. **49**(6), 1143–1151 (2003)
12. Hasan, K.M., Parker, D.L., Alexander, A.L.: Comparison of gradient encoding schemes for diffusion-tensor MRI. J. Magn. Reson. Imaging. **13**(5), 769–780 (2001)
13. Papadakis, N.G., Xing, D., Houston, G.C., Smith, J.M., Smith, M.I., James, M.F., Parsons, A.A., Huang, C.L.H., Hall, L.D., Carpenter, T.A.: A study of rotationally invariant and symmetric indices of diffusion anisotropy. Magn. Reson. Imaging. **17**(6), 881–892 (1999)
14. Chang, L.C., Koay, C.G., Pierpaoli, C., Basser, P.J.: Variance of estimated DTI-derived parameters via first-order perturbation methods. Magn. Reson. Med. **57**(1), 141–149 (2007)
15. Skare, S., Hedehus, M., Moseley, M.E., Li, T.Q.: Condition number as a measure of noise performance of diffusion tensor data acquisition schemes with MRI. J. Magn. Reson. **147**(2), 340–352 (2000)
16. Poupon, C., Mangin, J.F., Clark, C.A., Frouin, V., Regis, J., Le Bihan, D., Bloch, I.: Towards inference of human brain connectivity from MR diffusion tensor data. Med. Image. Anal. **5**(1), 1–15 (2001)
17. Basser, P.J., Pajevic, S., Pierpaoli, C., Duda, J., Aldroubi, A.: In vivo fiber tractography using DT-MRI data. Magn. Reson. Med. **44**(4), 625–632 (2000)
18. Conturo, T.E., Lori, N.F., Cull, T.S., Akbudak, E., Snyder, A.Z., Shimony, J.S., McKinstry, R.C., Burton, H., Raichle, M.E.: Tracking neuronal fiber pathways in the living human brain. Proc. Natl. Acad. Sci. USA **96**(18), 10422–10427 (1999)
19. Mori, S., van Zijl, P.C.M.: Fiber tracking: principles and strategies - a technical review. NMR Biomed. **15**(7–8), 468–480 (2002)
20. Behrens, T.E.J., Berg, H.J., Jbabdi, S., Rushworth, M.F.S., Woolrich, M.W.: Probabilistic diffusion tractography with multiple fibre orientations: What can we gain? Neuroimage **34**(1), 144–155 (2007)
21. Tuch, D.S.: Q-Ball imaging. Magn. Reson. Med. **52**(6), 1358–1372 (2004)
22. Tournier, J.D., Calamante, F., Connelly, A.: Robust determination of the fibre orientation distribution in diffusion MRI: Non-negativity constrained super-resolved spherical deconvolution. Neuroimage **35**(4), 1459–1472 (2007)
23. Tournier, J.D., Yeh, C.H., Calamante, F., Cho, K.H., Connelly, A., Lin, C.P.: Resolving crossing fibres using constrained spherical deconvolution: Validation using diffusion-weighted imaging phantom data. Neuroimage **42**(2), 617–625 (2008)
24. Basser, P.J., Mattiello, J., LeBihan, D.: Estimation of the effective self-diffusion tensor from the NMR spin-echo. J. Magn. Reson. B **103**(3), 247–254 (1994)
25. Momot, K.I., Kuchel, P.W.: PFG NMR diffusion experiments for complex systems. Concepts Magn. Reson. **28A**, 249–269 (2006)
26. Momot, K.I., Kuchel, P.W.: Convection-compensating diffusion experiments with phase-sensitive double-quantum filtering. J. Magn. Reson. **174**(2), 229–236 (2005)

27. Coremans, J., Luypaert, R., Verhelle, F., Stadnik, T., Osteaux, M.: A method for myelin fiber orientation mapping using diffusion-weighted MR-images. Magn. Reson. Imaging **12**(3), 443–454 (1994)
28. Press, W.H., Teukolsky, S.A., Vetterling, W.T., Flannery, B.P.: Numerical Recipes in Fortran. Cambridge University Press, New York (1992)
29. Lenglet, C., Campbell, J.S.W., Descoteaux, M., Haro, G., Savadjiev, P., Wassermann, D., Anwander, A., Deriche, R., Pike, G.B., Sapiro, G., Siddiqi, K., Thompson, P.M.: Mathematical methods for diffusion MRI processing. Neuroimage **45**(1), S111–S122 (2009)
30. Basser, P.J., Pajevic, S.: Statistical artifacts in diffusion tensor MRI (DT-MRI) caused by background noise. Magn. Reson. Med. **44**(1), 41–50 (2000)
31. Meder, R., de Visser, S.K., Bowden, J.C., Bostrom, T., Pope, J.M.: Diffusion tensor imaging of articular cartilage as a measure of tissue microstructure. Osteoarthr. Cartilage. **14**, 875–881 (2006)
32. de Visser, S.K., Crawford, R.W., Pope, J.M.: Structural adaptations in compressed articular cartilage measured by diffusion tensor imaging. Osteoarthr. Cartilage. **16**(1), 83–89 (2008)
33. de Visser, S.K., Bowden, J.C., Wentrup-Byrne, E., Rintoul, L., Bostrom, T., Pope, J.M., Momot, K.I.: Anisotropy of collagen fibre alignment in bovine cartilage: Comparison of polarised light microscopy and spatially-resolved diffusion-tensor measurements. Osteoarthr. Cartilage. **16**(6), 689–697 (2008)
34. Momot, K.I.: Diffusion tensor of water in model articular cartilage. Eur. Biophys. J. **40**(1), 81–91 (2011)

Index

G. Dougherty (ed.), *Medical Image Processing: Techniques and Applications*, Biological and Medical Physics, Biomedical Engineering, DOI 10.1007/978-1-4419-9779-1,